Quality Assurance of Postharvest Stored Products

Quality Assurance of Postharvest Stored Products

Edited by **Fernando Plath**

SYRAWOOD
PUBLISHING HOUSE

New York

Published by Syrawood Publishing House,
750 Third Avenue, 9th Floor,
New York, NY 10017, USA
www.syrawoodpublishinghouse.com

Quality Assurance of Postharvest Stored Products
Edited by Fernando Plath

International Standard Book Number: 978-1-68286-033-5 (Hardback)

Printed in the United States of America.

Contents

Preface

The main objective of postharvest treatment is to determine and maintain quality of the crop and to ensure its availability for future consumption. Once harvested, crops are subjected to various processes to assure fine quality of postharvest stored products. Postharvest handling prevents crop from deteriorating by using different techniques. The book provides an overarching account of current technologies to enhance shelf life of crops and to maintain their quality during storage, packing and transportation. It includes applications of postharvest physiology and is a great pick for students and researchers engaged in this field.

The researches compiled throughout the book are authentic and of high quality, combining several disciplines and from very diverse regions from around the world. Drawing on the contributions of many researchers from diverse countries, the book's objective is to provide the readers with the latest achievements in the area of research. This book will surely be a source of knowledge to all interested and researching the field.

In the end, I would like to express my deep sense of gratitude to all the authors for meeting the set deadlines in completing and submitting their research chapters. I would also like to thank the publisher for the support offered to us throughout the course of the book. Finally, I extend my sincere thanks to my family for being a constant source of inspiration and encouragement.

Editor

Optimization of the level of guar gum in low fat yak milk paneer

G. Kandeepan* and S. Sangma

National Research Center on Yak (ICAR), Dirang, West Kameng District, Arunachal Pradesh 790101, India.

Growing market for health promoting food products offers scope to develop dietary fiber enhanced low fat paneer. Therefore, a study was undertaken to compare sensory quality attributes of full fat, low fat and dietary fiber enhanced low fat paneer prepared from yak milk. Optimization of fat to 1% level significantly increased density but decreased casein in yak milk. Yield was significantly reduced in low fat fiber enhanced yak milk paneer. A well acceptable low fat paneer with improved body and texture can be prepared with 1% fat in yak milk. Guar gum as dietary fiber can be successfully incorporated in low fat yak milk paneer at the level of 0.15% with improved body and texture and juiciness.

Key words: Yak, milk, full fat, low fat, guar gum, dietary fiber, paneer, sensory quality.

INTRODUCTION

Yak *(Poephagus grunniens)* is a unique bovine species living at very high altitude of 3000 - 6000 m above mean sea level. It is well adapted to extreme cold, low atmospheric oxygen concentration, high solar radiation and poor feed resources. Despite the challenging living conditions, yak serves as the sole source of income to the yak herdsmen by means of its valuable products (Kandeepan and Sangma, 2009).

Yak milk is a highly nutritious product rich in fat, essential minerals and healthy poly unsaturated fatty acids like conjugated linoleic acid and omega 3 fatty acids (Neupaney et al., 2003; Or-Rashid et al., 2008; Peng et al., 2008). The composition of yak milk includes around 8.50% fat, 4.25% protein, 18.77% total solids and 0.87% total ash (Mondal and Pal, 1996; Sharma et al., 2006). Yak milk is creamy white in colour, having thick consistency and peculiar wet yak hair odour or sweat odour which is more prevalent in winter season. Preparation of paneer, an acid coagulated milk product, from yak milk is an excellent way to overcome this odour.

Traditionally, paneer is prepared from full fat milk resulting in rich flavour, soft consistency and lower sliceability. In India, around 3% of the total milk produced is processed into paneer (Aneja et al., 2002). Paneer is an ubiquitous product in India, having great demand among the consumers in national and international market. Paneer from yak milk needs special attention in this regard with emphasis on growing concerns of health related problems arising from the consumption of fat rich milk products. It is often blamed for its high content of fat and lack of dietary fiber, a much needed component for a healthy food. Consumption of fat rich food products increases the risk of arterial hypertension, coronary heart disease and obesity. Several clinical investigations have proved that dietary fibers have the potential to reduce blood low density lipoprotein cholesterol by 2 mg/dL (Brown et al., 1999), risk of diabetes mellitus type 2 (Willet et al., 2002), coronary heart disease (Bazzano et al., 2003), blood pressure by 2.8/1.1 mmHg (Streppel et al., 2005), obesity (Liu, 2003) and colorectal cancer (Schatzkin et al., 2007).

Guar gum, also called guaran, is a polysaccharide composed of galactose and mannose sugars. It is primarily the ground endosperm of guar beans. The guar seeds are dehusked, milled and screened to obtain the guar gum. It is typically produced as a free flowing, pale, off-white colored, coarse to fine ground powder. Guar gum is a viscous, soluble dietary fiber having better emulsifying, thickening and stabilizing properties which can also prevent the growth of ice crystals in frozen foods (Chaplin, 2009). It acts as a bulk forming laxative. It is claimed to be effective in promoting regular bowel movements and relieve constipation and chronic related

---

*Corresponding author. E-mail: drkandee@gmail.com.

functional bowel ailments such as diverticulosis, Crohn's disease, colitis and irritable bowel syndrome (Anonymous, 2009). The increased mass in the intestines stimulates the movement of waste and toxins from the bowel and colon. Guar gum dissolves in water and absorbs it to become a gelatinous, viscous substance which is fermented by bacteria in the colon into gases and physiologically active byproducts. Because it is soluble, guar gum is able to absorb toxic substances released by bacteria that cause infective diarrhea.

Guar gum is used in dairy to thicken the milk, yogurt, kefir and processed cheese products. It helps to maintain homogeneity and texture of ice creams and sherbets. But there is no evidence of any research intervention on the value addition of paneer from yak milk with dietary fiber for promoting it as healthy food. Poorest of the poor people, the yak herdsmen would also fetch a higher price for his product, if there is a shift in processing from full fat paneer to low fat paneer incorporated with dietary fiber. This would add variety to the category of milk products prepared from yak milk, offering wide choice of food available for the consumers in order to select according to their health status. But till date, there is no information available on the development and quality evaluation of dietary fiber incorporated low fat yak milk paneer. Therefore, an investigation was undertaken with an objective to optimize the level of guar gum to incorporate in low fat yak milk paneer.

MATERIALS AND METHODS

Milk collection

Yak milk was obtained from the Institute's farm of National Research Centre on Yak, Nyukmadung, Arunachal Pradesh, India. The milk was transported to the processing unit in aluminium cans. The filtered clean fresh milk was utilized for the experiment.

Cream separation

The cream of the yak milk was separated by using cream separator (Lakshmi Ball Bearing Cream Separator, RS-11, Chadha Electro Industries, Delhi, India). The discs were placed into the bowl spindle and the bowl was assembled. The bowl was placed onto the separator machine and covered with cream and skim milk outlets and float device. The milk to be separated was weighed and warmed to 40 °C. The representative sample was taken for testing. The machine was started and allowed to attain maximum speed. The separator was flushed with scalding water. After all water had run out, warm milk was put in the supply tank and the flow rate was maintained. The cream and skim milk were collected from their respective outlets into clean and dry containers of known tare weight. When all the milk had been separated, the system was flushed with hot water to rinse the cream spout. The weight of cream and skim milk obtained were recorded and representative samples were drawn for testing.

Standardization of fat percentage

The fat content of the milk for preparing low fat paneer was

optimized to 1% by Pearson square method (Thompkinson and Latha, 2006). A square was drawn and the desired fat required in the mix was placed in the centre. The fat percent of materials to be mixed was placed at the left hand corners of the square. The value in the centre was subtracted from the larger number at the left hand corner of the square. The remainder was placed at the diagonally opposite right hand corner. Similarly the smaller number was subtracted and the remainder was placed diagonally opposite at right hand corner. The numbers on the right hand side represented parts of each of original materials to be blended to make a product of fat percent given in the centre. The numbers at the right hand side were added which represented parts of finished product.

Preparation of guar gum

Gar gum solution was prepared by blending appropriate quantity of guar gum in 250 ml of boiled 1% citric acid solution in a mixer grinder (Khaitan home mate 807 JS mixer grinder, Khaitan Co., Kolkata, India) for 15 s pause 15 s at very low speed.

Preparation of paneer

Full fat paneer was prepared without reducing the level of fat but low fat paneer was prepared from the standardized level of 1% fat in yak milk. The standardized milk was heated to 90 °C for 15 s and cooled to 76 °C. Then 1% citric acid boiled and kept at 70 °C was mixed with the guar gum solution and added to the hot milk. The whole content was allowed to cool to 70 °C. The coagulum settled within 10 min. Then the coagulum was transferred to a muslin cloth or strainer. The whey was allowed to drain and coagulated mass was pressed with weight (about 5 times the weight of coagulum) for 10 min. The blocks were immersed in chilled water at 4 °C for 3 h. Excess water was drained and the product was packaged in low density polyethylene (LDPE) bags and stored at 4 °C.

Quality evaluation

The composition of yak milk was analyzed in a milk analyzer (Milkana Superior, Mayasan. A.S. Istanbul, Turkey). The yield of the paneer was recorded as the total quantity of milk taken upon the weight of paneer produced for each group.

Sensory evaluation

A trained panel was used to study the perceived differences between the different groups of paneer by scoring descriptive profile of sensory attributes. A ten member trained panelists consisting of scientists and research assistants of National Research Centre on Yak (ICAR), Dirang, Arunachal Pradesh, India. Panelists were trained according to guidelines for sensory analysis of milk products and well acquainted with different sensory attributes during their under graduate/post graduate/doctoral programme. They were briefly explained about the nature of the experiment without disclosing the identity of samples.

Samples were warmed (40 - 45 °C) using microwave oven (LG electronics India (P) Ltd., Mumbai) for 1 min and served to the panelists. Each panelist received two cubes from each sample of paneer in a statistically randomized order. Each session included samples from all groups of paneer. Panelists were provided with filtered water to cleanse their pallets between samples. Panelists evaluated samples for appearance, flavour, body and texture and juiciness using eight-point scales (1 = extremely poor to 8 = excellent). Panelists reported scores to the nearest half-point increment. Panelists' scores were averaged for statistical analysis.

Table 1. Comparison of the composition of full fat and low fat yak milk.

Parameters	Yak milk	
	Full fat	**Low fat**
Fat (%)	5.07 ± 0.09^a	1.03 ± 0.18^b
Solids not fat (%)	9.16 ± 0.20	9.39 ± 0.16
Protein (%)	3.47 ± 0.08	3.53 ± 0.06
Casein (%)	2.93 ± 0.04^a	1.31 ± 0.07^b
Density (g/cm^3)	1.0319 ± 0.77^b	1.0342 ± 0.83^a

n = 10 Means with different superscripts in the same row indicate significant difference (p < 0.05).

Table 2. Comparison of yield and sensory quality full fat, low fat and guar gum incorporated paneer from yak milk.

Parameters	Full fat paneer	Low fat paneer	Guar gum incorporated low fat paneer		
			0.10%	**0.15%**	**0.20%**
Yield (%) [#]	21.67 ± 0.56^a	15.94 ± 1.47^b	13.78 ± 0.12^c	13.54 ± 0.14^c	13.27 ± 0.01^c
Sensory attributes [##][*]					
Appearance	7.50 ± 0.06^a	7.25 ± 0.02^b	7.20 ± 0.03^b	7.15 ± 0.04^b	7.00 ± 0.04^c
Flavour	7.50 ± 0.06^a	6.80 ± 0.03^b	6.70 ± 0.03^{bc}	6.66 ± 0.02^{cd}	6.55 ± 0.04^d
Body and texture	6.50 ± 0.05^e	6.80 ± 0.03^c	6.90 ± 0.02^b	7.10 ± 0.04^a	6.60 ± 0.02^d
Juiciness	7.25 ± 0.03^a	6.45 ± 0.03^c	6.50 ± 0.03^c	6.75 ± 0.02^b	6.26 ± 0.03^d

#n = 5, ##n = 50. *Based on 8 point descriptive scale. Means with different superscripts in the same row indicate significant difference (p < 0.05).

Tests were conducted under white fluorescent lights in partitioned booths to isolate panelists.

Statistical analysis

The data were analyzed using SPSS (version 10.0 for Windows; SPSS, Chicago, Ill., U.S.A.) with randomized block design (Snedecor and Cochran, 1995). The data were subjected to analysis of variance, least significant difference and Duncan's multiple range test to compare the means to find the difference between groups. The smallest difference ($D_{5\%}$) for two means to be significantly different was p < 0.05.

RESULTS AND DISCUSSION

The result of the composition of full fat and low fat yak milk is shown in Table 1. It indicated that reducing the level of fat significantly (p < 0.05) reduces the level of casein in the yak milk. Optimization of the level of fat to 1% in yak milk significantly increased the density of the milk compared to the full fat yak milk. Decreasing the level of fat increased the level of total solids which contributed to the increase in density of the milk.

The result of the yield and sensory characteristics of full fat, low fat and guar gum incorporated paneer from yak milk is presented in Table 2. The physicochemical attributes of low fat paneer was significantly (p < 0.05) distinct compared to the full fat yak milk paneer

(Kandeepan et al., 2009). The yield of paneer was significantly (p < 0.05) different between full fat paneer, low fat paneer and guar gum incorporated low fat paneer. The yield did not differ significantly between the different levels of guar gum incorporation in low fat paneer. Decreasing the level of fat probably reduces the level of fat globules available for the casein molecule to bind during the coagulation by the addition of citric acid to the hot milk.

The low fat paneer from yak milk improved the body and texture of the product which was appreciated by the scores reported by sensory panel, as shown in Table 2. The improved body and texture scores offered very good sliceability to the paneer. However, the scores reported for the appearance, flavour and juiciness were significantly (p < 0.05) lower than the sensory scores of full fat paneer (Kandeepan et al., 2009). The decrease in appearance, flavour and juiciness scores was attributed to the lower level of fat present in the paneer. The appearance scores of low fat paneer did not differ significantly up to 0.15% incorporation of guar gum. Although, flavour scores differed significantly (p < 0.05) with the addition of guar gum in low fat paneer, the level of addition was significantly (p < 0.05) acceptable up to 0.15%. Incorporation of guar gum as dietary fiber significantly (p < 0.05) improved the body and texture of the low fat paneer, 0.15% level scoring highly significant value. This improved body and texture might be attributed to the better thickening property of the guar gum. The juiciness of low fat paneer was significantly (p < 0.05) improved with 0.15% level of addition of guar gum in low fat paneer.

Conclusion

Reducing the level of fat significantly increased the density but decreased the casein content in yak milk. A well acceptable low fat paneer can be prepared with 1% level of fat in yak milk. Low fat content of the yak milk significantly reduces the yield. Incorporation of guar gum in low fat yak milk paneer significantly improves the body and texture but reduces the flavour. Guar gum as dietary fiber can be successfully incorporated in low fat paneer at the level of 0.15% with improved body and texture and juiciness of low fat yak milk paneer.

ACKNOWLEDGMENT

The authors thank the Director, National Research Centre on Yak (ICAR), Dirang, for providing necessary facilities to accomplish this research.

REFERENCES

Aneja RP, Mathur BN, Chandan RC, Banerjee AK (2002). Technology of Indian Milk Products, Delhi, India: A Dairy India Publication. pp. 51-59.

Anonymous (2009). Wikipedia. http://en.wikipedia.org/wiki/Guar_gum.

Bazzano LA, He J, Ogden LG, Loria CM, Whelton PK (2003). Dietary fiber intake and reduced risk of coronary heart disease in US men and women: the National Health and Nutrition Examination Survey I Epidemiologic Follow-up Study. Arch. Int. Med. 163(16): 1897-1904.

Brown L, Rosner B, Willet WW, Sacks FM (1999). Cholesterol-lowering effects of dietary fiber: a meta-analysis. Am. J. Clin. Nutr. 69(1): 30-42.

Chaplin M (2009). Water Structure and Science: Guar Gum. London: London South Bank University website. http://www.lsbu.ac.uk/water/hydro.html.

Kandeepan G, Sangma S (2009). Technologies for processing traditional yak milk products. Beverage Food World. 36(8): 47-48.

Kandeepan G, Sangma S, Bhattacharya M, Ghosh MK, Konwar P, Ormilla K, Krishnan S, Jayakumar S, Chouhan VS, Ramesha KP (2009). Quality characteristics of traditional yak milk paneer. In: Programme and compendium of invited papers and abstracts, International conference on yak husbandry challenges and strategies, April 20-22, Dirang, Arunachal Pradesh, India: National Research Centre on Yak (ICAR). p. 164.

Liu S (2003). Relation between changes in intakes of dietary fiber and grain products and changes in weight and development of obesity among middle-aged women. Am. J. Clin. Nutr. 78(5): 920-927.

Mondal D, Pal RN (1996) Chemical composition of yak milk. Ind. J. Dairy Sci. XLIX: 12.

Neupaney D, Jin Bo Kim, Ishioroshi M, Samejima K (2003). Study on some functional and compositional properties of yak butter lipid. Anim. Sci. J. 74(5): 391-397.

Or-Rashid MM, Odongo NE, Subedi B, Karki P, McBride BW (2008). Fatty acid composition of yak (Bos grunniens) cheese including conjugated linoleic acid and trans-18:1 fatty acids. J. Agri. Food Chem. 56: 1654-1660.

Peng YS, Brown MA, Wu, JP, Wei LX, Wu JL, Sanbei DZ (2008). Fatty acid profile in milk fat from qinghai plateau yak at different altitudes and parities. . Anim. Sci. October 1, 2008.

Schatzkin A, Mouw T, Park Y, Subar AF, Kipnis V, Hollenbeck A, Leitzmann MF, Thompson FE (2007). Dietary fiber and whole-grain consumption in relation to colorectal cancer in the NIH-AARP Diet and Health Study. Am. J. Clin. Nutr. 85 (5): 1353-1360.

Sharma DK, Ghosh K, Raquib M, Bhattacharya M (2006). Yak products' profile: an overview. J. Food Sci. Technol. 43(5): 447-452.

Snedecor GW, Cochran WG (1995). Statistical Methods, 8th edn. New Delhi: Oxford and IBH Publishing Co.

Streppel MT, Arends LR, Veer PV, Grobbee DE, Geleijnse JM (2005). Dietary Fiber and Blood Pressure .A Meta-analysis of Randomized Placebo-Controlled Trials. Arch. Int. Med. 165: 150-156.

Thompkinson DK, Latha Sabikhi (2006). Laboratory Manual Market Milk. Karnal, India: National Dairy Research Institute. pp. 37-39.

Willet W, Manson J, Liu S (2002). Glycemic index, glycemic load, and risk of type 2 diabetes. Am. J. Clin. Nutr. 76: 274S-280.

"Effect of black bean (*Vigna mungo*) on the quality characteristics of oven-roasted chicken *seekh kababs*"

Z. F. Bhat[1]* and V. Pathak[2]

[1]Division of Livestock Products Technology Faculty of Veterinary Sciences and Animal Husbandry, Sher-e-Kashmir University of Agricultural Sciences and Technology of Jammu, R. S. Pura, Jammu, Jammu and Kashmir –181 102, India.
[2]Division of Livestock Products Technology UP Pt. Deen Dayal Upadhyaya Veterinary Science University Mathura (UP) -281001, India.

The study was aimed at optimizing the basic formulation and processing conditions for the preparation of chicken *seekh kababs* from spent hens meat by oven roasting method of cooking and their extension with black bean paste (hydrated 1:1 w/w). Three levels of black bean paste viz. 10, 15 and 20% were used as extender replacing lean meat in the formulation. The chicken *seekh kababs* formulated without black bean served as control and were compared with *kababs* extended with different levels of black bean for various physicochemical and sensory properties. pH, emulsion stability and cooking yield increased significantly ($P < 0.05$) with increase in the extension level with highest value for the *kababs* extended with 20% black bean. All the proximate parameters that is, moisture, protein and fat percent except ash content decreased significantly ($P < 0.05$) with the increasing extension level for both raw as well as cooked *kababs*. Scores for all the sensory parameters except for appearance decreased significantly ($P < 0.05$) with increase in the level of extension. All the sensory parameters of the *kababs* extended with 10% black bean were comparable with control. Sensory scores and physicochemical properties indicated that 10% black bean paste was optimum extension level for the formulation of extended chicken *seekh kababs*.

Key words: *Seekh kababs*, *vanaraja*, black bean, oven-roasting, quality parameters.

INTRODUCTION

Seekh kababs are one of the popular convenient ready to eat meat products usually prepared from lamb and beef, though particular styles of *seekh kababs* can be made from meat of other animals like chicken, fish etc. These can be prepared in versatile forms suitable to different food patterns and cooking styles with several types of seasonings and flavourings. These are easy to prepare, take least time for cooking, have a unique flavour and taste, and can be served to a large number of customers in a relatively short period of time. Thus, *seekh kababs* come under the category of fast foods. But the cost of these products is high to be affordable to all the sections of society. Thus, its economic formulation keeping the sensory attributes to acceptable limit is a challenge and as such the addition of low cost non-meat proteins (like black bean) to stretch the availability of *seekh kababs* is an important research area.

The increased concern for nutritional security of common mass demands a holistic approach to stretch the availability of quality protein sources by reducing the cost of formulated products. Poultry industry, a vibrant, organized and scientific sector now days, can play a key role in ensuring quality animal proteins at cheaper rate particularly through culled and spent hen meat. Processing of meat from spent hen to different value added products open the avenues for not only its judicious utilization but a readily accessible animal

---

protein sources for poor. Furthermore, the raising cost of broiler and mutton coupled with increased availability of spent hens has increased the development of meat products based on low cost meats and meat replacers.

Spent hens are old and culled chickens, which have completed their productive and reproductive phase of life and are considered as byproduct of egg industry (Mahapatra, 1992). Their meat is considered poor because of higher toughness and less juiciness which are due to high collagen content (Abe et al., 1996) and high degrees of cross linkages (Wenhen et al., 1973; Bailey, 1984) as compared to broiler meat. Problem of poor utilization can be resolved by development of further processed convenience products (Kondaiah, 1990; Chowdhuri et al., 1992) such as sausages, patties, kababs, rolls, steaks, nuggets, blocks etc. But the products prepared from spent hen meat have comparatively poor sensory properties and lower yield. The emulsion with inferior emulsifying capacity due to high proportion of connective tissue and less salt soluble proteins (Huspeth and May, 1969), higher cooking loss because of high fat content and poor water binding capacity (Acton and Dick, 1978; Buyck et al., 1982), low emulsion stability due to low concentration of salt soluble proteins (Hargus et al., 1970) are the shortcomings of using spent hens meat that can be overcome by suitable food additives or extenders like starch and milk proteins (Chung et al., 1989; Tarte et al., 1989).

Non-meat proteins from a variety of plant sources such as soy proteins (Gujral et al., 2002; Pietrasik and Duda, 2000; Porcella et al., 2001), buck wheat protein (Bejasano and Corke, 1998), samh flour (Elgasim and Al-wesali, 2000), common bean flour (Dzudie et al., 2002) and bengal gram and green gram (Modi et al., 2003; Bhat and Pathak, 2009) and corn flour (Serdarouglu and Degirmencioglu, 2004) have been used as binders and extenders in comminuted meat products. Stability, yield, textural palatability and cost of meat products are the major criteria for non-meat proteins (Roberts, 1974). Legumes provide energy, proteins, minerals, vitamins and the most important that is, dietary fibre required for human health. Several studies have proved that dietary fibers have the potential to reduce blood low density lipoprotein cholesterol (Brown et al., 1999), risk of diabetes mellitus type 2 (Willet et al., 2002), coronary heart disease (Bazzano et al., 2003), blood pressure (Streppel et al., 2005), obesity (Liu, 2003) and colorectal cancer (Schatzkin et al., 2007). Inclusion of legumes in daily diet has many physiological effects in controlling and preventing various metabolic diseases such as coronary heart disease and colon cancer (Tharanathan and Mahadevamma, 2003).

A very few workers have attempted the still inconclusive studies on the legumes as extenders in seekh kababs. Thus, the present study was envisaged to evaluate the effects of black bean on the quality characteristics of chicken seekh kababs.

MATERIALS AND METHODS

Source of chicken meat

Vanaraja birds (irrespective of sex) of the age group of over 80 weeks were purchased from State Animal Husbandry Department. The birds were slaughtered using ritual Halal method. The body fat was trimmed and deboning of dressed chicken was done manually removing all tendons and separable connective tissue. The lean meat was packed in polythene bags and frozen at −20°C until use.

Condiments and refined wheat flour

Onion, garlic and ginger in the ratio of 3:2:1 were ground in a mixer to the consistency of fine paste. Refined wheat flour was purchased from local market and used.

Spice mixture

The spice mix formula used by Kumar and Sharma (2005) was followed and is presented in Table 1. The spices were purchased from local market. After removal of extraneous matter, all spices were dried in an oven at 50°C for overnight and then ground in grinder to powder. The coarse particles were removed using a sieve (100 mesh) and the fine powdered spices were mixed in required proportion to obtain spice mixture for chicken kababs. The spice mixture was stored in plastic airtight container for subsequent use.

Extender

Black bean also referred to as urad dal or bean, black matpe bean, black lentil, or white lentil (Vigna mungo), is a bean grown in southern Asia. The black bean was obtained from local market and converted to paste form after overnight soaking (1:1 w/w hydration). The paste of pulse was incorporated at 10, 15 and 20% levels in the formulation replacing lean meat.

Fat

Refined cottonseed oil of brand name 'Ginni' was purchased from local market and used. The approximate composition of the oil is presented in Table 2.

Preparation of seekh-kababs from meat of spent hen

Lean meat from spent hen was cut into smaller chunks and minced in a Sirman mincer (MOD-TC 32 R10 U.P. INOX, MARSANGO, ITALY) with 6 mm plate followed by common grind size, the 4 mm plate. The common salt, vegetable oil, refined wheat flour (maida), nitrite, sodium tripolyphosphate, spice mixture and condiment mixture were added to weighed meat according to formulation. Meat emulsion for chicken kababs was prepared in Sirman Bowl Chopper (MOD C 15 2.8G 4.0 HP, MARSANGO, ITALY). Minced meat was blended with salt, sodium tripolyphosphate and sodium nitrite for 1.5 min. Water in the form of crushed ice was added and blending continued for 1 min. This was followed by addition of refined vegetable oil and blended for another 1 to 2 min. This was followed by addition of spice mixture, condiments and other ingredients and again mixed for 1.5 to 2 min to get the desired emulsion. Aliquots of raw emulsions from various treatments under each trial of an experiment were collected in plastic bottles for analysis. The various ingredient used in the formulation of the chicken seekh-kababs are presented in the Table 3.

Table 1. Composition of spice mixture.

Ingredient	Percent (%)
Aniseed	12
Bay leaves	2
Black pepper	12
Cardamom	5
Cinnamon	5
Cloves	2
Colored chilli	1
Coriander	20
Cumin seed	15
Mace	2
Nutmeg	2
Red chilli	12
Thymol	10

Table 2. Composition of refined cottonseed oil (per 100 g).

Ingredients	Percent (w/w)
Energy	900 k. cal
Carbohydrate	0
Proteins	0 g
Cholesterol	0 g
Saturated fatty acids	24 g
Mono-unsaturated fatty acids	22 g
Poly-unsaturated fatty acids	54 g
Trans fatty acids	0 g

Above values are approximate.

Table 3. Formulation of *kababs* from meat of spent hens.

Ingredients	Percent (w/w)
Lean meat	67.7
Added water	10.0
Vegetable oil	9.0
Condiment mixture	5.0
Refined wheat flour	4.0
Spice mixture	2.0
Table salt	1.5
Monosodium Glutamate	0.5
Sodium Tripolyphosphate	0.3
Sodium nitrite	120 ppm

Moulding of *seekh kababs*

Kababs were moulded on steel skewers. The steel skewers of 10 mm diameter and of length sufficient to fit in the hot air oven were used for the purpose and typically, the skewer was a round rod with one end pointed and the other blunt. Holding the skewer in one hand, an accurately weighed quantity (60 g) of meat mix/emulsion, in the form of a ball, was taken in the other hand, pierced through the pointed end and pressed on to middle of the skewer. With the help of moistened palm and fingers, it was gently spread evenly and moulded into a cigar shaped *kabab*. The length of the *kabab* was determined by the graduated scale and averaged 18 cm.

Cooking of *seekh kababs*

Skewers with raw *kababs* on them were placed longitudinally on the two edges of a perforated oven tray in a convection oven (YORCO SALES PVT. LTD. INDIA, MODEL-YS1-431, S.NO. 02B2843). The molded raw *kababs* were smeared with vegetable oil and cooked in a preheated hot air oven at 180 ± 2°C for a total time of about 12 min. The internal temperature of *kababs* was monitored by a thermometer and cooked to an internal temperature of 78 ± 2°C. The *kababs* were removed from the skewers, cooled to room temperature and weighed. Pooled sample of each treatment was assigned for analysis.

Analytical procedures

The pH of raw mix/emulsion soon after its preparation and cooked *kababs* was determined by the method of Keller et al. (1974) using a digital meter (SYSTRONICS DIGITAL pH METER 803, SERIAL NO. 603). Emulsion stability of meat emulsion was determined as per procedure described by Townsend et al. (1968). Proximate composition viz. moisture, fat, ash and crude protein content of chicken *kababs*, raw and cooked were determined by standard methods described by AOAC, 1995.

Cooking yield

The weight of each *kabab* was recorded before and after cooking. The cooking yield was calculated and expressed as percentage by a formula:

Cooking yield percent = (Weight of cooked *kababs* x 100) / Weight of raw *kababs*

Sensory evaluation

The sensory evaluation of the product was carried for attributes, namely appearance, flavour, juiciness, texture and the overall acceptability of samples by a panel of trained members composed of scientists and research scholars of the 'division' based on an 8-point hedonic scale, wherein 8 denoted "extremely desirable" and 1 denoted "extremely undesirable" (Seman et al., 1987). The panels were trained for four basic tastes, that is recognition and threshold test and hedonic tests routinely performed in the 'division'. Panelists were seated in a room free of noise and odours and suitably illuminated. Coded samples for sensory evaluation were prepared and served warm to panelists. Water was provided for oral rinsing between the samples.

Statistical analysis

Means and standard errors were calculated for different parameters. Factorial design of experiment was fallowed. Analysis of variance was performed as per Snedecor and Cochran (1980). In significant effects, least significant differences were calculated at

Table 4. Effect of black bean paste extension on pH, emulsion stability and proximate composition of raw chicken *seekh-kababs*. (Mean ±SE)[*]

Parameters	Levels of black bean extension (%)			
	0	10	15	20
pH	6.09[a] ± 0.01	6.14[ab] ± 0.03	6.16[ab] ± 0.05	6.19[b] ± 0.01
Emulsion stability (%)	86.61[a] ± 0.29	91.32[b] ± 0.28	92.83[c] ± 0.31	93.41[c] ± 0.28
Moisture (%)	63.25[a] ± 0.37	62.63[ab] ± 0.25	61.92[ab] ± 0.67	61.36[b] ± 0.31
Protein (%)	15.62[a] ± 0.52	14.87[ab] ± 0.24	14.04[bc] ± 0.13	13.51[c] ± 0.50
Fat (%)	14.10[a] ± 0.10	13.56[ab] ± 0.21	13.20[bc] ± 0.29	12.61[c] ± 0.37
Ash (%)	2.19[a] ± 0.02	2.51[b] ± 0.04	2.79[c] ± 0.04	2.86[c] ± 0.03
Moisture: Protein	4.07[a] ± 0.15	4.22[ab] ± 0.10	4.40[ab] ± 0.04	4.58[b] ± 0.16

*Mean ±SE with different superscripts in a row differs significantly (P < 0.05), n = 6 for each treatment.

appropriate level of significance for a pair wise comparison of treatment means.

RESULTS

The mean values of various parameters namely pH, emulsion stability and proximate composition of raw *seekh-kababs* from meat of spent hens extended with 0, 10, 15 and 20% levels of hydrated black bean (1:1 w/w) paste are presented in Table 4.

pH and emulsion stability of raw chicken *seekh kababs*

The mean values of pH of the raw *seekh kababs* ranged from 6.09 to 6.19. An increase in pH with increase in the level of extension was recorded which was evident from significant higher value (P < 0.05) at 20% level (6.19) of extension as compared to control (6.09) while slightly higher values were observed at 10% (6.14) and 15% (6.16) levels which were comparable to the control and 20% levels.

The mean values of the emulsion stability ranged from 86.61 to 93.41%. A significantly higher (P < 0.05) value for emulsion stability was recorded at all extension levels as compared to control whereas at 15% (92.83%) extension level it was comparable to 20% (93.41%) level. The mean value of emulsion stability for control *kababs* was recorded as 86.61%.

Proximate composition of raw chicken *seekh kababs*

The mean values of the moisture percentage of raw *seekh kababs* ranged from 61.36 to 63.25%. A gradual decrease in moisture was recorded and was significantly low (P < 0.05) at 20% (61.36%) level as compared to control, whereas at 10 (62.63%) and 15% (61.92%) extension levels it was comparable to others.

The mean values of the protein percentage of raw *seekh kababs* ranged from 13.51 to 15.62%. Protein percentage showed a significant decrease (P < 0.05) in alternate succession with highest value for control (15.62%) and lowest for *kababs* extended with 20% (13.51%) black bean paste.

The mean values of the fat percentage of raw *seekh kababs* ranged from 12.61 to 14.10%. Fat percentage also showed a significant decrease in alternate succession with highest value for control (14.10%) and lowest for *kababs* extended with 20% (12.61%) black bean paste.

The mean values of the ash percentage of raw *seekh kababs* ranged from 2.19 to 2.86%. Ash content of the *kababs* increased significantly (P < 0.05) with increase in levels of black bean in the formulation with highest value for *kababs* extended with 20% (2.86%) black bean paste and lowest for control (2.19%). However, at 15% (2.79%) it was comparable to 20% level.

Moisture to protein ratio showed an increase along with level of extension which was evident from significant higher value (P < 0.05) at 20% level of extension as compared to control while slightly higher values were observed at 10 and 15% levels which were comparable to the control and 20% levels.

Oven roasted chicken *seekh kababs*

The mean values of various parameters namely pH, cooking yield and proximate composition of cooked *seekh kababs* from meat of spent hens extended with 0, 10, 15 and 20% levels of hydrated black bean paste (1:1 w/w) are presented in Table 5.

pH and cooking yield of oven roasted chicken *seekh kababs*

The mean values of pH of the oven roasted *seekh kababs* ranged from 6.14 to 6.30. A significant (P < 0.05)

Table 5. Effect of black bean paste extension on physicochemical properties of oven roasted chicken *seekh-kababs*. (Mean ±SE)*.

Parameters	Levels of black bean extension (%)			
	0	10	15	20
pH	$6.14^a \pm 0.01$	$6.23^{ab} \pm 0.03$	$6.27^b \pm 0.02$	$6.30^b \pm 0.08$
Cooking yield (%)	$82.73^a \pm 0.63$	$85.69^b \pm 0.61$	$87.68^c \pm 0.52$	$88.90^c \pm 0.60$
Moisture (%)	$58.89^a \pm 0.76$	$58.83^a \pm 0.47$	$57.89^{ab} \pm 0.57$	$56.82^b \pm 0.54$
Protein (%)	$19.21^a \pm 0.55$	$18.59^{ab} \pm 0.62$	$17.46^b \pm 0.46$	$15.95^c \pm 0.15$
Fat (%)	$14.58^a \pm 0.46$	$14.13^{ab} \pm 0.50$	$13.55^{ab} \pm 0.40$	$12.99^b \pm 0.13$
Ash (%)	$2.31^a \pm 0.05$	$2.65^b \pm 0.06$	$2.88^c \pm 0.03$	$2.99^c \pm 0.06$
Moisture: Protein	$3.07^a \pm 0.08$	$3.19^a \pm 0.15$	$3.33^{ab} \pm 0.09$	$3.56^b \pm 0.05$

*Mean ±SE with different superscripts in a row differs significantly ($P < 0.05$), n (Physicochemical parameters) = 6.

influence on pH and emulsion stability was recorded. The pH of cooked *kababs* at 15% (6.27) and 20% (6.30) extension levels were significantly higher ($P < 0.05$) than control (6.14), whereas at 10% level (6.23) pH value was comparable to all.

The mean values of cooking yield of the oven roasted chicken *seekh kababs* ranged from 82.73 to 88.90%. A significantly higher ($P < 0.05$) value for cooking yield was recorded at all extension levels as compared to control, however at 15% (87.68%) extension level it was comparable to 20% level (88.90%).

Proximate composition of oven roasted chicken *seekh kababs*

The mean values of the moisture percentage of oven roasted *seekh kababs* ranged from 56.82 to 58.89%. A gradual decrease in moisture was recorded with increase in levels of lean replacement and was significantly low ($P < 0.05$) at 20% level (56.82%) as compared to control (58.89%), whereas at 15% (57.89%) extension level it was comparable to all others.

The mean values of the protein percentage of oven roasted *seekh kababs* ranged from 15.95 to 19.21%. Protein percentage of *kababs* showed a gradual decrease with increase in level of black bean paste and was significantly low ($P < 0.05$) at 15 and 20% level as compared to control, whereas at 10% (18.59%) extension level it was comparable to control (19.21%) and 15% (17.46%).

The mean values of the fat percentage of oven roasted *seekh kababs* ranged from 12.99 to 14.58%. Fat percent showed a significant decline ($P < 0.05$) at 20% (12.99%) extension level as compared to control (14.58%) while the other variants were comparable with both control and 20%.

The mean values of the ash percentage of oven roasted *seekh kababs* ranged from 2.31 to 2.99%. Ash percent showed a significant increase ($P < 0.05$) at all

extension levels as compared to control whereas at 15% level (2.88%) it was comparable to 20% level (2.99%).

Moisture to protein ratio showed an increase along with level of extension which was evident from significant higher value ($P < 0.05$) at 20% level of extension as compared to control. Slightly higher values were observed at 10 and 15% levels which were comparable to the control whereas at 15% level it was comparable to all.

Sensory attributes of oven roasted chicken *seekh kababs*

The mean values of various sensory parameters of oven roasted *seekh kababs* from meat of spent hens extended with 0, 10, 15 and 20% levels of hydrated black bean paste (1:1 w/w) are presented in Table 6.

Table 6 revealed that extension with hydrated black bean paste had a significant influence ($P < 0.05$) on flavour, juiciness, texture and overall acceptability. The mean values of appearance scores of oven roasted *seekh kababs* from meat of spent hens ranged from 6.85 to 7.09. The appearance scores showed a declined trend with increase in percent extension level, though the decline was non-significant.

Flavour, juiciness and overall acceptability of *kababs* showed similar pattern. The scores were significantly lower ($P < 0.05$) at 20% extension level (6.38, 6.42 and 6.38) as compared to control (6.92, 6.92 and 7.04) and 10% level (6.71, 6.78 and 6.80); whereas these were comparable to scores at 15% level (6.59, 6.54 and 6.64). The scores at 10% extension were comparable to control while at 15% level these were comparable to 10% as well as 20% levels.

The mean values of texture scores of the oven roasted *seekh kababs* from meat of spent hens ranged from 6.33 to 6.97. The texture scores were significantly lower ($P < 0.05$) at 15 (6.54) and 20% (6.33) levels as compared to control (6.97) while at 10% level (6.64) it was comparable

Table 6. Effect of black bean paste extension on sensory attributes of oven roasted chicken *seekh-kababs*. (Mean ±SE)*.

Sensory attributes	Levels of black bean extension (%)			
	0	10	15	20
Appearance	7.09 ± 0.11	6.95 ± 0.07	6.92 ± 0.11	6.85 ± 0.10
Flavour	6.92[a] ± 0.09	6.71[ab] ± 0.10	6.59[bc] ± 0.09	6.38[c] ± 0.13
Juiciness	6.92[a] ± 0.08	6.78[ab] ± 0.09	6.54[bc] ± 0.09	6.42[c] ± 0.12
Texture	6.97[a] ± 0.11	6.64[ab] ± 0.13	6.54[b] ± 0.11	6.33[b] ± 0.16
Overall acceptability	7.04[a] ± 0.10	6.80[ab] ± 0.13	6.64[bc] ± 0.10	6.38[c] ± 0.14

[a]Mean ±SE with different superscripts in a row differs significantly (P < 0.05). Mean values are scores on 8 point descriptive scale where 1- extremely poor and 8- extremely desirable, n = 21 for each treatment.

to control and 15% level.

DISCUSSION

pH and emulsion stability of raw chicken *seekh kababs*

An increase in pH along with level of extension was recorded which was evident from significant (P < 0.05) higher value at 20% level of extension as compared to control while slightly higher values were observed at 10 and 15% levels which were comparable to the control and 20% levels. Gradual increase in pH with increase in extension level was expected because of neutral nature of extender and was in agreement with findings of Huang et al. (1996), Prabhakara and Janardhana (2000), Kumar and Sharma (2006) and Bhat and Pathak (2009). A significantly (P < 0.05) higher value for emulsion stability was recorded at all extension levels as compared to control whereas at 15% extension level it was comparable to 20% level. The increase in emulsion stability with increase of extension level could be attributed to gelatinizing property of increasing starch component on heating, which stabilized the emulsion (Comer, 1979). Gelatinization of this starch improves the binding properties of meat proteins (Puolanne and Ruusunen, 1983) and was in agreement with the results reported by Sharma et al. (1998), Bond et al. (2001), Kumar and Sharma (2006) and Bhat and Pathak (2009). Improvement of binding properties in meat products by addition of protein and starches of vegetable origin have also been reported by Price and Schweigert (1971); Sharma et al. (1988); Comer et al. (1986) indicated about the possible interaction between soluble meat and vegetable proteins and stated that fillers appeared to increase fat agglomeration while improving stability. It may also be attributed to high emulsifying ability of black bean paste. Shabakov et al. (1983) reported that pea flour bound both fat and water and formed stable dense foam.

Proximate composition of raw chicken *seekh kababs*

A gradual decrease in moisture was recorded and was

significantly (P < 0.05) low at 20% level as compared to control, whereas at 10 and 15% extension levels it was comparable to others. It may be due to less moisture content in hydrated black bean paste than that of lean chicken meat. Similar findings were reported by Kumar and Sharma (2005, 2006) and Bhat and Pathak (2009) in chicken patties and chicken *seekh kababs* respectively. Protein percentage showed a significant (P < 0.05) decrease in alternate succession. It may be due to lower protein content of hydrated black bean paste than that of chicken meat and is in agreement with the findings of Kumar and Sharma (2005, 2006) and Bhat and Pathak (2009). Fat percentage also showed a significant (P < 0.05) decrease in alternate succession. This may be attributed to the dilution effect caused by incorporation of black bean paste which is particularly low in fat content. Similar observation was reported by Kumar and Sharma (2006). Ash percent showed a significant (P < 0.05) decline at all extension levels as compared to control whereas at 15% level it was comparable to 20% level. It may be attributed to the declining trend of moisture with increasing level of extension, which results into an increasing dry matter content. Similar findings were observed by Bhat and Pathak (2009). Moisture to protein ratio showed an increase along with level of extension which was evident from significant (P < 0.05) higher value at 20% level of extension as compared to control while slightly higher values were observed at 10 and 15% levels which were comparable to the control and 20% levels.

pH and cooking yield of oven roasted chicken *seekh kababs*

A significant (P < 0.05) influence on pH and cooking yield was recorded. The pH of cooked *kababs* at 15 and 20% extension levels were significantly (P < 0.05) higher than control, whereas at 10% level pH value was comparable to all. Gradual increase in pH with increase in extension level was expected because of neutral nature of extender and was in agreement with findings of Huang et al., (1996) and Prabhakara and Janardhana (2000), Kumar and Sharma (2005, 2006) and Bhat and Pathak (2009). A significantly (P < 0.05) higher value for cooking yield was recorded at all extension levels as compared to control whereas at 15% extension level it was comparable to 20% level. Increase in cooking yield along with increase in level of extender was as per the findings of Huang et al. (1996, 1999), Sharma et al. (1998) and Bond et al. (2001), Kumar and Sharma (2005, 2006) and Bhat and Pathak (2009). But here increase was not exactly as a result of increased moisture percentage but due to higher moisture and fat retention of chicken *kababs* extended with black bean paste. This observation is in agreement with Mcwatters (1977) and Mitsyk and Mikhailovskii (1981). Grinding of black bean might have enhanced the absorptivity by increasing surface area, lowering drip losses and increasing cooking yield as suggested by Shanner and Baldwin (1979). Higher yields in comminuted meat products have also been reported by Comer (1979), Padda et al. (1989) and Minerich et al. (1991).

Proximate composition of oven roasted chicken *seekh kababs*

A gradual decrease in moisture was recorded and was significantly (P < 0.05) low at 20% level as compared to control, whereas at 15% extension level it was comparable to others. Similar finding was observed with the incorporation of chickpea paste by Abo-Bakr (1987), Nag (1998), Kumar and Sharma (2005, 2006) and Bhat and Pathak (2009). It may be due to less moisture content in black bean paste than that of lean chicken meat. Protein percentage showed a gradual decrease and was significantly (P < 0.05) low at 15 and 20% level as compared to control, whereas at 10% extension level it was comparable to control and 15%. Similar finding was observed with the incorporation of chickpea paste by Nag (1998), with barley flour and pressed rice flour by Kumar and Sharma (2005, 2006) in chicken patties and with green gram paste by Bhat and Pathak (2009) in chicken *seekh kababs*. The probable reason may be due to lower protein content of black bean paste than that of chicken meat. Fat percent showed a significant (P < 0.05) decline at 20% extension level as compared to control while others had comparable values. Similar finding was observed with the incorporation of chickpea paste by Nag (1998), Kumar and Sharma (2005, 2006) and Bhat and Pathak (2009) and may be attributed to the dilution effect caused by incorporation of black bean paste which is particularly low in fat content. Ash percent showed a significant (P < 0.05) increase at all extension levels as compared to control whereas at 15% level it was comparable to 20% level. This may be attributed to the declining trend of moisture with increasing percentage of extension, resulting to increasing dry matter content with level of extension. Jindal and Bawa (1998) also reported an increased ash content of cooked sausages with

increase in soy flour levels. Similar findings were reported by Nayak and Tanwar (2004), Kumar and Sharma (2005, 2006) and Bhat and Pathak (2009). Moisture to protein ratio showed an increase along with level of extension which was evident from significant (P < 0.05) higher value at 20% level of extension as compared to control. Slightly higher values were observed at 10 and 15% levels which were comparable to the control whereas at 15% level it was comparable to all.

Sensory attributes of oven roasted chicken *seekh kababs*

Extension with black bean paste had a significant (P < 0.05) influence on flavour, juiciness, texture and overall acceptability. The appearance scores showed a declined trend with increase in percent extension level, though the decline was non-significant. Decline in appearance could be attributed to dilution of meat pigment. Zyl and Zayas (1996), Kumar and Sharma (2005, 2006) and Bhat and Pathak (2009) reported similar results. Flavour, juiciness and overall acceptability of *kababs* showed similar pattern. The scores were significantly (P < 0.05) lower at 20% extension level as compared to control and 10% level, whereas these were comparable to scores at 15% level. The scores at 10% extension were comparable to control while at 15% level these were comparable to 10% as well as 20% levels. Flavour score deceased as a result of dilution of meaty flavour with increase in extension level. Padda et al. (1989) also observed a decline in flavour scores of goat meat balls extended with roasted Besan. Kumar and Sharma (2005, 2006) and Bhat and Pathak (2009) also presented similar findings in the flavour of extended meat products. As the moisture retention increased with increase in extension it might have led to less preference by sensory panelists and therefore, a lower score for juiciness. Huang et al. (1999) reported similar results for juiciness in beef patties. The decrease in juiciness with increase in extension level has also been reported by Shaner and Baldwin (1979), Kumar and Sharma (2005, 2006) and Bhat and Pathak (2009). The texture scores were significantly (P < 0.05) lower at 15 and 20% levels as compared to control while at 10% level it was comparable to control and 15% level. The decrease in texture scores at higher levels of extender may be due to replacement of structural meat proteins by extender as reported by Verma et al. (1984). Such a decline in texture was also supported by findings of Huang et al. (1996), Kumar and Sharma (2005, 2006) and Bhat and Pathak (2009). A declining trend in overall acceptability was reflective of change in scores of flavour and texture with increased extension levels and similar findings were reported by Nag (1994), Kumar and Sharma (2005, 2006) and Bhat and Pathak (2009). The sensory of *kababs* for all attributes at 10% extension were quite comparable to control. Hence, 10% extension with black bean paste was taken as optimum extension level for the formulation of extended chicken *seekh kababs*.

Conclusion

The chicken *seekh kababs* from meat of spent hens can be successfully extended with black bean. Three levels viz. 10, 15 and 20% of black bean paste were used for extension of chicken *seekh kababs* and on the basis of analysis of different physicochemical and sensory parameters, 10% level of extension was adjudged as optimum for oven roasting method of cooking. Chicken *seekh kababs* of very good acceptability and nutritive value could be prepared by extension with black bean paste substituting lean chicken meat from spent hen in formulation. There is also a significant reduction in cost of the products developed without compromising with the quality which will ensure the nutritional security of the people in developing countries especially the rural one. Besides ensuring the nutritional security, it will also provide a potential outlet for the poultry industry byproduct (spent hens) and thereby increasing the profit margins of the poultry industry. Further research should be focused on the use of higher amounts and different legumes in *kababs* particularly *seekh kababs*.

REFERENCES

Abe Hirta A, Kimura T, Yamuchi K (1996). Effects of collagen on the toughness of meat from spent laying hens. J. Jpn. Soc. Food Sci. Technol., 43(7): 831-834.

Abo Bakr TM (1987). Nutritional evaluation of sausages containing chick peas and faba beans as meat proteins extenders. Food Chem., 23: 143-150.

Acton JC, Dick RL (1985). Functional properties of raw materials. Meat Ind., 31: 32-36.

AOAC (1995). Official methods of analysis. 16th Edn. Association of official Agricultural Chemists, Washington, DC.

Bailey AJ (1984). The chemistry of intra molecular collagen. The Royal Society of Chemistry, Burlington House. Recent Adv. Chem. Meat. 22-47.

Bazzano LA, He J, Ogden LG, Loria CM, Whelton PK (2003). Dietary fiber intake and reduced risk of coronary heart disease in US men and women: The National Health and Nutrition Examination Survey Epidemiologic Follow-up Study. Arch. Int. Med., 163(16): 1897-1904.

Bejasano FP, Corke H (1998). Amarantus and buck wheat protein concentrate effects on an emulsion-type meat product. Meat Sci., 50(30): 343-353.

Bhat ZF, Pathak V (2009). Effect of mung bean (*Vigna radiata*) on quality characteristics of oven roasted chicken *seekh kababs*. Fleischwirtschaft Int., 6: 58-60.

Bond JM, Marchello MJ, Slanger WD (2001). Physical-chemical and self-life characteristics of low-fat ground beef patties formulated with waxy hull-less barley. J. Muscle Foods, 12(1): 53-69.

Brown L, Rosner B, Willet WW, Sacks FM (1999). Cholesterol-lowering effects of dietary fiber: A meta-analysis. Am. J. Clin. Nutr., 69(1): 30-42.

Buyck MJ, Seideman SC, Quenzer NM, Donnelly LS (1982). Physical and sensory properties of chicken patties made with various levels of fat and skin. J. Food Protect., 45: 214-217.

Chowdhuri J, Kumar S, Keshari RC (1992). Physicochemical and organoleptic properties of chicken patties incorporating texturised soy proteins. In. Proc. National Meet of Food Scientists and Technologists, CFTRI, Mysore (10th April). M.F.P.I., p. 53.

Chung S, Bechtel P, Villota R (1989). Production of meat based intermediate moisture snack foods by twin-screw extrusion. 49th Annual Meeting, Institute of Food Technologists, Chicago, IL. June 25-29.

Comer FW (1979). Functionality of fillers in comminuted meat products. Can. Inst. J. Food Sci.Technol., 12: 157-165.

Comer FW, Chew N, Lovelock L, Allan-Wojtas P (1986). Comminuted meat products: Functional and microstructural effects of fillers and meat ingredients. Can. Inst. J. Food Sci.Technol., 19(2): 68-74.

Dzudie T, Joel Scher J, Hardy J (2002). Common bean flour as an extender in beef sausages. J. Food Eng., 52(2): 143-147.

Elgasim EA, Al-Wesali MS (2000). Water activity and hunter colour values of beef patties extended with samh (*Mesembryanthemum forsskalei* Hochst) flour. Food Chem., 69: 181-185.

Gujral HS, Kaur A, Singh N, Sodhi SN (2002). Effect of liquid whole egg, fat and textured soy protein on the textural and cooking properties of raw and baked patties from goat meat. J. Food Eng., 53: 377-385.

Hargus GL, Froning GW, Mebus CA, Neelakantan S Hartung TE (1970). Effect of processing variables on stability and protein extractability of turkey meat emulsion. J. Food Sci., 35: 682-692.

Huang-Jen-Chieh, Zayas JF, Bowers J (1996). Functional properties of sorghum flour as an extender in ground beef patties. IFT Annual Meeting: Book of abstracts, ISSN 1082-1236, pp. 63-64.

Huang-Jen-Chieh, Zayas JF, Bowers J (1999). Functional properties of Sorghum flour as an extender in ground beef patties. J. Food Qual., 22(1): 51-61.

Huspeth JP, May KN (1969). Emulsifying capacity of salt soluble protein of poultry meat. II. Heart, gizzard, and skin from broilers, turkeys, hens and ducks. Food Technol., 23: 3733-3774.

Jindal V, Bawa AS (1988). Utlization of spent hens and soy flour in the preparation of sausages. Indian J. Meat Sci. Technol., 1: 23-27.

Keller JE, Skelley GC, Acton JC (1974). Effect of meat particle size and casing diameter on summer sausage properties during. J. Milk Food Technol., 37: 297-300.

Kondaiah N (1990). Poultry meat and its place in market. Poultry Guide. 27: 41-45.

Kumar RR, Sharma BD (2005). Evaluation of the efficacy of pressed rice flour as extender in Chicken Patties. Indian J. Poult. Sci., 40(2): 165-168.

Kumar RR, Sharma BD (2006). Efficacy of Barley Flour as Extender in Chicken Patties from Spent hen meat. J. Appl. Anim. Res., (30): 53-55.

Liu S (2003). Relation between changes in intakes of dietary fiber and grain products and changes in weight and development of obesity among middle-aged women. Am. J. Clin. Nutr., 78(5): 920-927.

Mahapatra CM (1992). Poultry Products Technology—Prospects and Problems. Poult. Guide, 29: 69-70.

McWaters KH (1977). Performance of defatted pea nut, soybean and field pea meals as extenders in ground beef patties. J. Food Sci., 42: 1492-1495.

Minerich PL, Addis PB, Epley RJ, Bingham C (1991). Properties of wild rice/ground beef mixtures. J. Food Sci., 56: 1154-1157.

Mitsyk VE, Mikhailovskii VS (1981). Minced meat semi-manufactured products incorporating pea-meal and sunflower seed meal protein. Tovarovedenie, 14: 44-46. (c.f. FSTA, 14: 7 s 1311).

Modi VK, Mahendrakar NS, Narasimha Rao D, Sachindra NM (2003). Quality of buffalo meat burger containing legume flours as binders. Meat Sci., 66: 143-149.

Nag S, Sharma BD, Kumar S (1998). Quality attributes and shelf life of chicken nuggets extended with rice flour. Indian J. Poult. Sci., 33(2): 182-186.

Nayak NK, Tanwar VK (2004). Effect of tofu addition on physico-chemical and storage properties of cooked chicken meat patties. Indian J. Poult. Sci., 39(2): 142-146.

Padda GS, Sharma N, Bisht GS (1989). Effect of some vegetative extenders on organoleptic and physico-chemical properties of goat meat balls. Indian J. Meat Sci. Technol., 2: 116-122.

Pietrasik Z, Duda Z (2000). Effect of fat content soy protein carragenan mix on the quality characteristics of communted scalded sausages.Meat Sci., 56: 181-188.

Porcella MI, Sanchez G, Vaudonga SR, Zanelli ML, Descalzo AM, Meichtri LH (2001). Soy protein isolate added to vacuum packaged chorizos effect on drip loss, quality characteristics and stability during storage. Meat Sci., 57: 437-443.

Prabhakara Reddy K, Janardhana Rao B (2000). Effect of binders and precooking meat on quality of chicken loaves. J. Food Sci. Technol., 37(5): 551-553.

Price JF, Schweighert BS (1971). In: Science of Meat and Meat Products. Freeman and Co., San Franscisco.

Puolanne EJ, Ruusunen M (1983). Effect of potato flour and dried milk on the water binding capacity and consistency of Brushwurst sausages. Fleischwirtschaft (Abst.), 63: 631-633.

Kumar RR, Sharma BD (2005). Evaluation of the efficacy of pressed rice flour as extender in Chicken Patties. Indian J. Poult. Sci., 40(2): 165-168.

Roberts LH (1974). Sausage emulsion–functionality of non-meat proteins during emulsification. In proceedings, Meat Industry Research Conference, American Meat Institute, Washington, DC.

Schatzkin A, Mouw T, Park Y, Subar AF, Kipnis V, Hollenbeck A, Leitzmann MF, Thompson FE (2007). Dietary fiber and whole-grain consumption in relation to colorectal cancer in the NIH-AARP Diet and Health Study. Am. J. Clin. Nutr., 85(5): 1353-1360.

Seman DL, Moody WG, Fox JD, Gay N (1987). Influence of hot and cold deboning on the palatability, textural and economic traits of restructured beef steaks. J. Food Sci., 52: 879-882, 889.

Serdarouglu M, Degirmencioglu O (2004). Effects of fat level (5, 10 and 20%) and corn flour (0, 2 and 4%) on some properties of Turkish type meatballs (koefte). Meat Sci., 68(2): 291-296.

Shabakov MS, Starodubtseva LA, Vakhrameev VK (1983). Some functional properties of proteinaceous pea meal. Pishchevaya Tekhnologiya, (c.f. FSTA, 18: 2 J 146), 6: 40-42.

Shaner KM and Baldwin RE (1979). Sensory properties, proximate analysis and cooking losses of meat loaves extended with chickpea meal or textured soy protein. J. Food Sci., 44: 1191-1193.

Sharma BD, Padda GS, Keshri RC and Sharma N (1988). Acceptability studies of different stuff binders in chevon samosas. Indian J. Sci. Technol., 1: 32-36.

Sharma BD, Kumar S, Nanda PK (1998). Development of low cost high value formulations for convenience meat products utilizing non-conventional extenders. Annual Report, Indian Vety. Research Institute, Izatnagar, Bareilly (U.P.). Sub Project-III, Div of L.P.T., pp. 65-66.

Snedecor GW, Cochran WG (1980). In: Statistical Methods. 7th Edn. Oxford and IBH Publishing Co., Calcutta.

Streppel MT, Arends LR, Veer PV, Grobbee DE, Geleijnse JM (2005). Dietary Fiber and Blood Pressure .A Meta-analysis of Randomized Placebo-Controlled Trials. Arch. Int. Med., 165: 150-156.

Tarte R, Molin RA, Koymazadeh M (1989). Development of beef/corn extruded snack model product. Proc. 50th Annual Meeting, Institute of Food Technologists, Chicago, IL. June 25-29.

Tharanathan RN, Mahadevamma S (2003). Grain legumes-A boon to human nutrition. Trends Food Sci. Technol., 14: 507-518.

Townsend WE, Witnauer LP, Riloff JA, Swift CE (1968). Comminuted meat emulsions. Differential thermal analysis of fat transition. Food Technol., 22: 319-323.

Verma MM, Ldward DA, Lawrie RA (1984). Lipid oxidation and Metmyoglobin formation in sausages containing chickpea flour. Meat Sci., 11: 171-189.

Wenham LN, Fairbain K, McLeod W (1973). Eating quality of mutton compared with lamb and its relationship to freezing practices. J. Anim. Sci., 36: 1081.

Willet W, Manson J, Liu S (2002). Glycemic index, glycemic load, and risk of type 2 diabetes. Am. J. Clin. Nutr., 76: 274S-280S.

Zyl HV, Zayas JF (1996). Effect of three levels of sorghum flour on the quality characteristics of frankfurters. IFT Annual Meeting: Book of Abstracts, ISSN 1082-1236, p. 64.

3

Ethylene and ethephon induced fruit ripening in pear

Dhillon W. S.[1]* and Mahajan B. V. C.[2]

[1]Department of Horticulture, Punjab Agricultural University, Ludhiana, India.
[2]Punjab Horticultural Postharvest Technology Centre, Punjab Agricultural University campus, Ludhiana, India.

The fruits of 'Patharnakh' pear were harvested at physiological maturity and subjected to different treatments of ethephon for proper ripening, both in the solution form (500, 1000 and 1500 ppm) and as gas application (100 ppm). After treatment, the fruits were kept at 20°C and at ambient temperature for 4, 8, 12 and 16 days. The ripening treatments were better at 20°C as compared to ambient temperature. Among different solution treatments of ethephon, colour development and fruit ripening, which is judged on the basis of fruit firmness and chemical composition of the fruit, was better at 1000 ppm. The ethylene gas application was also equally comparable with this treatment. The optimum ripening in fruits and acceptable quality was achieved after 8 days of ripening period in 'Patharnakh' at 20°C temperature with 1000 ppm ethephon solution, and 100 ppm ethylene gas treatments.

Key words: Cultivar, ethephon, ethylene, fruit firmness, Patharnakh.

INTRODUCTION

The consumption of 'Patharnakh' cv. of pear (*Pyrus pyrifolia* (Burm) Nakai) as table fruit is very low due to more gritty cells in pulp and hard texture of the fruits. The hardness in pear fruit is due to low activity of enzymes responsible for degrading cell wall polysaccharides and hydrochloride soluble pectin into sugars and water soluble pectin, respectively (Ning et al., 1997). This hardy nature of fruits and its organoleptic quality can be improved by following suitable ripening techniques. Pear fruits fail to ripen until they are either exposed to a critical period of chilling temperature (Blankenship and Richardson, 1985) or exposed to exogenous application of ethylene enhancing chemicals, which trigger the ripening process. Proper ripening governs the post-harvest dessert quality of pear fruits.

The changes in cell wall composition which accompany the softening of ripening fruit apparently result from the action of enzymes produced by the fruit (Pressey, 1977). Initiation of ripening activities of climacteric fruit is controlled by the threshold level of internal ethylene concentration (Hansen, 1966). Exogenous ethylene application to immature (Lieberman et al., 1977) or mature (Wang and Mellenthin, 1972) 'd Anjou' pears before completion of the cold requirement can induce ripening and softening without involvement of the respiratory climacteric. Ethylene regulates fruit ripening by coordinating the expression of genes that are responsible for a variety of processes, including a rise in respiration, autocatalytic ethylene production and changes in color, texture, aroma and flavor (Oetiker and Yang, 1995). The temperate pears like Gebhard red strain, harvested at commercial maturity with flesh firmness of 64.5 N, did not ripen normally at 20°C even though the chilling requirement had been met by storage at -1°C (Honma et al., 1997). The fruit did not ripen without the exogenous application of ethylene. Their study showed that 3 day treatment with 100 $\mu l/^{-1}$ ethylene readily induced pulp tissue to convert 1-aminocyclopropane 1- carboxylic acid (ACC) to ethylene. ACC synthase activity was induced only by ethylene treatment, and did not increase until the fruit had been transferred to 20°C for 3 days. This problem of incomplete ripening also exists in hard pear varieties grown in sub- tropics. This offer an opportunity to

*Corresponding author. E-mail: wasakhasingh@yahoo.com.

examine the effect of exogenous ethylene as specific ripening characteristics in a mature fruit of hard pear variety cultivated in sub -tropics. The study was conducted to investigate the effects of ethylene gas, and ethephon - an ethylene releasing chemical; ripening temperature (at ambient, and at 20°C), and duration of time, on softening and ripening of 'Patharnakh' pears.

MATERIALS AND METHODS

The fruits of 'Patharnakh' hard pear were picked at physiological maturity in last week of July during 2006 and 2007. The data presented in this paper is based upon the average values of these two years. The fruit samples were collected at random from all the sides including internal and peripheral areas of the tree. The fruits were subjected to different treatments of ethephon for proper ripening, both in solution form (500, 1000 and 1500 ppm), and ethylene (100 ppm) as gas. The fruits were dipped in required concentration for five minutes and then dried under shade. The fruits were exposed to ethylene gas (100 ppm) for 24 h in fruit ripening chamber. After the treatment, the fruits were kept at 20°C and at ambient temperature for 4, 8, 12 and 16 days. The control fruits were also kept at same environments for comparison. Each treatment was replicated three times. The experiment was laid- out in "completely randomized block design" (Sharma, 1998).

The observations on different parameters were recorded after each interval of ripening period. The color of the fruit was recorded/measured visually, as well as with Hunter Color Lab and the results were expressed as L*, a*, b* (Hunter, 1975). The physiological loss in fruit weight (PLW) was calculated on initial weight basis after every ripening interval. Organoleptic evaluation of the fruits was done by five judges on the basis of Hedonic scale (1 to 9 points), on the basis of general appearance, taste and texture (Amerine et al., 1965). A thin layer of 1 cm^2 from two sides at the shoulder end of the fruit was removed to measure the flesh firmness with the help of IRC-FT 327 penetrometer. 8 mm probe of the penetrometer was pushed gently into the flesh and the puncture resistance measured in lbs force. The total soluble solids(TSS) were determined from fresh strained thoroughly stirred juice of fruits on each sampling date with the help of a hand refractometer (Erma made in Japan). The readings were corrected at 20°C and expressed as percentage soluble solids. The volume was made to 100 ml by adding distilled water. Out of this, 10 ml was taken and titrated against 0.1 N sodium hydroxide solution, using phenolphthalein as an indicator. The results were expressed as percentage Malic Acid (AOAC, 2000). The reducing sugars and titratable acidity were determined by the standard procedure (AOAC, 2000).

RESULTS, DISCUSSION AND CONCLUSION

Certain chemicals, especially ethylene, are known to influence coloring of various fruits. The color of the fruit increased gradually and uniformly in case of ethylene 100 ppm and ethephon 1000 ppm treated fruits. The color values recorded with Hunter Color Lab in terms of L*, a*, b* were 55.78, -4.18, 25.76 for control; 58.58, 3.68, 28.47 for ethylene 100 ppm, and 58.00, 3.10, 28.30 for ethephon 1000 ppm, respectively. In the present studies, b values stands for yellow color of the fruit, which clearly

shows that there was dramatic improvement in color with ethephon 1000 ppm and ethylene 100 ppm as gas treatments over control (Table 1). The fruits with these treatments were complete yellow in color after 8 days of ripening at 20°C. However, the color turned out to be deep yellow after 12 days of ripening and dull yellow after 16 days. The color in untreated fruits was light to dull yellow under both ambient and 20°C ripening environments. The change in color during ripening may be due to the synthesis of mainly carotenoids accompanied by the simultaneous loss of chlorophyll (Reyes and Paul, 1995). Exposures of fruits to gas ethylene or ethephon solution have been reported to improve their color and quality during storage and marketing (Kulkarni et al., 2004).

The PLW was significantly higher at ambient temperature as compared to 20°C under all the ripening treatments (Table 2). It also increased significantly after each storage interval from 4 to 16 days under both the ripening environments. After 4 days of ripening period, the lowest PLW was recorded under 1000 ppm ethephon solution treatment at 20°C temperature. However, the high PLW was recorded under 1500 ppm ethephon treatment at ambient temperature. The loss of more than 5% moisture leads to shriveling of fruits and all the treatments showed PLW beyond this limit after 4 days at ambient ripening storage. All the treatments also showed higher PLW more than 5% after 16 days of storage at 20°C temperature, except 500 ppm ethephon treatment which recorded the values very close to it, that is, 4.99%. The PLW was with in permissible limits up to 12 days of ripening storage at 20°C under all the treatments. The loss in weight was less under control treatment as compared to ethephon treatments up to 8 days of storage indicating that the ripening process of the fruits was not started properly. The fruits under ethephon treatments started ripening of fruits immediately after treatment, hence the loss was higher. But, the loss was suddenly on higher side after 12 days of storage in control fruits. Continuous processes of respiration and transpiration have resulted in weight loss during ripening at both the ripening environments. In earlier studies too, the loss in weight of fruit during storage both at ambient and in cold room increased with the enhancement of storage days in pear (Dhillon et al., 2005$_b$).

The organoleptic rating increased up to 8 days of storage under all the treatments at both the ripening environments (Table 3). Thereafter, it starts declining. The organoleptic rating was 7.96, 8.05 and 7.95 after 4 days under 500, 1000 ppm and fogging treatments (100 ppm), which rose to 8.13, 8.18 and 8.10 after 8 days, respectively. The highest score of 8.18 was recorded after 8 days at 20°C ripening treatment at ethephon 1000 ppm. The next best treatment was recorded to be fogging (100 ppm). The organoleptic rating score was only 6.63 and 6.00 after 4 days in control under 20°C and ambient

Table 1. Effect of ethephon and ethylene gas on color development (fruit color) during ripening in pear.

Treatments	Temperature	Storage days				
		0	4	8	12	16
Ethephon 500 ppm	20°C	Green	Green yellow	Green yellow	Yellow	Dull yellow
	Ambient	Green	Green yellow	Green yellow	Yellow	Dull yellow
Ethephon 1000 ppm	20°C	Green	Yellow green	Yellow	Deep yellow	Dull yellow
	Ambient	Green	Green yellow	Yellow	Deep yellow	Dull yellow
Ethephon 1500 ppm	20°C	Green	Yellow green	Dull yellow	Dull yellow	Dull yellow
	Ambient	Green	Yellow green	Yellow	Deep yellow	Dull yellow
Ethylene gas 100 ppm	20°C	Green	Green yellow	Yellow	Deep yellow	Dull yellow
	Ambient	Green	Green yellow	Yellow	Deep yellow	Dull yellow
Control	20°C	Green	Green	Light green	Dull yellow	Dull yellow
	Ambient	Green	Green	Light green	Dull yellow	Dull yellow

Table 2. Effect of ethephon and ethylene gas on physiological loss in weight (PLW) during ripening in pear.

Treatments	Temperature	Storage days				
		4	8	12	16	Mean
Ethephon 500 ppm	20°C	0.64	1.23	2.45	5.02	2.34
	Ambient	3.40	6.98	13.01	20.43	10.95
Ethephon 1000 ppm	20°C	0.64	1.22	3.54	5.14	2.63
	Ambient	4.15	7.45	14.63	27.32	13.39
Ethephon 1500 ppm	20°C	0.67	1.28	3.92	5.62	2.87
	Ambient	4.48	7.93	15.68	28.76	14.21
Ethylene gas 100 ppm	20°C	0.70	1.34	2.61	5.22	2.47
	Ambient	3.27	6.32	14.00	26.98	12.64
Control	20°C	0.48	0.95	2.85	5.26	2.38
	Ambient	2.81	5.90	12.52	26.23	11.86
Mean		2.12	4.06	8.32	15.60	

C D (0.05) Treatments = 0.03, storage days = 0.05, treatment × storage days = 0.14.

temperature treatments, respectively. Under control, the highest score achieved was 6.93 at 20°C temperature after 8 days, which were significantly lower than ethephon treatments at 20°C. The score decreased significantly after 16 days under all the treatments at both the ripening environments, thereby indicating that the fruits show over ripening. At this stage, the fruits also lost their optimum firmness. The fruits having score above 7.8 were excellent in texture, flavor and taste. The fruit having score of 7.4 to 7.8 had acceptable eating quality.

However, the fruit quality was not to the desired level of those fruits having scored less than 7.4.

The increase in organoleptic rating mainly associated with improvement in fruit color, increase in TSS, decrease in acidity and fruit firmness. The organoleptic score is the balance between sugars and acids. The pear fruits fail to soften until they are exposed to a critical period of chilling temperature which is responsible for biosynthesis of ethylene, thereby triggering the ripening process (Blankenship and Richardson, 1985). Ethylene is

Table 3. Effect of ethephon and ethylene gas on organoleptic rating (1 to 9 point Hedonic scale) during ripening in pear.

Treatments	Temperature	Storage days				
		4	8	12	16	Mean
Ethephon 500 ppm	20 °C	7.96	8.13	7.98	7.57	7.91
	Ambient	6.47	6.81	6.58	6.13	6.50
Ethephon 1000 ppm	20 °C	8.05	8.18	7.98	7.68	7.97
	Ambient	6.38	6.76	6.65	6.20	6.50
Ethephon 1500 ppm	20 °C	7.57	7.71	7.50	7.05	7.46
	Ambient	6.08	6.64	6.08	6.00	6.20
Ethylene gas 100 ppm	20 °C	7.95	8.10	7.90	7.25	7.80
	Ambient	6.63	6.95	6.45	6.06	6.52
Control	20 °C	6.63	6.93	6.70	6.13	6.59
	Ambient	6.00	6.30	6.10	6.00	6.10
Mean		6.97	7.25	6.99	6.61	

C D (0.05) Treatments = 0.03, storage days = 0.05, treatment × storage days = 0.09.

Table 4. Effect of ethephon and ethylene gas on fruit firmness (lb) during ripening in pear.

Treatments	Temperature	Storage days				
		4	8	12	16	Mean
Ethephon 500 ppm	20 °C	15.08	14.29	12.07	10.62	13.01
	Ambient	15.45	15.03	10.90	10.41	12.95
Ethephon 1000 ppm	20 °C	14.85	13.15	12.02	10.44	12.61
	Ambient	14.19	14.52	10.88	9.79	12.34
Ethephon 1500 ppm	20 °C	14.93	13.08	9.58	7.29	11.22
	Ambient	14.69	12.38	11.77	6.33	11.29
Ethylene gas 100 ppm	20 °C	15.10	13.44	12.64	9.79	12.74
	Ambient	15.57	13.93	12.13	9.96	12.89
Control	20 °C	16.93	15.74	11.76	10.13	13.64
	Ambient	17.73	13.23	10.41	8.26	12.41
Mean		15.45	13.88	11.41	9.30	

C D (0.05) Treatments = 0.03, storage days = 0.05, treatment × storage days = 0.09.

produced in chilled fruits upon rewarming at a specific temperature (Lilievre et al., 1997). This activity was enhanced by the exogenous application of ethephon in present studies, which is evident from the softening of fruit with ethephon treatments and subsequent ripening at 20 °C. The decrease in organoleptic rating after certain period of ripening might be associated with increase in some biochemical changes. The juicy and buttery texture of ripened pear fruits also indicates the involvement of cell substances and then degradation by enzymes (pectinase and polygalacturonase) during ripening process (Chen et al., 1981).

The firmness of the fruit decreased significantly under all the treatments at both the ripening environments with the increase in ripening interval from 4 to 16 days (Table 4). The fruits showed highest firmness after 4 days under

Table 5. Effect of ethephon and ethylene gas on total soluble solids (TSS%) during ripening in pear.

| Treatments | Temperature | Storage days | | | | |
		4	8	12	16	Mean
Ethephon 500 ppm	20°C	12.83	13.55	12.05	11.14	12.39
	Ambient	12.15	12.33	11.65	11.05	11.79
Ethephon 1000 ppm	20°C	13.12	14.02	12.08	11.73	12.73
	Ambient	12.62	12.77	10.73	11.08	11.80
Ethephon 1500 ppm	20°C	13.05	14.04	11.94	11.55	12.65
	Ambient	12.87	12.70	11.83	11.84	12.31
Ethylene gas 100 ppm	20°C	13.68	14.04	11.93	11.20	12.71
	Ambient	12.13	11.56	10.93	11.00	11.40
Control	20°C	11.08	11.90	10.73	9.05	10.69
	Ambient	11.00	11.10	10.23	8.95	10.32
Mean		12.45	12.80	11.41	10.86	

C D (0.05)Treatments = 0.04, storage days = 0.06,Treatment ×storage days = 0.12.

control treatment at ambient temperature. The fruits recorded above 14.5 lb force were recorded under 1000 ppm ethephon treatment at ambient temperature after 4 days. However, the firmness below 14.5 lb force was recorded under treatment of ethephon 500 ppm at 20°C, ethephon 1500 ppm and exposure to 100 ppm ethylene gas at both the ripening environments and in control at ambient temperature after 8 days of ripening period. All the treatments under both the ripening temperatures showed considerably low fruit firmness after 12 days which was less than ideal fruit firmness of 13.0 lb force. Below 13.0 lb force showed over ripening of the fruits. Considerable reduction in fruit firmness was noted after 16 days. In general, the fruit firmness was lower under all the ethephon treatments as compared to control after 4 days, however, the fruit firmness suddenly decreased under control fruits at ambient temperature after 8 days and thereafter.

Of course, the fruit firmness decreased under control fruits at ambient temperature, but the fruits did not develop the other quality parameters properly. The fully ripened 'Patharnakh' fruit exhibited flesh firmness in range of 13.5 to 15.0 lb/ inch² (Dhatt et al., 2005). The 'conference' pear fruits depicted a decrease in flesh firmness when chilled at -1°c and ripened at 20°C (Barkley et al., 1982). The softening of flesh could be due to the degradation of soluble pectin by high activity of endopolyglacturonase in fruits (Martin-Cabrejas et al., 1994). The change in fruit firmness was also attributed to change in the turgor of the cells and changes in the composition of cell wall pectin's and lipo protein

membrane bordering the cells (Chenn et al., 1991). The cortical tissues associated with swelling of parenchyma cell walls and dissolution of pectin polysaccharides were responsible for decrease in fruit firmness during ripening (Martin-Cabrejas et al., 1994). The decrease in flesh hardness was also associated with high cellulose activity during fruit ripening in pear (Ning et al., 1997). Similarly, the fruit firmness in 'Patharnakh' pear fruit decreased during ripening at 20°C after chilling the fruit at 0-1°c (Dhillon et al., 2005[b]).

A significant increase in TSS was observed under all the treatments up to 8 days of ripening period and decreased thereafter (Table 5). The highest level of TSS (14.04%) was noted under fogging 100 ppm, and ethephon 1500 ppm at 20°C treatments, closely followed by ethephon 1000 ppm at 20°C. This treatment holds better TSS level even after 12 and 16 days of ripening period. All the ethephon treatments improved the TSS in pear fruits significantly over control at both the ripening environments. The highest level of TSS under control at 20°C was recorded after 8 days which were significantly less than ripening treatment of ethephon even after 4 days. A similar pattern to that of TSS was observed in reducing sugars (Table 6). These sugars increased initially (up to 8 days) and decreased thereafter under both the environments in all the fruit ripening treatments. All the ethephon treatments improved reducing sugars in the fruit over control. The level of reducing sugars was higher in 1000 ppm ethephon and 100 ppm ethylene gas treatments when compared with 500 and 1500 ppm ethephon treatments. The increase in soluble solids and

Table 6. Effect of ethephon and ethylene gas on juice acidity (%) during ripening in pear.

Treatments	Temperature	Storage days				
		4	8	12	16	Mean
Ethephon 500 ppm	20°C	0.431	0.325	0.345	0.348	0.362
	Ambient	0.464	0.353	0.368	0.370	0.389
Ethephon 1000 ppm	20°C	0.423	0.317	0.340	0.353	0.358
	Ambient	0.438	0.334	0.346	0.364	0.370
Ethephon 1500 ppm	20°C	0.424	0.324	0.330	0.353	0.358
	Ambient	0.433	0.329	0.338	0.355	0.364
Ethylene gas 100 ppm	20°C	0.405	0.318	0.310	0.333	0.341
	Ambient	0.443	0.338	0.341	0.351	0.368
Control	20°C	0.480	0.355	0.340	0.378	0.388
	Ambient	0.528	0.378	0.370	0.378	0.413
Mean		0.447	0.337	0.343	0.358	

C D (0.05) Treatments = 0.003, storage days = 0.005, treatment × storage days = 0.010.

Table 7. Effect of ethephon and ethylene gas on reducing sugars (%) during ripening in pear.

Treatments	Temperature	Storage days				
		4	8	12	16	Mean
Ethephon 500 ppm	20°C	5.29	6.03	5.26	4.95	5.38
	Ambient	5.26	5.83	5.22	4.91	5.31
Ethephon 1000 ppm	20°C	5.44	6.03	5.37	5.03	5.47
	Ambient	5.22	5.78	5.21	4.90	5.28
Ethephon 1500 ppm	20°C	5.68	6.00	5.36	5.05	5.52
	Ambient	5.14	5.85	5.21	4.95	5.29
Ethylene gas 100 ppm	20°C	5.55	6.13	5.59	5.22	5.62
	Ambient	5.39	5.89	5.42	5.11	5.45
Control	20°C	5.32	5.83	5.24	4.78	5.29
	Ambient	5.09	5.61	4.98	4.60	5.07
Mean		5.34	5.90	5.29	4.95	

C D (0.05)Treatments = 0.05, storage days = 0.07,treatment × storage days = 0.15.

sugars upon ripening could be due to hydrolysis of starch and organic compounds (Sinha et al., 1983; Lelievre et al., 1997). Also an increase in TSS was observed in fruits chilled at 0°C and subsequently ripened at 20°C in Baggugosha (Singh, 1999) and in Punjab Beauty (Dhillon et al., 2005[a]) pear fruits. The decrease in TSS level after 8 days of storage for ripening at ambient and at 20°C temperatures might be due to the inter conversion of some of the sugars into volatile organic acids. Such findings have been reported in grapes by Peynaud and Ribbereau (1971) during cold storage studies.

The juice acid content decreased significantly under all the ethephon treatments with the prolongation of ripening period from 4 to 8 days and starts increasing thereafter (Table 7). In control fruits, the acid content decreased up to 12 days and slightly increased after 16 days of ripening interval under both 20°C and ambient ripening temperatures. The acid content under all the treatments was significantly lower at 20°C when compared with ambient temperature. The lowest acidity level was, however, recorded under ethephon 1000 ppm at 20°C after 8 days, while highest (0.528%) under control fruits at ambient temperature after 4 days. The reduction in acid content was pronounced under ethephon treatments over control after 4 days, which narrowed down later on. In general, when the values were compared with those values obtained after 4 days, the acid content in pear juice decreased with the prolongation of ripening period. This decrease might be due to the utilization of available organic acids at a faster rate in the respiration during ripening. This process might have been triggered with the exogenous application of ethephon. The conversion of organic acids into soluble sugars and long chain polysaccharides may also leads to decrease in acids (Lelievre et al., 1997). Similar results were also reported by Mahajan et al. (2008) in guava fruits, who recorded a decrease in acid content during ripening and storage. The increase in acid content in juice after 8 days in ethephon ripening treatments and after 12 days in control under both the environments might be associated with the increase in weight loss. In over all, the fruits treated with ethephon 1000 ppm or fogging with ethephon at 100 ppm and ripened at 20°C temperature for 8 days exhibited best quality.

REFERENCES

Amerine MA, Pangborn RM, Roessler EB (1965). Principals of sensory evaluation of food. Academic Press, London.

AOAC (2000). Official Methods of Analysis. 15th Edition, Association of Official Analytical Chemists. Benjamin Franklin Station, Washington DC.

Barkley IM, Knee M, Casimir MA (1982). Fruit softening. 1. Changes in cell wall composition and endo-polygalacturonase in ripening pears. J Exp. Bot., 33: 1248-1255.

Blankenship SM, Richardson DG (1985). Development of ethylene biosynthesis and ethylene induced ripening in 'Anjou' pears during the cold requirement for ripening. J Amer Soc Hort Sci 110: 520-523.

Chen PM, Spotts RA, Mellenthin WM (1981). Stem end decay and quality of low oxygen stored d' Anjou pears. J Am. Soc. Hort. Sci., 106(6): 695-698.

Dhatt AS, Mahajan BVC, Bhatt AR (2005). Effect of pre and post harvest calcium treatments on storage life of Asian pear. Acta Horticult., 696: 497-501.

Dhillon WS, Mahajan BVC, Dhatt AS, Sandhu AS (2005[a]). Changes in ripening behaviour of pear cv. Punjab Beauty. Indian J. Hort., 62(2): 193-195.

Dhillon WS, Mahajan BVC, Dhatt AS, Sandhu AS (2005[b]). Waxing and storage studies in soft-pear cv. Punjab Beauty. Acta Horticult., 696: 485-492.

Hansen E (1966). Postharvest physiology of fruit. Annu. Rev. Plant Physiol., 17: 459-480.

Honma Y, Ikema M, Toma C, Ehara M, Iwanaga M, Chen PM, Varga DM, Facteau TJ (1997). Promotion of ripening of 'Gebhard' red d' Anjou pears by treatment with ethylene. Postharvest Biotech., 12: 213-220.

Hunter S (1975). The measurement of appearance. John Wiley and Sons, New York, pp. 304-305.

Kulkarni SG, Kudachikar B, Rawana KVR (2004). Studies on effect of ethereal dip treatment on ripening behavior of mango. J. Food Sci. Technol., 41: 216-220.

Lieberman M, baker JE, Sloger M (1977). Influence of plant hormones on ethylene production in apple, tomato and avocado slices during maturation and senescence. Plant Physiol., 60: 214-217.

Lelievre JM, Tichit L, Dao P, Fillion L, Nam YW, Pech JL, Latche A (1997). Effects of chilling on the expression of ethylene biosynthetic gene in ' Passe-crassane' pears (Pyrus communis) fruits. Plant Mol. Biol., 33: 847-855.

Mahajan BVC, Singh G, Dhatt AS (2008). Studies on ripening behavior and quality of winter guava with ethylene gas and ethephon treatments. J. Food Sci. Technol., 45: 81-84.

Martin-Cabrejas MA, Waldran KW, Selvendaran RR, Parket ML, Maotes GK (1994). Ripening related changes in cell wall of 'Spanish' pear. Physiol. Pl., 91: 671-679.

Ning B, Kubu Y, Inaba A, Nakamura R (1997). Softening characteristics of Chinese pear 'YaLi' fruit with special relation to change in cell wall polysaccharides and degrading enzymes. Scientific Reports of Faculty of Agriculture, Okayama University, 86: 71-78.

Oetiker JH, Yang SF (1995). The role of ethylene in fruit ripening. Acta Horticult., 398: 212-216.

Pressey R (1977). Enzymes involved in fruit softening. In: Ory RL, Angelo Aj St. (Eds). Enzymes in food and beverage processing. ACS Symp. Ser. 47. Amer Chem Soc Washington, DC, pp. 172-191.

Reyes MU, Paul RF (1995). Effect of storage temperature and ethylene treatment on guava fruit ripening. Post harvest Biol. Technol., 6: 357-365.

Sharma JR (1998). Statistical and biometrical techniques in plant breeding. New Age International (P) Limited Publishers, New Delhi.

Singh A (1999). Standardization of ripening technique in 'Baggugosha' pear. M Sc Thesis, Punjab Agric Univ, Ludhiana.

Sinha MM, Tripathi SD, Tiwari JP, Mishra RS (1983). Effect of Alar and CCC on flowering and fruiting in peach cv. Alexandria. Punjab Hort. J., 23: 43-46.

Wang CY, Mellenthin WM (1972). Internal ethylene levels during ripening and climacteric in d' Anjou pears. Plant Physiol., 50: 311-312.

4

Utilization of clove powder as phytopreservative for chicken nuggets preparation

Devendra Kumar* and V. K. Tanwar

Department of Livestock Products Technology, College of Veterinary and Animal Science, G.B. Pant University of Agriculture and Technology, Pantnagar-263145, India.

An investigation was undertaken to explore the possibilities of utilization of clove powder as phyto-preservative in chicken nuggets prepared from spent hen meat and then storage study was conducted at 4±1°C, to assess the effect of test ingredient (clove powder) on sensory and microbiological qualities of optimized product in comparison with its control counterpart. Addition of clove powder in standard formulation caused a slight change in emulsion stability, cooking yield, water activity and composition of chicken nuggets. The findings showed antimicrobial effect of clove powder as microbial load in optimized preparation was found to be significantly (P≤0.05) lower throughout the storage period when compared with control preparation. Sensory scores of optimized and control preparation were found to be acceptable even after 15 days of storage.

Key words: Clove, phyto-preservative, chicken nugget, storage stability.

INTRODUCTION

Utilization of spices in various forms like powder, extract or essential oils has been well documented for inhibiting the growth of many spoilage bacteria and fungi in foods (Meena and Sethi, 1997; Subbulakshmi and Naik, 2002; Rajkumar and Berwal, 2003). Clove (*Eugenia caryophyllus*) is reported to have antibacterial (Everting and Deibel, 1992; Suresh et al., 1992), antimycotic (Karapinar, 1990; Illicim et al., 1998) and yeast inhibitory (Farag et al., 1989) activity. Such properties of clove may be attributed to its 2-methoxy-4-(2-propenyl) phenol content (Beuchat and Golden, 1989 and Suresh et al., 1992).

The basic approach in developing convenience meat products for the Indian market must be governed by consideration of cost. Hence, the choice of meat falls on the low-graded meat like that of culled and spent hens, which can be upgraded with functional food additives into good quality value added products. Lipid oxidation is one of the major causes of deterioration in the quality of meat and meat products. It may also decrease the nutritional

value by forming potentially toxic products during cooking and processing (Shahidi et al., 1992; Maillard et al., 1996). Clove also possesses good antioxidant activity (Gulcin et al., 2004). Therefore, an investigation was undertaken to evaluate the effect of incorporating clove powder on physico-chemical, sensory and storage stability of chicken nuggets prepared from spent hen meat.

MATERIALS AND METHODS

Procurement of the materials

The spent hens were procured from Instructional Poultry Farm (IPF), Govind Ballabh Pant University of Agriculture and Technology, Pantnagar. The birds were slaughtered and dressed, manually de-boned, packed in low density polyethylene (LDPE) bags and stored overnight at 4 ± 1°C in refrigerator. The ingredients of spice mix were procured from Pantnagar local market, cleaned and dried in oven at 50°C for 2 h; grounded and sieved through 100 mesh and the fine powder obtained was stored for subsequent use. The condiment mix contained onion, garlic and ginger; prepared afresh in appropriate ratio as fine paste. Clove buds, refined soybean oil, refined wheat flour, table salt, sugar, skim milk powder and eggs were procured from Pantnagar local market. Clove buds were cleaned and ground to fine powder and stored in air-tight

---

*Corresponding author. E- mail: drdevvet24@rediffmail.com, drdev24@gmail.com.

containers at room temperature (24 ± 2 ℃) for subsequent use.

Preparation of chicken nuggets

Meat was cut into small pieces and grounded twice in a meat mincer (Hobart®, USA) with 5 mm plate followed by 3 mm plate. Emulsion of each formulation was prepared using bowl chopper (Hobart®, USA). All the nugget formulations consisted of spent hen meat 60%, vegetable oil (fat) 10%, ice flakes 10%, refined wheat flour 2.5%, skim milk powder 2%, whole egg liquid 5%, table salt 2%, sugar 1%, sodium tri-polyphosphate (STPP) 0.25%, condiments 5%, spices mix 1.5% and sodium nitrite 150 ppm. The prepared emulsion was tightly packed in oil coated metallic mold fitted with lids and steam cooked for 45 min at $34.47\ KN/m^2$ pressure. Subsequently, the cooked product was cooled, weighed and removed carefully from the moulds. The meat block thus obtained was carefully sliced and cut in nugget size pieces (4×1.5×1.5 cm). The nuggets were packed in sterilized LDPE bags and stored at refrigerated temperature (4±1 ℃) for analysis.

Optimization of clove powder

The meat emulsion for control product consisted of basic formulation given earlier without test ingredient. Clove powder was added to the earlier formulation at three levels – 0.1, 0.2 and 0.3% (w/w) of meat emulsion. The preliminary trials were conducted to access the best level of incorporation of clove powder into chicken nuggets, on the basis of sensory evaluation by semi-trained sensory panel of 11 panelists. On the basis of sensory evaluation, chicken nuggets containing 0.1% clove powder were selected for further study.

Physico-chemical characteristics

The emulsion stability was determined by the method of Baliga and Madaiah (1970) with minor modifications (50 g emulsion was taken in LDPE bags). The pH value was recorded by using a digital pH meter (ECI Ltd, India) and water activity by using water activity meter (Rotronic Hygrolab 3). Thiobarbituric acid (TBA) value (mg malonaldehyde/kg of sample) was estimated as per procedure given by Tarladgis et al. (1960). Moisture, protein, fat and total ash content of chicken nuggets were determined following AOAC (1984) procedures. Total Plate count (TPC), total lipolytic, coliform and yeast and mold counts were done by following standard methods of APHA (1992). The sensory quality of samples was evaluated using 8 point hedonic scale (Keeton et al., 1984) using semi-trained sensory panel of 11 panelists.

Storage study

Storage study of product was conducted by keeping the products at 4±1 ℃ for 15 days. Sensory evaluation, pH and TBA values estimation and microbiological study were conducted on both preparations, control as well as the optimized one, after every 5 days of interval. Data were recorded and statistically analysed to evaluate the stability of optimized product in comparison to control preparation.

Statistical analysis

Statistical analysis of the data was done using ANOVA technique according to the method described by Snedecor and Cochran (1994) on completely randomized design (CRD). Average of three

Table 1. Effect of clove powder[1] on physico-chemical properties of chicken nuggets   (Mean ± SE).

Parameters	Control	Optimized product
Emulsion stability	94.375±0.311	94.478±0.301
Cooking yield	95.835±0.232	96.083±0.098
Water activity	0.972±0.0026	0.974±0.0026
Moisture	62.031±0.192	62.037±0.238
Protein	15.354±0.422	15.173±0.405
Fat	13.093±0.310	12.783±0.125
Total ash	2.764±0.041	2.730±0.033

[1]Clove powder is added in the formulation at 0.1% level (w/w) of meat emulsion.

replicates was used in calculations.

RESULTS AND DISCUSSION

Physico-chemical characteristics

Table 1 indicated that incorporation of clove powder in chicken nugget formulation caused slight increase in emulsion stability, moisture content and water activity, whereas a slight decrease in cooking yield, protein, fat and total ash content were observed after incorporation. However, statistically all these changes were non-significant.

Effect on sensory parameters

Sensory evaluation of all prepared samples presented in Table 2 indicated that, chicken nuggets containing clove powder scored significantly (P<0.05) higher for appearance/colour, flavour, juiciness and overall acceptability than the control preparation while texture scores remained unaffected statistically. The better appearance/colour scores of optimized nuggets might be due to a slight reddish tint imparted by clove powder. Decrease in flavour scores might be due to development of oxidative rancidity and microbial deterioration in products during storage. All sensory scores decreased significantly (P<0.05) with the advancement of storage period. Biswas et al. (2006) also reported that all the sensory quality values decreased significantly with the advancement of storage period.

Effect on pH and TBA value

A non-significant effect on pH value was observed in the nuggets after clove powder incorporation, but a significantly (P≤0.05) increasing trend in pH value was observed in both preparations with the advancement of

Table 2. Effect of clove powder[1] and storage period on quality attributes of chicken nuggets (Mean ± SE).

Product (Treatments)	Storage period (No. of days)				Treatment mean
	0	5	10	15	
Appearance/Colour					
Control	6.812±0.148	6.720±0.149	6.020±0.092	4.940±0.248	6.123±0.423b
Optimized	7.236±0.113	7.112±0.090	6.850±0.079	6.252±0.085	6.862±0.219a
Flavor					
Control	6.878±0.126	5.600±0.169	3.652±0.202	3.532±0.092	4.915±0.808b
Optimized	7.374±0.112	5.402±0.101	4.880±0.145	4.528±0.149	5.546±0.689a
Texture					
Control	.630±0.119	6.500±0.106	6.234±0.085	5.450±0.101	6.203±0.264b
Optimized	6.470±0.078	6.324±0.049	6.180±0.068	5.980±0.129	6.238±0.105b
Juiciness					
Control	6.378±0.077	6.250±0.081	6.044±0.118	5.820±0.122	6.123±0.122b
Optimized	6.580±0.076	6.390±0.058	6.160±0.046	6.000±0.069	6.264±0.114a
Overall acceptability					
Control	7.100±0.127	6.460±0.147	5.340±0.136	4.220±0.082	5.780±0.635b
Optimized	7.404±0.084	6.850±0.135	6.080±0.051	5.780±0.073	6.437±0.302a
pH					
Control	6.128±0.0058B	6.160±0.0094A	6.234±0.0074C	6.262±0.0073D	6.196±0.027b
Optimized	6.128±0.0058B	6.158±0.0066A	6.223±0.0176C	6.246±0.0097D	6.189±0.024b
TBA					
Control	0.347±0.003	0.724±0.030	1.097±0.047	1.387±0.038	0.889±0.226a
Optimized	0.344±0.004	0.380±0.012	0.575±0.022	0.677±0.020	0.494±0.079b
Total plate count					
Control	2.545±0.131	3.388±0.187	4.285±0.152	6.436±0.099	3.913±0.944a
Optimized	2.382±0.136	2.474±0.135	2.735±0.128	3.05±0.092	2.660±0.149b
Lipolytic count					
Control	2.1152±0.137	2.725±0.132	3.167±0.119	3.577±0.091	2.896±0.313a
Optimized	2.0891±0.119	2.143±0.152	2.255±0.129	2.436±0.099	2.231±0.076b
Coliform count					
Control	ND	ND	1.814±0.155	2.5478±0.070	1.090±0.647a
Optimized	ND	ND	ND	1.390±0.049	0.347±0.347b
Yeast and mold count					
Control	ND	ND	1.164±0.061	2.068±0.105	0.808±0.501a
Optimized	ND	ND	0.680±0.280	1.18±0.092	0.465±0.287b

*: Mean values bearing same superscripts column-wise and row-wise (alphabets) do not differ significantly (P<0.05). [1]Clove powder is added in the formulation at 0.1% level (w/w) of meat emulsion.

storage period (Table 2). The increase in pH value during storage period suggests that, there was significant breakdown of meat protein on storage of the product. The increase in pH value during storage of meat was also reported earlier (Yadav and Sanyal, 1999). Chicken nuggets containing clove powder (optimized product) maintained significantly (P≤0.05) lower TBA values throughout the storage period. Lower TBA values in optimized product were due to the anti-oxidant properties of clove powder. The antioxidant effect of clove was also reported earlier (Gulcin et al., 2004; Shobana and Naidu, 2000).

Effect on microbial count

An increasing trend in microbial counts was observed in both preparations with the advancement of storage period (Table 2), but chicken nuggets containing clove powder maintained a significantly (P≤0.05) lower microbial counts (TPC, coliforms, lipolytic, yeast and molds) throughout the storage period than control preparation. Lower microbial count observed in nuggets containing clove powder suggested antimicrobial properties of cloves, which were also reported earlier by Suresh et al. (1992), Yadav (2005) and Rajkumar and Berwal (2003).

Conclusion

From the present study, it can be concluded that in preparation of chicken nuggets, clove powder can be used at 0.1% level (w/w) of meat emulsion with beneficial effect on physico-chemical and sensory qualities of the product. This product can be stored at refrigeration temperature (4±1 °C) for 15 days in LDPE bags with good acceptability.

ACKNOWLEDGEMENT

Authors are grateful to the Director of Research and Dean, Faculty of Veterinary Science, G.B. Pant University of Agriculture and Technology, Pantnagar for providing necessary facilities to carry out the above research work.

REFERENCES

AOAC (1984). Official Methods of Analysis. 14th edn. Association of Official Analytical Chemists. Washington, D.C.

APHA (1992). Compendium of Methods for the Microbiological Examination of Foods. 4th edn. C. Vandergrant and D.F. Splittstoesser. American Public Health Association, Washington, D.C, pp. 919-927

Baliga BR, Madaiah N (1970). Quality of sausages emulsion prepared from mutton. J. Food Sci., 35(4): 383-385.

Beuchat LR, Golden DA (1989). Antimicrobials occurring naturally in foods. Food Technol., 43: 134-142.

Biswas S, Chakraborty A, Sarkar, S (2006). Comparision among the qualities of patties prepared from chicken broiler, spent hen and duck meats. J. Poultry Sci., 43: 180-186.

Everting WT, Deibel KE (1992). Sensitivity of Listeria monocytogens to spices at two temperatures. J. Food Safety. 12: 129-131.

Farag RS, Daw ZY, Hewedi FM, El-Baroty GSA (1989). Antibacterial activity of some Egyptian spices essential oils. J. Food Protection, 52: 665-667.

Gulcin I, At G, Beydemir U, Elmasta M, Lu ORK (2004). Comparison of antioxidant activity of clove_ (Eugenia caryophylata Thunb) buds and lavender (Lavandula stoechas L.). Food Chem., 87(3): 393-400.

Illicim A, Digrak M, Bagei E (1998). The investigation of antimicrobial effect of some plant extracts. Turkish J. Bio., 22: 119-125.

Karapinar M (1990). Inhibitory effect of anethole and euginol on growth and toxin production of Aspergillus parasitius. Intern. J. Food Microbio., 10: 193-200.

Keeton JT, Foegeding EA, Patina AC (1984). A comparison of non meat products, sodium tripolyphosphate and processing temperature effects on physical and sensory properties of frank furthers. J. Food Sci., 49: 1462-1474.

Maillard MN, Soum MH, Boivin P, Berset C (1996). Antioxidant activity of barley and malt- Relationship with phenolic content. Lesbensm Wissen, 29: 238-244.

Meena MR, Sethi V (1997). Role of spices and their essential oils as preservatives and antimicrobial agents – A review. Indian Food Packer, 25: 38-45.

Rajkumar R, Berwal JS (2003). Inhibitory effect of clove on toxigenic molds. J. Food Sci. Technol., 40(4): 416-418.

Shahidi F, Janitha PK, Wanasundara PD (1992). Phenolic antioxidants. Crit. Rev. Food Sci. Nutr., 32: 67-103.

Shobana S, Naidu KA (2000). Antioxidant activity_ of selected Indian spices . Prostaglandins, Leukotrienes and Essential Fatty Acids, 62(2): 107-110.

Snedecor GW, Cochran WG (1994). Statistical methods. First east west press edition, New Delhi.

Subbulakshmi G, Naik M (2002). Nutritive value and technology of spices: current status and future perspectives. J. Food Sci. Technol., 39: 319-344.

Suresh P, Ingel VK, Vijayalakshmi N (1992). Antibacterial activity of eugenol in comperision with other antibiotics. J. Food Sci. Technol., 29(4): 254-257.

Tarladgis BG, Watts BM, Younathan MT, Dugan LR (1960). A distillation method for the quantitative determination of malonaldehyde in rancid foods. . J. Am. Oil Chem. Soc., 37: 44-48.

Yadav AS (2005). Recent advances in bio- and phyto-preservation of meat. In: Recent Advances in Poultry and Egg Processing and Quality Assessment of Poultry Products,2005. Aug. 29- Sept. 7, CARI Izatnagar, India, pp. 75-79.

Yadav PL, Sanyal MK (1999). Development of livestock products by combination preservation technique. In: Proceedings of National Seminar on Food Preservation by Hurdle Technology and Related Areas, DFRL, Mysore, India, pp. 104-109.

5

Effect of vacuum tumbling time on physico-chemical, microbiological and sensory properties of chicken tikka

Sanjay Kumar Bharti[1], Anita B.[1], Sudip Kumar Das[1]* and Subhasish Biswas[2]

[1]Department of Livestock Products Technology, College of Veterinary and Animal Sciences, G. B. Pant University of Agriculture and Tech, Pantnagar– 263145, India.
[2]Department of Livestock Products Technology, West Bengal University of Animal and Fishery Sciences, Kolkata – 700037, India.

Chicken tikka prepared by using vacuum tumbler (tumbling time 15, 30 and 45 mins) were stored at refrigerated temperature (4±1°C) and subjected to comparative studies on 0, 7 and 14 day. Product having more tumbling time showed significant (P<0.05) higher values for cooking yield, pH, moisture and ash%, lower values for cholesterol (mg/100 g), and protein%. The thio-barbituric acid (TBA) value (mg malonaldehyde/kg) showed a non-significant (P>0.05) higher value with control, however, a significant increasing trend was noticed throughout the storage period. Microbiological study revealed that the microbial count (cfu/g) in total plate count and lipolytic count were significantly lower (P<0.05) in product having more tumbling time and in all groups of chicken tikka, the microbial count increased significantly with advancement of storage period. All the sensory attributes viz. appearance and color, texture, juiciness and overall acceptability increased significantly with increase in tumbling time.

Key words: Vacuum tumbling, chicken tikka, sensory property, meat processing.

INTRODUCTION

Chicken tikka is a Pakistani dish made by baking small pieces of chicken which have been marinated in spices and yogurt. It is traditionally made on skewers in a tandoor (Indian clay oven) and is usually boneless. Marination is used to improve both sensory (flavor, color moisture and texture) as well as functional properties of meat (water-holding capacity, stability and cooked yield). Marinades are preliminary a mixture of salt, organic acids, nitrites and spices in a solution in which meat is soaked. Skinless and boneless meat are marinated in a tumbler (massager), operated in a static, vacuum or high pressure to improve marinade absorption and uniformity (Sams, 2001). The agitation, which can be applied for one to several hours (slow or intermittent), helps disrupt some of the tissue structure, assists in distributing the brine solution and develops a protein exudate that will later serve as "glue" to bind the meat chunks during cooking. Operating under vacuum helps in removing the air bubbles from the exudate and might also assists in protein extraction (Barbut, 2005). The results of tumbling meat have been studied by several investigators. According to Treharne (1971), tumbling is defined as the massaging of meat surfaces; however, many meat processors now make a distinction between "Tumbling" and "Massaging". Tumbling involves the physical process of meat rotating in drum, falling and making contact with metal walls and paddles. This process involves a transfer of kinetic energy and consequently causes alteration in muscle tissue.

In contrast, the process of "Massaging" is considerably less rigorous. It usually involves a stationary drum with paddles rotating around a vertical axle. This process does not involve free falling of meat contents. Consequently, the process mainly involves muscle tissue rubbing other muscle tissue and the smooth surface of the drum. Theoretically, this result is less transfer of kinetic energy and therefore less heat rise in the product. Tumbling has many beneficial effects, some of which are due to the formation of a protein exudate. According to

---
*Corresponding author. E-mail: sudipvet@gmail.com

Table 1. Composition of spice and condiment mix.

Ingredient	%
Garlic paste	10.8
Ginger paste	2.7
Onion paste	27.0
Red chilly powder	2.1
Coriander powder	4.0
Garam masala powder	2.7
Curd	40.9
Lime	5.4
Salt	4.0
Sodium nitrite	150 ppm
STPP(Sodium tri-poly phosphate)	0.4

Rust and Olson (1973), this protein exudate acts as a sealer when the protein is denatured during thermal processing. Vartorella (1975) and Krause (1976) added that this sealer helps hold in juices during smoking and cooking, and results in increased yields, increased juiciness, and improved slicing characteristics of the finished product. Other benefits of tumbling include improved tenderness and more uniform cured meat color (Krause, 1976).

To provide ready to cook, value added products of poultry and seafood, at either the food processing plant or the supermarket/butcher shop vacuum tumbling is an important method. The vacuum causes the product to absorb more marinade, which makes the product juicier and faster cooking.

The tumbling massages the product, which makes it tenderer. With all of the aforementioned physical changes caused by tumbling or massaging, it seems logical that there must be several significant changes that occur due to alteration in tumbling time. In view of the above facts the present study was envisaged with the following objectives:

1. Determination of influence of vacuum tumbling time on eating quality characteristics of prepared chicken tikka and selecting the best on the basis of physico-chemical, microbiological and sensory evaluation;
2. Assessment of physico-chemical, microbiological and sensory properties of chicken tikka during storage.

MATERIALS AND METHODS

The experiment was conducted in the Department of Livestock Products Technology, College of Veterinary and Animal Sciences, G.B. Pant University of Agriculture and Technology, Pantnagar. Poultry of nearly 1.5 kg body weight were procured from instructional poultry farm (IPF) Nagla, Govind Ballabh Pant University of Agriculture and Technology, Pantnagar. A total of 14 birds were utilized for the study. Birds used in this study were

randomly selected from their respective flocks at IPF, Nagla. The birds were slaughtered at the Department of Livestock Products Technology following standard protocol and allowed to bleed for 180 s, in bleeding cones.

The birds were dressed as per approved scientific methods and manually de-boned. The slaughtering technique and transportation method was approved by the Ethical committee involving concerned Head of the Department and other members. The meat obtained was washed with clean water. Boneless lean meat was collected and stored. On the basis of literature available and various preliminary trials, four groups of chicken tikka were prepared. that is, one control and three treatments (vacuum tumbled for 15, 30 and 45 min respectively). Spice mix, condiment mix, curd, oil, lemon, table salt and packaging materials were procured from local market. All the chemical and media used in the study were of analytical grade and were obtained from standard firms (Hi media, Merck; India).

Preparation of chicken tikka

Boneless lean meat was cut into small cubes of nearly one inch size with the help of deboning knife. Spice and condiment mix was prepared as per the formulation given in Table 1. The paste along with condiment paste and spice mix was properly mixed with finely beated curd. Later chicken pieces were added to the above mix. The prepared mix was weighed equally in four parts and vacuum tumbled in a vacuum tumbler (Promarks Vac- TM50) for 15, 30, 45 min. A control was also prepared along with tumbled treatments.

Tumbled and non tumbled meat was cooked in a preheated oven to maximum 240°C under smokeless, moderate and uniform heat. The temperature was maintained in the oven throughout the cooking period of about 20 min. to permit thorough and uniform cooking. During cooking the meat pieces were turned over once to avoid drying, charring or blistering. Cooking was done till the meat attained a golden brown colour and was fully cooked. The chicken tikka were packed aerobically in adequate number of LDPE bags, sealed in a packaging machine and stored at refrigeration temperature (4±1°C). The samples of chicken tikka were analyzed for physico-chemical, microbiological and sensory characteristics at regular intervals of 0, 7 and 14 days or till spoilage, whichever was earlier.

Analytical procedure

Physico chemical characteristics

Cooking yield: To determine the cooking yield tumbled and non tumbled chicken tikka were weighed before and after cooking. The cooking yield was calculated as:

$$\text{Cooking yield (\%)} = \left[\frac{\text{Weight of cooked chicken tikka}}{\text{Weight of raw chicken tikka}}\right] \times 100$$

pH determination

For determination of pH, representative samples of 10 g of patties from each treatment were homogenized for 30 s with 100 ml distilled water using a blender. The pH of prepared homogenates was recorded by using a digital pH meter (WTW®, Germany, Model 330i fitted with Sen Tix sp electrode) by immersing the electrode of pH meter into aliquot of the sample (Egbert et al., 1992). The pH meter was calibrated with known buffers of pH 7 and 4.01 before use every time.

Cholesterol content

Total cholesterol was determined as per Zaltkis et al. (1953) with little modifications as described by Rajkumar et al. (2004). Lipid extract was prepared by mixing one gram of sample with 10 ml of freshly prepared 2:1 Chloroform: Methanol solution and homogenizing it in a blender. Homogenate was filtered using Whatman filter paper No. 42 and 5 ml of filtrate was added with equal quantity of distilled water, mixed and centrifuged at 3000 rpm for 7 min. Top layer (methanol) was removed by suction. Volume of bottom layer (Chloroform) having cholesterol was recorded. The O.D. of standard and sample against blank was taken at 560 nm. Total cholesterol mg percent was recorded as follows:

$$\frac{O.D. \text{ of sample}}{O.D. \text{ of standard}} \times \frac{Vol. \text{ of Choloform(ml)}}{Weight \text{ of the sample taken (g)}} \times conc. \text{ of standard} = Cholesterol \text{ in (mg)}\%$$

Proximate analysis

Moisture, fat, protein and total ash contents of cooked patties were determined by the procedure given by AOAC (1984).

Storage study

Thiobarbituric acid (TBA) value

TBA was estimated as per procedure given by Tarladgis et al. (1960).

Microbiological analysis

Total plate count, coliform count, lipolytic count and yeast and mold count were determined as per the procedure described by APHA (1992).

Sensory evaluation

The sensory qualities of samples were evaluated by meat descriptive analysis method. The patties were oil fried and served warm to panelists for sensory evaluation. The sensory quality of samples was evaluated using 8 point descriptive scale (Keeton et al., 1984) where 8 denoted extremely desirable and 1 denoted extremely poor. A sensory panel of seven judges were requested to evaluate the product for different quality attributes viz: appearance, flavor, texture, juiciness and overall acceptability.

Statistical analysis

Statistical analysis of the data obtained, was done using ANOVA technique according to the method described by Snedecor and Cochran (1994) by completely randomized design (CRD). Further, to determine the significance between treatments, Tukey's HSD test was conducted by a SPSS® – 16 software package.

RESULTS AND DISCUSSION

The abbreviations used in results for treatments and control are as follows for control: Con; for vacuum tumbling 15 min: VT15, for vacuum tumbling 30 min: VT30 and for vacuum tumbling 45 min: VT45. Data presented in Table 2 indicate that cooking yield of chicken tikka of group VT45 was significantly (P<0.05) higher than the Con, VT15 and VT30 groups. This result is in agreement with the findings of Dzudie and Okubanjo,(1999) who reported that the product tumbled for a longer time had a lower cooking loss, when compared to those cooked for a short time due to increased amount of extractable soluble proteins. Muller (1991) also reported higher product yield due to tumbling as compared to non-tumbled control. Increased tumbling time provides better chances for migration of curing solution in increased ionic strength and pH, which in turn enhance the product yield. Ghavimi et al. (1986) observed insignificant difference between product yield from vacuum and aerobically tumbled meats. These data agree with the report of Rust and Olson (1973) who felt that the exudates of myofibrillar protein seals moisture in the product as it coagulates on and immediately below the surface.

The pH of group VT45 was observed to be 6.247±0.014, which was found to be highest among all the different test groups (Table 2). The observations are same as those of Froning and Sackett (1985) who reported that pH of the cooked muscle was significantly higher in tumbled carcasses as compared to the non tumbled treatments. It might be due to increased penetration of curing solution in treatment as compared to the control samples. Plimpton et al. (1991) also reported higher pH in tumbled samples. Cassidy et al. (1978) stated that besides enhanced distribution, salt might retard enzyme action necessary for glycolysis and production of lactic acid. More tumbling time causes rapid migration of curing solution in the muscle tissue. Ledward (1979) also found that ham tumbled under vacuum had comparatively higher cure absorption. The increment of pH could be attributed to the modification of meat protein conformation during thermal denaturation (Ang and Hamm, 1982).

The Mean±SE for cholesterol level for VT45 was 40.091±1.382 (mg/100 g) which differs significantly from Con, VT15 and VT30 but later two groups that is VT15 and VT30 did not differ significantly from each other (Tables 2). Cholesterol levels were found to be slightly higher in mechanically deboned broilers as compared to hand-deboned ones (Sharma et al., 2002). Moisture content of group VT45 was significantly (P<0.05) higher than the Con but no significant difference between VT15, VT30 and VT45 groups was seen (Table 3). It might be due to comparatively high levels of salt soluble proteins, which form a seal upon heating, thus retaining higher

Table 2. Effect of different treatments on cooking yield (%), pH, cholesterol (mg/100 g) and TBA values (mg malonaldehyde/kg) of chicken tikka (Mean±SE).

Treatment / Parameter	Control	Vacuum tumbling 15 min	Vacuum tumbling 30 min	Vacuum tumbling 45 min
Cooking Yield (%)	$65.702^a \pm 0.089$	$71.044^b \pm 1.007$	$78.862^c \pm 0.332$	$80.752^d \pm 0.385$
pH	$6.002^a \pm 0.020$	$6.115^b \pm 0.010$	$6.177^c \pm 0.009$	$6.247^d \pm 0.014$
Cholesterol (mg/100 g)	$64.915^a \pm 1.160$	$55.712^b \pm 1.282$	$52.629^{bc} \pm 1.163$	$40.091^d \pm 1.382$
TBA (mg malonaldehyde/kg)	0.362 ± 0.004	0.354 ± 0.004	0.350 ± 0.005	0.346 ± 0.004

*Mean values bearing same or no superscript row wise do not differ significantly (P<0.05).

Table 3. Effect of different treatments on proximate composition (%) of chicken tikka (Mean±SE).

Treatment / Parameter	Control	Vacuum tumbling 15 min	Vacuum tumbling 30 min	Vacuum tumbling 45 min
Moisture (%)	$60.426^a \pm 0.241$	$62.256^b \pm 0.436$	$62.757^{bc} \pm 0.568$	$63.300^{bcd} \pm 0.455$
Protein (%)	28.435 ± 0.892	$26.820^a \pm 0.585$	$26.149^{ab} \pm 0.249$	$25.766^{abc} \pm 0.197$
Fat (%)	10.806 ± 0.016	10.775 ± 0.015	10.759 ± 0.010	10.746 ± 0.017
Ash (%)	0.956 ± 0.012^a	0.987 ± 0.018^a	1.014 ± 0.008^{ab}	1.045 ± 0.007^{bc}

*Mean values bearing same or no superscript row wise do not differ significantly (P<0.05).

moisture during cooking (Rust and Olson, 1973). Vacuum tumbled meat block had slightly higher moisture than aerobically tumbled product, which could also be accounted for the same reason.

Protein content of group VT45 was significantly (P<0.05) lower than the Con, VT15 and VT30 groups (Table 3). Con. and VT15 groups did not differ significantly but VT30 significantly differed from Con. These results are in agreement with that of (Theno et al., 1978). Retention of higher amount of moisture in the tumbled samples decreased protein percentage. The ether extract of different groups did not exhibited significant difference from each other as shown in Table 3 and it ranged from 10.746 to 10.806%. Similar were the findings of Theno et al. (1978) who reported that there were no significant (P<0.05) differences in the fat percentage of tumbled and control samples. The Mean±SE for ash for VT45 was 1.045±10.007% which differs significantly from Con. VT15 and VT30 groups did not differ significantly but VT30 significantly differed from Con. Babji et al. (1982) reported that tumbling disrupted muscle cells, thus facilitating the diffusion of curing ingredients into the meat. This could explain the higher ash content of the tumbled product. In contrast, Katsaras and Budras (1993) found no significant difference in chemical composition of the tumbled salted and non-tumbled salted turkey breast meat.

There was no significant (P<0.05) difference in the TBA value of the chicken tikka of different test groups. But it was found to be non-significantly lower in VT45 group as compared to Con, VT15 and VT30 groups (Table 2).

Total plate count of group VT45 was significantly (P<0.05) lower than Con, VT15 and VT30 groups (Table 4). VT15 and VT30 groups did not differ significantly but VT15 differed significantly from Con. Because the meat surface is destructured by tumbling, the transfer surface area is likely to increase. Consequently, tumbling would enhance water and acid transport between meat and solution. Marinade penetration and diffusion are therefore accelerated, thus decreasing the microbial count (Ghavimi et al., 1986, Pohlman et al., 2002). Ockerman et al. (1978) reported significant decontamination effect by vacuum tumbling by using antimicrobial agents such as trisodium phosphate and cetylpyridium chloride.

The lipolytic count of VT45 group was found to be lowest 2.847±0.001 log cfu/g, as shown in Table 4, and it differs significantly from the lipolytic count of other three groups. Tumbling, which has been shown to help in the distribution of curing agents and marinade solution during processing in several studies (Leak et al., 1984 and Kamchorn et al., 1983) should increase the distribution of antimicrobial agents. In the current study also, at a given storage time and enhanced tumbling time, fewer total microorganisms were detected. The coliform colonies and yeast and mold colonies were not detected on 0 day in all groups that is Con, VT15, VT30 and VT45 group.

Appearance and color score from sensory panel of group VT45 was significantly (P<0.05) higher than the Con, VT15 and VT30 groups (Table 5). Color and appearance of the chicken tikka were improved significantly due to tumbling. Similar finding was also reported by Theno et al. (1978). This might be due to

Table 4. Effect of different treatments on microbiological quality (log cfu/gm) of chicken tikka (Mean±SE).

Treatment Parameter	Control	Vacuum tumbling 15 min	Vacuum tumbling 30 min	Vacuum tumbling 45 min
TPC	$3.398^a \pm 0.000$	$3.383^b \pm 0.001$	$3.375^{bc} \pm 0.001$	$3.368^{cd} \pm 0.001$
Lipolytic	$2.943^a \pm 0.000$	$2.896^b \pm 0.000$	$2.876^c \pm 0.001$	$2.847^d \pm 0.001$
Coliforms	ND	ND	ND	ND
Yeast and Molds	ND	ND	ND	ND

*Mean values bearing same or no superscript row wise do not differ significantly (P<0.05).

Table 5. Effect of different treatments on sensory attributes (scores) of chicken tikka (Mean±SE).

Treatment Parameter	Control	Vacuum tumbling 15 min	Vacuum tumbling 30 min	Vacuum tumbling 45 min
Appearance/Color	6.408 ± 0.015^a	6.531 ± 0.039^b	6.656 ± 0.042^c	6.777 ± 0.047^d
Flavor	6.019 ± 0.108	6.170 ± 0.069	6.270 ± 0.058	6.394 ± 1.102
Texture	6.015 ± 0.027^a	6.400 ± 0.061^b	6.764 ± 0.006^c	7.019 ± 0.011^d
Juiciness	5.996 ± 0.038^a	6.398 ± 0.032^b	6.824 ± 0.027^c	7.229 ± 0.019^d
Overall acceptability	6.118 ± 0.031^a	6.437 ± 0.032^b	6.746 ± 0.044^c	6.991 ± 0.016^d

*Mean values bearing same or no superscript row wise do not differ significantly (P<0.05).

increased penetration of curing solution containing nitrite in the tumbled product. Weiss (1973) reported that vacuum treatment allowed meat system to make more efficient use of nitric oxide formed from nitrite. A significant improvement in internal color development is of particular interest and agrees with the reports of Theno et al., 1976 and Treharne (1971). The effect of tumbling on the rate and uniformity of diffusion of curing ingredients probably accounts for the color development, but the disruption of sarcolemma as suggested by Krause (1978) could increase the accessibility of myoglobin to the nitrite, thus helping to explain the significant improvement in color development. There was no significant (P<0.05) difference observed in the flavor score of chicken tikka due to different treatments. But it was found to be lower in Con as compared to VT45 (Table 5). Harmon et al. (1992) and Acton (1972) reported that tumbling significantly improved flavor scores. However, in this study no significant differences were observed.

The mean value score for texture of chicken tikka differs significantly (P<0.05) from each other. The texture of chicken tikka having tumbling time of 45 min was found to be highest (7.019±0.011), as shown in Table 5, and it differs significantly from the texture of other three groups. Improvement in tenderness in tumbled product might be due to cellular disruption and myofibrillar fragmentation of the muscle tissue (Babji et al., 1982). During tumbling, enhanced extraction of salt soluble proteins and mechanical action of friction allowed protein-to-protein interaction during cooking (Pietrasik and Shand, 2005)

resulting in improved textural characteristics. Acton (1972) also observed better texture and improved sliceability due to tumbling. Udayasaha et al. (1999) suggested that the beneficial effect of brine addition could be due to moisture enhancement, which would improve the textural characteristics of the final product.

The juiciness score of group VT45 was observed to be highest and of group Con to be lowest among the entire different test groups (Tables 5). The results confirm the finding of Theno et al. (1978) who reported that juiciness scores of tumbled product were significantly higher than control. It could be due to higher retention of moisture by the extracted salt soluble proteins. However, irrespective of time, juiciness scores of aerobically tumbled and vacuum tumbled products are different and comparable within their comparative groups. The overall mean value of acceptability score was superior for VT45 followed by for VT30, VT15 and Con (Table 5). Tumbling significantly improved the overall acceptability of product.

A similar finding was reported by Theno et al. (1978) who found improved overall acceptability of restructured buffalo meat slices. Vacuum tumbling for 3 h enhanced the overall acceptability of product to a great extent. The results of sensory agree with the statements of Theno et al. (1976), who illustrated that tumbling had a significant influence on external appearance, internal color, sliceability, taste and aroma. The most dramatic effects were in slice ability and yield. Based on the result of physico-chemical, microbiological and sensory analysis, the product VT45 was selected for storage studies along

Table 6. Effect of treatment and control on pH and TBA values (mg malonaldehyde/kg) of chicken tikka during storage.

Storage period	Treatment		Day mean
	Control	Vacuum tumbled (VT45)	
	pH		
0 day	6.022 ± 0.020^{c}	6.152 ± 0.040^{ec}	6.087 ± 0.030
7 day	6.085 ± 0.018^{b}	6.212 ± 0.040^{db}	6.148 ± 0.029
14 day	6.145 ± 0.014^{a}	6.267 ± 0.038^{da}	6.206 ± 0.026
Treatment mean	6.084 ± 0.017	6.210 ± 0.026	
	TBA		
0 day	0.362 ± 0.004^{c}	0.351 ± 0.004^{e}	0.357 ± 0.004
7 day	0.630 ± 0.211^{b}	0.591 ± 0.003^{db}	0.610 ± 0.107
14 day	1.296 ± 0.011^{a}	0.909 ± 0.018^{da}	1.102 ± 0.014
Treatment mean	0.762 ± 0.075	0.617 ± 0.008	

*Mean values bearing same or no superscript row wise do not differ significantly (P<0.05).

with control.

Evaluation of effect of treatment (VT45) and control on storage stability

A significant (P<0.05) effect of treatment as well as storage period on TBA values of chicken tikka during the observation period and the interaction between treatment and storage period was also found to be highly significant (P<0.05). The initial TBA (mg malonaldehyde/kg) value for Con group was found to be 0.362±0.004 and for VT45 it was 0.351±0.004. These values increased significantly (P<0.05) during refrigerated storage reaching 1.296±0.001 for Con and 0.909±0.018 for VT45 on 14th day of observation. The overall mean values of TBA for Con and VT45 groups were significantly (P<0.05) different, which were found to be 0.762±0.075, and 0.617±0.008 respectively (Table 6). With the advancement of storage period there was increase in the TBA values of the treatments, this might be due to the increased lipid oxidation and production of volatile metabolites in the presence of oxygen during storage and aerobic packaging (Kumar and Sharma, 2004).

A significant (P<0.05) effect of treatment as well as storage period on pH values of chicken tikka was observed during the observation period and the interaction between treatment and storage period was not found to be significant (P<0.05) (Table 6). The overall mean value of VT45 group was significantly (P<0.05) higher than the Con group. A significantly (P<0.05) increasing trend was also observed in pH with the advancement of storage period. The overall mean value for the test group was 6.152±0.040 on 0th day and increased to 6.267±0.038 on 14th day of observation (Table 6). Increasing trend observed in pH during storage may be attributed to proteolysis, due to bacterial growth. The breakdown products of proteins contributed to

increase in the pH of product. However, Cremer and Chipley (1997) reported an increase in pH during storage of low fat pork patties attributed to proteolysis due to bacterial growth.

A highly significant (P<0.01) increasing trend was also observed in TPC with the advancement of storage period (Table 7). The overall mean value for test group was 3.368±0.001 log cfu/g on 0th day and increased to 5.027±0.005 log cfu/g on 14th day of observation. The products under treatment maintained lower TPC values than the control throughout the storage period and were within the limit as proposed by (Froning et al., 1971) for cooked meat products that is, 5.33×10^{3} cfu/g for total plate count. Tandon, 1974 also reported that microbial loads in turkey frankfurters showed some increase in total counts during refrigeration storage. Hobbs (1983) reported a bacterial count of 7.8×10^{7}/g in chicken sausages stored at 5°C for 7 days. The lipolytic count of Con and VT45 group were 2.944±0.001 and 2.846±0.001 log cfu/g respectively on 0th day of storage which increased significantly to 4.007±0.031 log cfu/g in Con and 3.971±0.014 log cfu/g in VT45 group respectively on 14th day of observation (Table 7). A significant (P<0.05) increasing trend was also observed with the advancement of storage period. The overall mean value for the test group increased to 3.971±0.014 log cfu/g on 14th day of observation.

The coliforms were not detected on 0th day in Con and VT45 group On 7th day coliform count were observed in both Con and VT45 group but Con showed a significantly (P<0.05) higher value that is, 1.316±0.001 log cfu/g than the treatment (Table 7). The higher coliform count observed in the processed treatments compared with fresh might be caused by contamination due to either the equipment or the ingredients. Coliforms are common contaminants in fresh and processed foods that are often detected in food processing area (Jay, 1992; Anand et al., 1992). The yeast and mold colonies were not

Table 7. Effect of treatment and control on microbiological quality (log cfu/gm) of chicken tikka during storage.

Storage period	Treatment		Day Mean
	Control	Vacuum Tumbled (VT45)	
Total plate count			
0 day	3.398 ± 0.001[c]	3.368 ± 0.001[f]	3.383 ± 0.001
7 day	4.266 ± 0.001[bc]	4.190 ± 0.023[e]	4.228 ± 0.008
14 day	5.178 ± 0.002[ad]	5.027 ± 0.005[d]	5.102 ± 0.003
Treatment Mean	4.281 ± 0.001	4.195 ± 0.009	
Lipolytic count			
0 day	2.944 ± 0.001[cf]	2.846 ± 0.001[f]	2.896 ± 0.001
7 day	3.609 ± 0.005[be]	3.511 ± 0.004[e]	3.560 ± 0.004
14 day	4.077 ± 0.031[ad]	3.971 ± 0.014[d]	4.028 ± 0.022
Treatment Mean	3.543 ± 0.012	3.442 ± 0.006	
Coliform count			
0 day	ND	ND	ND
7 day	0.682 ± 0.001[be]	0.631 ± 0.001[e]	0.656 ± 0.001
14 day	1.239 ± 0.001[ad]	1.180 ± 0.155[d]	1.209 ± 0.078
Treatment Mean	0.640 ± 0.001	0.603 ± 0.001	
Yeast and Mold count			
0 day	ND	ND	ND
7 day	0.682 ± 0.001[be]	0.631 ± 0.001[e]	0.656 ± 0.001
14 day	1.239 ± 0.001[ad]	1.180 ± 0.155[d]	1.209 ± 0.078
Treatment Mean	0.640 ± 0.001	0.603 ± 0.001	

*Mean values bearing same or no superscript row wise do not differ significantly (P<0.05).

detected on 0th day in all groups but on 7th day onward it was detected in Con and VT45 group. The overall mean value of yeast and mold count for VT45 group was found to be 1.180±0.155 log cfu/g and for Con group it was 1.239±0.001 log cfu/g (Table 7). There was significant increase in yeast and mold count from 7th day to 14th day of observation period. Sharma (1999) reported yeast and mold counts of chicken products to be very low at the time of preparation of product but, increased as high as 3.82 log cfu/gm during refrigeration storage. Biswas et al., 2006 reported that fungi got upper hand over bacteria in meat and meat products.

A significant declining trend for appearance/color score was observed with the advancement of storage period (Table 8). The overall mean for appearance/color score was found to be 6.240±0.023 for Con group at 14th day and for treatment it was 6.616±0.050. Jacobson and Koehler, 1970 reported that all the sensory quality values decreased significantly with the advancement of storage period. A declining trend for flavor score was observed in both groups with the advancement of storage period. The overall mean value for VT45 group was found to be 6.394±0.063 on 0th day and a drastic decrease in flavor score was observed on 14th day of observation (Table 8).

Nath (1992) also stated that pronounced flavor changes were observed in refrigerated samples at 4 °C. The TBA values correlated significantly with flavor changes indicating that oxidative changes occurred as flavor deterioration during refrigerated storage.

A significant decreasing trend in the mean value for texture during storage period was observed which were 7.019±0.011 on 0th day and 6.092±0.011 on 14th day of observation (Table 8). A significant reduction in the juiciness score was observed with the advancement of storage period. On 0th day of observation the overall mean value for juiciness in Con was 5.996±0.038 and 7.229±0.019 for VT45 group and on 14th day of observation it was reduced to 5.481±0.038 and 6.493±0.020 respectively (Table 8). The overall acceptability no doubt, decreased during storage, for both samples. These findings are in agreement with those of Mandal (1993.). On going through the above findings, it can be concluded that for preparation of chicken tikka, with the use of vacuum tumbling, the product vacuum tumbled for 45 min improved the physico-chemical, microbiological and sensory quality of the product and it can be stored at refrigeration temperature for 14 day in LDPE bags with good overall acceptability.

Table 8. Effect of treatment and control on sensory attributes of chicken tikka during storage.

Storage period	Treatment		
	Control	Vacuum tumbled (VT45)	Day mean
Appearance/color			
0 day	6.411 ± 0.015^a	6.777 ± 0.047^{da}	6.594 ± 0.031
7 day	6.341 ± 0.023^b	6.716 ± 0.053^{db}	6.528 ± 0.038
14 day	0.240 ± 0.023^c	6.616 ± 0.050^{cc}	6.428 ± 0.036
Treatment Mean	6.330 ± 0.020	6.703 ± 0.050	
Flavor			
0 day	6.019 ± 0.109^a	6.394 ± 0.063^{da}	6.206 ± 0.086
7 day	5.636 ± 0.305^b	6.012 ± 0.063^{ea}	5.824 ± 0.184
14 day	4.896 ± 0.109^c	5.270 ± 0.062^{fc}	5.083 ± 0.085
Treatment Mean	5.517 ± 0.174	5.892 ± 0.062	
Texture			
0 day	6.015 ± 0.027^a	7.019 ± 0.011^{da}	6.517 ± 0.019
7 day	5.925 ± 0.251^a	6.674 ± 0.012^{ea}	6.299 ± 0.131
14 day	5.340 ± 0.251^b	6.092 ± 0.011^{fb}	5.716 ± 0.131
Treatment Mean	5.760 ± 0.176	6.595 ± 0.011	
Juiciness			
0 day	5.996 ± 0.038^a	7.229 ± 0.019^{da}	6.613 ± 0.028
7 day	5.635 ± 0.038^b	6.870 ± 0.279^{eb}	6.252 ± 0.158
14 day	5.481 ± 0.038^c	6.718 ± 0.020^{ec}	5.987 ± 0.029
Treatment Mean	5.704 ± 0.038	6.846 ± 0.106	
Overall acceptability			
0 day	6.118 ± 0.031^a	6.991 ± 0.016^{da}	6.554 ± 0.023
7 day	5.635 ± 0.034^b	6.521 ± 0.023^{eb}	6.252 ± 0.028
14 day	5.342 ± 0.037^c	6.227 ± 0.022^{fc}	6.106 ± 0.029
Treatment Mean	5.699 ± 0.034	6.579 ± 0.020	

*Mean values bearing same or no superscript row wise do not differ significantly ($P<0.05$).

REFERENCES

Sams AR (2001). Poultry meat processing edition, Boca Raton, Florida, USA, CRC Press.

Barbut S (2005). Poultry products processing. First Indian reprint. CRC press LLC.

Treharne T (1971). Growing interest in Meat tumbling. Food manufacture, October, pp. 35-39.

Rust RE, Olson DG (1973). Meat curing principles and modern practice. Koch supplies, Inc., Kanas city.

Vartorella T (1975). Canned pork. M.Sc. Dissertation. The Ohio State Uni. Columbus, OH. USA.

Krause R (1976). Influence of tumbling and sodium triopolyphosphate on quality, yield, and cure distribution in hams. M.Sc. Dissertation. The Ohio State University, Columbus, OH. USA.

Egbert WR, Huffman DL, Chen, CM, Jones WR (1992). Microbial and oxidative changes in low fat ground beef during simulated retail distribution. J. Food Sci., 57(6): 1269-1274, 1293.

Zaltkis A, Zak B, Boyle HJ, Mich D (1953). A new method for direct determination of serum cholesterol. J. Lab. Clin. Med., 41: 486-936.

Rajkumar V, Agnihotri MK, Sharma N (2004). Quality shelf life of vacuum and aerobic packed chevon patties under refrigeration. Asian Aust. J. Anim. Sci., 17(4): 548-553.

AOAC (1984). Official Methods of Analysis. 16th edn. Association of official Analytical Chemists, Will Behington, D.C

Tarladgis BG, Watts BM, Younathan MT, Dugan LR (1960). A distillation method for the quantitative determination of malonaldehyde in rancid foods. J. Am. Oil Chem. Soc., 37: 44-48.

APHA (1992). Compendium of Methods for Methods for the Microbiological Examination of Foods 2nd edn. (ed. M. L. Speak). Am. Pub. Hlth. Assoc., Will Behington, D.C.

Keeton JT, Foegeding EA, Patina AC (1984). A comparison of non meat products, sodium tripolyphosphate and processing temperature effects on physical and sensory properties of frank furthers. J. Food Sci., 49: 1462-1474.

Snedecor GW, Cochran WG (1994). Statistical methods. First East West press edition, New Delhi.

Dzudie T, Okubanjo A (1999). Effect of rigor state and tumbling time on quality of goat hams. J. Food Engr., 42: 103-107.

Muller WB (1991). Effect of method of manufacture. Fleischwirtschaft, 71: 8, 10-18, 60.

Ghavimi B, Roigers RW, Althan TG, Ammerman GR (1986). Effects of non-vacuum, vacuum and nitrogen back flush tumbling on various characteristics of restructured cured beef. J. Food Sci., 51: 1166-1168.

Froning GW, Sackett B (1985). Effect of salt and phosphates during tumbling of turkey breast muscle on meat characteristics. Poultry Sci., 64: 1328-1333.

Plimpton RF, Perkins CJ, Sefton TL, Cahill VR (1991). Rigor Condition, Tumbling and Salt level Influence on Physical, Chemical and Quality Characteristics of Cured, Boneless Ham. J. Food Sci., 56: 1514-1528.

Cassidy RD, Ockerman HW, Krol B, Van RPS, Plimpton RF, Cahill VR (1978). Effect of tumbling method, phosphate level and final cook temperature on the histological characteristics of tumble porcine muscle tissue. J. Food Sci., 43: 1514-1518.

Ledward DA (1979). Meat. Effect of heating on foodstuffs. In R.J. Priestley (ed.). London: Applied Science Publishers, pp. 121-157.

Ang CYW, Hamm D (1982). Proximate analysis, selected vitamin and mineral and cholesterol content of mechanically deboned and hand-deboned broiler parts. J. Food Sci., 47: 885-888.

Sharma BD, Kumar S, Nanda PK (2002). Optimization of short term tumbling schedule for the processing of cured and restored buffalo meat blocks. Indian J. Anim. Sci., 72(8): 684-688.

Theno DM, Siegel DG, Schmidt GR (1978). Meat massaging: Effect of salt and phosphate on the ultra structure of cured porcine muscle. J. Food Sci., 43: 488-492.

Babji AS, Froning GW, Ngoka DA (1982). The effect of short term tumbling and salting on the quality of turkey breast meat. Poult. Sci., 61: 300-303.

Katsaras K, Budras KD (1993). The relationship of the microstructure of cooked ham to its properties and quality. Lebensmitted Wissenschaft und Technologie, 26: 229-234.

Pohlman FW, Stivarius MR, McElyea KS, Waldroup AL (2002). Reduction of E.coli, Salmonella, Typhimurium, Coliform, aerobic bacteria and improvement of ground beef color using trisodium phosphate or centylpyridium chloride before grinding. Meat Sci., 60: 349-356.

Ockerman HW, Plimpton JR, Cahill VR, Parrett NA (1978). Influence of short term tumbling, salt and phosphate on cured canned pork. J. Food Sci., 43: 878-881.

Leak FW, Kemp JD, Langlois BE, Fox JD (1984). Effect of tumbling and tumbling time on quality and micro flora of dry cured hams. J. Food Sci., 49: 695-698.

Kamchorn T Sebranek WG, Topel DG, Rust RE (1983). Use of vacuum during formation of meat emulsion. J. Food Sci., 48: 1039-41.

Weiss JM (1973). Ham tumbling and massaging. A special report printed from Western Meat Industry. Oct, 23.

Theno DM, Siegal DG, Schmidt GR (1976). Microstructure of sectioned and formed hams. University of Illinos, Urbana, IL. J. Anim. Sci., May, Abstract 40.

Krause RJ, Ockrman HW, Krol B, Moerman PC, Plimpton RF (1978). Influence of tumbling, tumbling time, trim and STPP on quality and yield of cured hams. J. Food Sci., 43: 853-855.

Harmon CJ, Means WJ, Kemp, JD (1992). Blind, sensory and chemical properties of restructured dry cured hams. J. Food Sci., 48: 1039-1040.

Acton JC (1972). Effect of heat processing on extractability of salt soluble proteins, tissue binding strength and cooking losses of poultry loves. J. Food Sci., 37: 244-46.

Pietrasik Z, Shand PJ (2005). Effect of mechanical treatments and moisture enhancement on the processing characteristics and tenderness of beef semi-membranous roast. Meat Sci., 71(3): 498-505.

Udayasaha MTRK, Kowala BN (1999). Effect of EDTA and ascorbic acid drip treatment on storage stability of goat meat at refrigeration temperature. J. Food Sci. Technol., 36 (2): 180-183.

Kumar M, Sharma BD (2004). The storage stability and textural physico-chemical and sensory quality of low fat ground pork patties with carrageenan as fat replacer. Int. J. Food Sci. Technol., 39: 31-42.

Cremer ML, Chipley JR (1997). Satellite food service system: Time and temperature and microbiological and sensory quality of precooked frozen hamburger patties. J. Food Prot., 40: 603-607.

Froning GW, Arnold RC, Mandigo RW, Neth CE, Hartung TE (1971). Quality and storage stability of frankfurters containing 15% mechanically deboned turkey meat. J. Food Sci., 36(7): 974-978.

Tandon PL (1974). Studies on the effect of two methods of processing on keeping quality of chicken sausage at different storage temperatures. M.V.Sc. Dissertation. Agra Univ., Agra. India.

Hobbs G (1983). Microbial spoilage of fish. In: Food Microbiology; Advances and prospects, (Ed. Roberts, T.A. and Skinner, F.A.) New York, Academic Press, p. 217.

Jay JM (1992). Modern Food Microbiology. 4th edn. New York, Van Nostrand Reinhold, p. 690.

Anand SK, Pandey NK, Mahapatra CM (1992). Changes in microflora during preparation and storage of chicken tandoori. Indian J. Poult. Sci., 27(1): 51-53.

Sharma BD (1999). Meat and meat products technology, Jaypee Brothers Publishers, pp. 85-87.

Biswas S, Chakraborty A, Sarkar S (2006). Comparison among the qualities of patties prepared from chicken broiler, spent hen and duck meats. J. Poult. Sci., 43: 180-186.

Jacobson M, Koehler HH (1970). Development of rancidity during short time storage of cooked poultry meat. J. Agric. Food Chem., 18: 1069.

Nath RL (1992). Effects of addition of chicken fat and cooking methods on the quality of chicken patties. M.V.Sc Dissertation. IVRI Deemed University, Izatnagar, India.

Mandal PK (1993). Effect of cooking method on the quality of chicken meat balls. MVSc Dissertation. IVRI Deemed University, Izatnagar, India.

6

Sennoside contents in Senna (*Cassia angustifolia Vahl.*) as influenced by date of leaf picking, packaging material and storage period

Anubha Upadhyay, Yogendra Chandel, Preeti Sagar Nayak* and Noor Afshan Khan

Department of Crop and Herbal Physiology, College of Agriculture, Jawaharlal Nehru Krishi Vishwa Vidyalaya (JNKVV), Jabalpur (M.P.) 482004, India.

A field experiment was conducted in the Senna (*Cassia angustifolia Vahl.*) during the Rabi season of 2008 to 2009 in the Dusty Area of the Department of Crop and Herbal Physiology, Jawaharlal Nehru Krishi Vishwa Vidyalaya. The experiment was carried out in a complete randomized design asymmetrical factorial with three treatments and three replications. *Cassia angustifolia* leaves were picked at three different stages of growth period viz. 90, 110 and 130 DAS (days after sowing) with different type of packaging material and storage period. The Sennoside A and B were determined by High Performance Thin Layer Chromatography. The Sennoside A and B content were found to be highest in leaf picking stage of 90 DAS followed by 130 and 110. Maximum contents of Sennoside A and B were estimated in black polythene followed by aluminum foil bag and transparent polythene.

Key words: Senna, *Cassia angustifolia,* Sennoside A and B, leaf picking stage, HPTLC.

INTRODUCTION

Cassia angustifolia (family Caesalpiniacea) popularly known as Senna, is a valuable plant drug in Ayurveda and modern system of medicine for the treatment of constipation (Atal and Kapoor, 1982; Das et al., 2003; Martindale, 1977; Sharma, 2004). Cultivation of Senna does not require much expenditure on irrigation, manuring, pesticides, protection and other pre- and post harvest care. This makes the plant ideal crop for acid regions where water provision, wasteland development, desertification control and sand dune stabilization are the major challenges (Tripathi, 1999). Senna is a sun-loving crop and requires bright sunshine for its successful growth. The crop is raised from seed and has a hard and tough seed-coat, ascertain amount of abrading of its surface is necessary to induce quick germination. The Sennosides had been extracted from Senna leaves, stems, pods, buds and flowers but no Sennosides were found in the seeds.

Sennoside levels in native Senna plants were similar to those found growing in other countries (Babash et al., 1985). The laxative property of Senna is based on two glycosides viz. Sennoside A and sennoside B whereas Sennosides C and D have also been reported in the plant. These Sennosides are the aloe-emodin diantraone diglucosides. Apart from Sennosides, the pod and leaf also contain glycosides of anthraquinones, rhein and chrysophenic acid. Recently, two naphthalene glycosides have also been isolated from leaves and pods (Gupta, 2008). The flavanoids that are reported from this plant are Kaenferol, Kaempterin and isorhamnetin. It also contains beta-sitosterol (0.33%). From the germplasm collected in Tamil Nadu, Andhra Pradesh and Gujarat, India the superior accession 'Sona' was derived, following evaluation and selection for leaf and pod yield. Cultivar Sona (released in 1997) is 115 to 120 cm tall, leafy (mean leaf area 11.28 cm^2), highly branched, flowering after 100 to 115 days, large number of pods plant^{-1} (5 cm long and 1.25 to 1.75 cm wide). Dry leaf and pod yield is 11 and 4 quintals ha^{-1}, respectively, while sennoside content average 3.5% (Singh et al., 1997). The deflowering

*Corresponding author. E-mail: preetisagarnayak@rediffmail.com.

Table 1. Effect of storage material and duration on the content of Sennoside A and B in Senna leaves picked at 90 DAS.

Storage period	Packaging materials					
	Sennoside A (%)			Sennoside B (%)		
Month	P_1	P_2	P_3	P_1	P_2	P_3
S_0 (0) Initial content		0.126			0.069	
S_1 (1)	0.123	0.117	0.113	0.067	0.065	0.064
S_2 (2)	0.113	0.110	0.107	0.063	0.061	0.060
S_3 (3)	0.100	0.095	0.088	0.057	0.054	0.055
S_4 (4)	0.088	0.077	0.081	0.048	0.046	0.047
S_5 (5)	0.065	0.060	0.058	0.043	0.041	0.042
S_6 (6)	0.054	0.045	0.045	0.036	0.034	0.035

P1- Black polythene bags, P2- Aluminum foil bags, P3- Transparent polythene bags.

increased Sennoside A and B concentration (percent dry weight) in leaves by 25%, total leaf dry mass by 63% and harvest index by 22% with the result that the Sennoside A and B yield (grams) per plant doubled in response to flowering. During the day time, net photosynthesis remained consistently lower in the deflowered plants. Youngest leaves had the greatest Sennoside A and B concentration. A clone raised from cuttings of one seedling had lower Sennoside A: B ratio than the plants raised from the seedlings (Ratnayaka et al., 2002). In the traditional medicine, *C. angustifolia* is used for its purgative properties. A new anthraquinone glycoside (emodin 8-O-sophoroside), and seven known glycosides, were isolated from the leaves of *C. angustifolia* and their structures were elucidated by spectral analysis (Kinjo et al., 1994). The *Cassia-angustifolia* is used in traditional medicine for its cathartic properties and in the treatment of tumours. Kaempferol 3-O- beta -glucoside and isorhamnetin 3-O- beta -glucosides were newly isolated in its leaves (Singh et al., 1995). No study was carried out about the growth stages on which the crop should be harvested to obtain the maximum sennoside content and how long its storage durability. Therefore, the said study was undertaken to evaluate the effect of different growth stages on sennoside content of Senna and after that its post-harvest technology. The main objectives were to observe the sennoside content in Senna leaves at different dates of leaf picking and the impact of packaging materials and storage periods.

MATERIALS AND METHODS

The experiment was conducted in the Department of Crop and Herbal Physiology at the dusty area, Jawaharlal Nehru Krishi Vishwa Vidyalaya Jabalpur, Madhya Pradesh during the year 2008 to 2009. The experiment was carried out in a complete randomized design (CRD) asymmetrical factorial with three treatments and three replications. The sowing date was 3rd July, 2008. The Leaf - picking stages (day after sowing) were 90 days (G_1), 110 days (G_2) and 130 days (G_3). The

packaging materials (storage containers) were black polythene (P_1), aluminum bag (P_2) and transparent polythene (P_3). The storage time (duration after packaging) were having one month intervals up to six month (S_1 to S_6). The Sennoside A and B (%) were estimated by High Performance Thin Layer Chromatography (Rajpal, 2002). The statistical analysis of data taken on different variables was carried out through CRD asymmetrical factorial design to know the degree of variation among the treatments.

RESULTS AND DISCUSSION

Initial observation before packaging

The initial content of Sennoside A and B in Senna leaves were estimated before packaging to be 0.126 and 0.069%, 0.112 and 0.064%, 0.114 and 0.066% at 90%, 110% and 130% DAS respectively.

Storage and leaf picking effect on Sennoside A and B

The analysis of Sennoside was carried out in different types of packaging materials and storage periods. The results are presented in Tables 1 to 4 shows there were significant differences in the Sennoside levels in the different containers. The highest content of Sennoside A and B were 0.123 and 0.067%, 0.110 and 0.063%, 0.113 and 0.064% in leaves picked at 90%, 110% and 130% DAS respectively in black polythene bag (P_1). During the storage minimum content of Sennoside A and B that is, 0.113 and 0.064%, 0.100 and 0.060%, 0.108 and 0.062% were recorded in transparent polythene bag (P_3) at 90%, 110% and 130% DAS respectively. Hence the Sennosides content were highest in black polythene bag (P_1), followed by aluminium foil bag (P_2) and the least in transparent polythene bag (P_3). After one month of storage, Sennoside A and B

Table 2. Effect of storage material and duration on the content of Sennoside A and B in Senna leaves picked at 110 DAS.

Storage period (Month)	Packaging materials					
	Sennoside A (%)			Sennoside B (%)		
	P_1	P_2	P_3	P_1	P_2	P_3
S_0 (0) Initial content		0.112			0.064	
S_1 (1)	0.110	0.105	0.100	0.063	0.061	0.060
S_2 (2)	0.101	0.095	0.090	0.057	0.054	0.053
S_3 (3)	0.095	0.091	0.087	0.054	0.052	0.050
S_4 (4)	0.080	0.074	0.068	0.045	0.043	0.042
S_5 (5)	0.061	0.056	0.054	0.041	0.040	0.039
S_6 (6)	0.050	0.044	0.042	0.034	0.033	0.031

P_1- Black polythene bags, P_2- Aluminum foil bags, P_3- Transparent polythene bags.

Table 3. Effect of storage material and duration on the content of Sennoside A and B in Senna leaves picked at 130 DAS.

Storage period (Month)	Packaging materials					
	Sennoside A (%)			Sennoside B (%)		
	P_1	P_2	P_3	P_1	P_2	P_3
S_0 (0) Initial content		0.114			0.066	
S_1 (1)	0.113	0.110	0.108	0.064	0.062	0.062
S_2 (2)	0.102	0.100	0.097	0.060	0.057	0.055
S_3 (3)	0.095	0.089	0.087	0.055	0.053	0.051
S_4 (4)	0.086	0.084	0.070	0.047	0.044	0.043
S_5 (5)	0.064	0.061	0.052	0.042	0.041	0.040
S_6 (6)	0.052	0.041	0.040	0.035	0.033	0.032

P_1- Black polythene bags, P_2- Aluminum foil bags, P_3- Transparent polythene bags.

Table 4. Interaction of packaging material and storage period on content of Sennoside A and B.

Treatments	Sennoside A (%)			Sennoside B (%)		
	SEd	SEm	CD 5%	SEd	SEm	CD 5%
G	0.0002	0.0002	0.0004	0.0002	0.0001	0.0004
P	0.0002	0.0002	0.0004	0.0002	0.0001	0.0004
GXP	0.0004	0.0003	0.0007	0.0003	0.0002	0.0007
S	0.0003	0.0002	0.0006	0.0003	0.0002	0.0005
GXS	0.0005	0.0004	0.0011	0.0005	0.0003	0.0009
PXS	0.0005	0.0004	0.0011	0.0005	0.0003	0.0009
GXPXS	0.0009	0.0007	0.0018	0.0008	0.0006	0.0016

G - Growth stages, P - Packaging materials, S - Storage period.

contents were maximum in leaves picked at 90, 110 and 130 DAS respectively in P_1 and were found at par with P_2 followed by P_3. At two months, of storage Sennoside A and B contents were found maximum with (0.113 and 0.063%, 0.101 and 0.057%, 0.102 and 0.060%) in leaves picked at 90, 110 and 130 DAS respectively in P_1 and were statistically at par with P_2 and P_3. Minimum Sennoside A and B were found observed in transparent polythene bags with values of 0.107 and 0.060%, 0.090 and 0.053%, 0.097 and 0.055%) for leaves picked at 90, 110 and 130 DAS respectively.

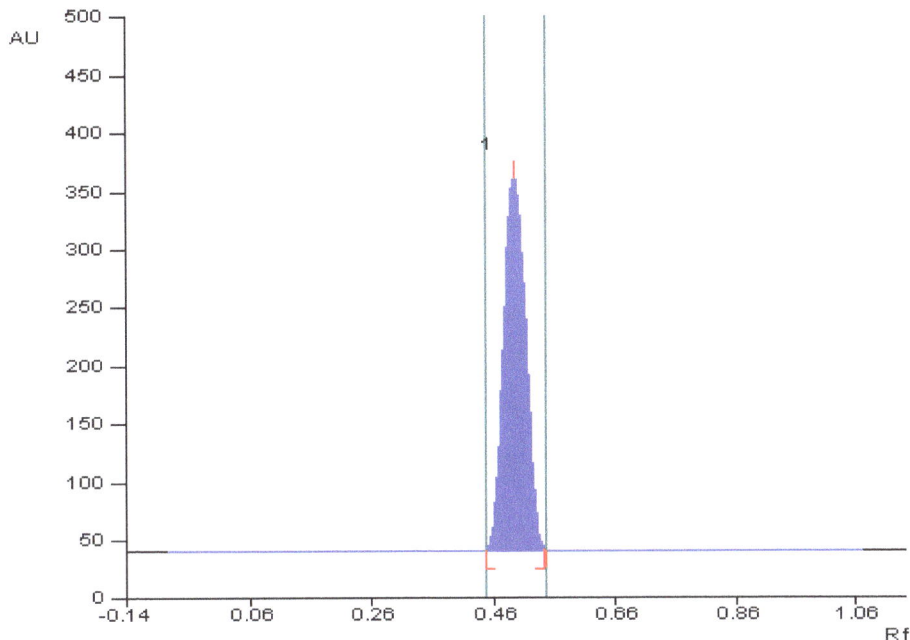

Figure 1. Chromatogram of Sennoside A standard.

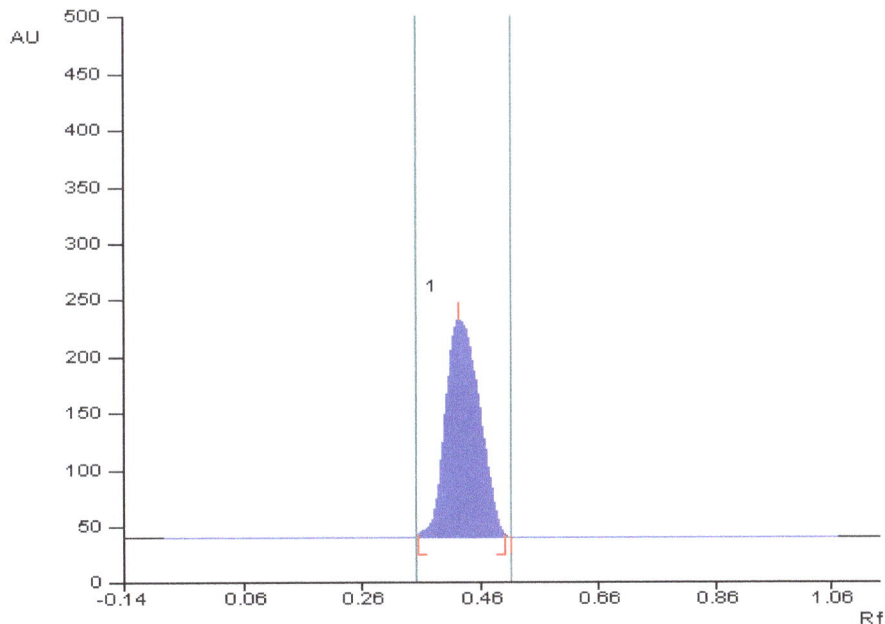

Figure 2. Chromatogram of Sennoside A content in leaf picked at 90 DAS and stored in black polyethylene bag at the 1st month.

At three month of storage, the significantly highest Sennoside A and B contents that is, 0.100 and 0.057%, 0.095 and 0.054%, 0.095 and 0.055% at 90, 110 and 130 DAS respectively were estimated in black polythene bags and least were (0.088 and 0.055%, 0.087 and 0.050%, 0.087 and 0.051%) at 90, 110 and 130 DAS respectively noted in transparent polythene bags P_3 (Figures 1 to 6).

At four month of storage, the maximum contents of Sennoside A and B were recorded to be that is, 0.088 and 0.048%, 0.080 and 0.045%, 0.086 and 0.047% at 90, 110 and 130 DAS respectively in

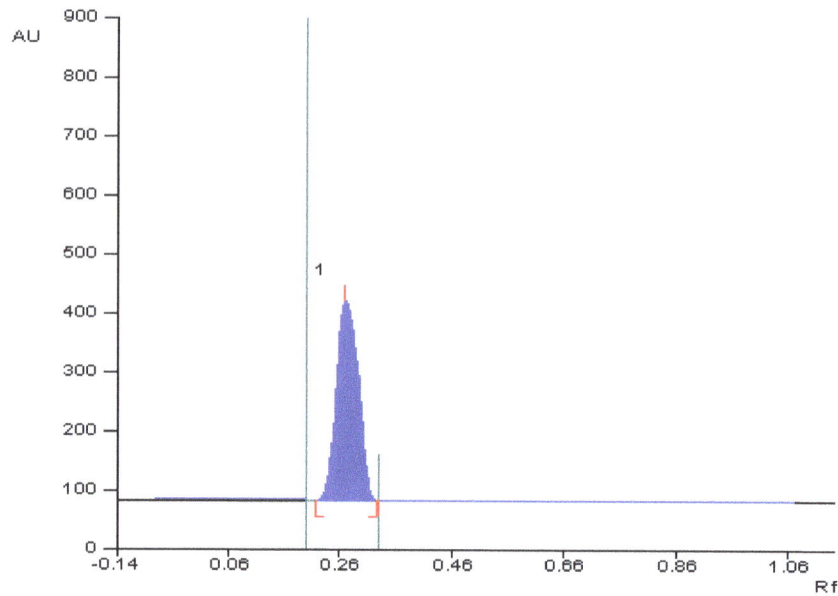

Figure 3. Chromatogram of Sennoside A content in leaf picked at 90 DAS and stored in a black polyethylene bag at the 6th month.

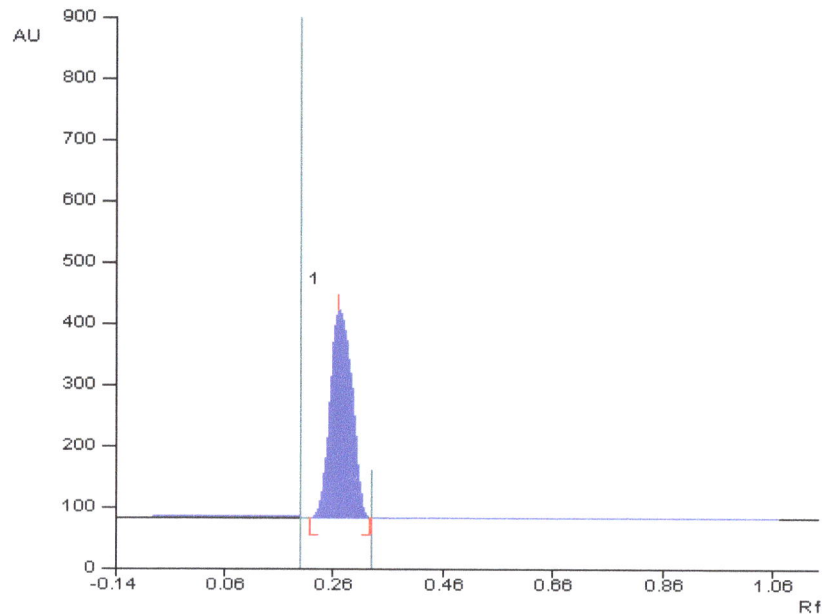

Figure 4. Chromatogram of Sennoside B standard.

black polythene P_1. The least contents were observed in transparent polythene bags (0.081 and 0.047%, 0.068 and 0.042%, 0.070 and 0.043%) in leaves picked at 90, 110 and 130 DAS.

At the fifth and sixth months, of storage period, the trend continued with leaves stored in black polythene bags having the maximum Sennoside A and B levels whereas the lowest was found in the transparent polythene bags. Though these values were lower than what was obtained in the previous months.

The results revealed that Sennoside A and B contents were significantly reduced in different storage containers that is, P_1, P_2 and P_3 (6.17 and 10.46%). Furthermore, it was found that the Sennoside A and B were significantly reduced with

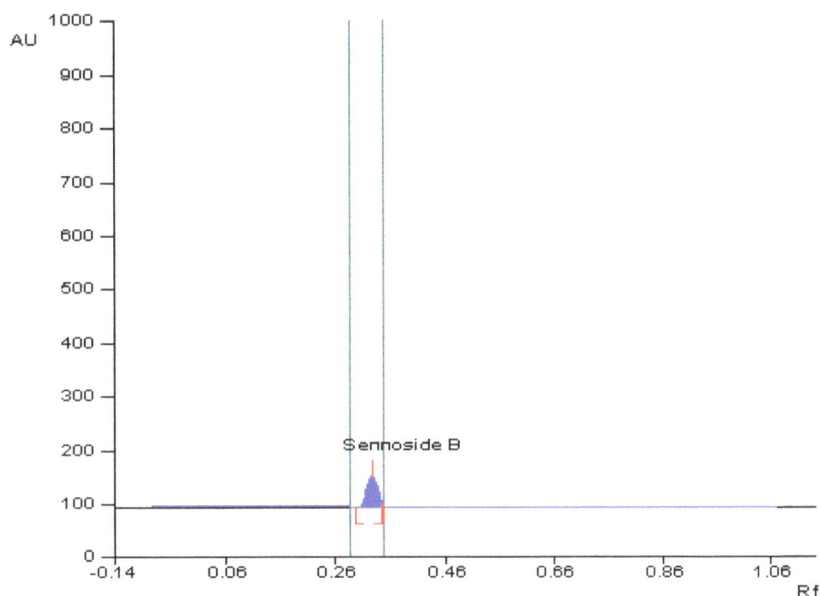

Figure 5. Chromatogram of Sennoside B content in leaf picked at 90 DAS and stored in a black polythene bag at the 1st month.

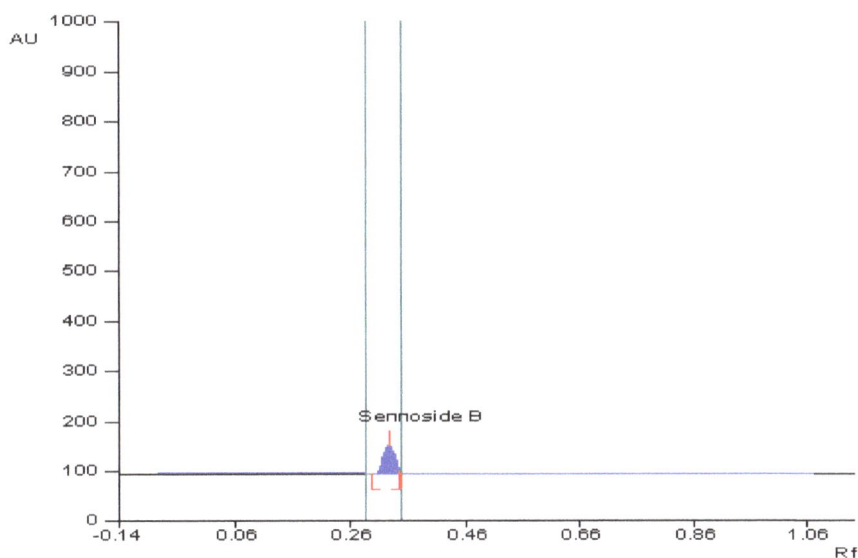

Figure 6. Chromatogram of Sennoside B content in leaf picked at 90 DAS and stored in a black polythene bag at the 6th month.

increasing storage period from the first month till the sixth month. The Sennoside A and B in all three leaves picking (90 DAS, 110 DAS and 130 DAS) were packed in different three packaging materials that is, black polythene bags (P_1), aluminum foil bags (P_2) and transparent polythene bags (P_3) and stored for six month. Maximum content of Sennosides A and B was obtained in freshly harvested Senna leaves but was found to decline significantly for rest of the storage period up to six months. The losses of Sennosides A and B in storage were found to be lowest in black polythene bags as compared to aluminum foil and transparent polythene bags. The storage of leaf should be protected from light and moisture (European Pharmacopoeia, 2005). Srivastava, et al. (1980) have reported that the glycosides is more in the leaves than in pods and it decreases with the age during the growth of plant, thus harvesting of the leaf crop between 40 and 70 days improved biological activity to the maximum in the crude drug. It is also found that prolonged exposure of growing Senna plants to cycle light and darkness affects the biosynthesis of anthracene derivatives in their leaves. They also found that in the absence of light (after 48 h), the conversion of non-rhein to rhein is suppressed as indicated by high proportion of non-rhein glycosides to total glycosides. Pareek, et al. (1983) has reported that total Sennosides content in stored produce fall after one year. Similar decrease was reported by Anonymous (2006) of total alkaloid content during storage in medicinal roots. Storage methods have also been reported to have a significant effect on saponin content. Decrease in saponin contents was observed at the later period of storage. Significantly higher saponin contents were retained due to storage in polythene bags. Similarly, Adom et al. (1996) reported that the powdered dry okra kept in polythene package survived storage best. So the package material and storage time significantly affected the quality parameters.

CONCLUSIONS

1) The Sennoside content in leaves was found highest in those picked at 90 DAS and was lowest at 110 days after sowing. So it is suggested that the senna leaves should be harvested at 90 days after sowing to get maximum Sennosides contents.

2) The best packaging material for storage of Senna leaves was found to be black polyethylene bags, then aluminum foil bags and transparent polyethylene bags. The reduction of Sennoside was found to be lowest in black polyethylene bags. Further, the storage of the leaves should be done in black polythene or in any dark container so that the deterioration of Sennoside should be minimum.

3) In conclusion, it is suggested that the Senna leaves should be harvested at 90 days after sowing to get maximum Sennosides contents. Secondly, dark-coloured packaging material should be used to store in order to minimize the degradation of Sennosides.

REFERENCES

Adom KK, Dzogbefia VP, Ellis WO, Simpson BK (1996). Solar drying of okra-effects of selected package materials on storage stability. Food Res. Int., 29(7): 589–593.

Anonymous (2006). Biennial report (2004-2005 and 2005-2006). Post harvest degradation of total alkaloid content in roots of Ashwagnadha. All India networking research project on medicinal and aromatic plants. CRMAP Anand, Gujrat, India, AUAC, pp. 350-351.

Atal CK, Kapoor BM (1982). Cultivation and utilization of Medicinal Plants. RRL, Jammu Tawi, India, p. 8.

Babash TA, Vandyshev VV, Perelson ME (1985). Presence of Sennosides in Senna and Rhubarb cultivated in the USSR. Khimiko-Farmatsevticheskii-Zhurnal, 19(6): 718-720.

Das PN, Purohit SS, Sharma AK, Kumar T (2003). A handbook of Medicinal Plants. Agrobios (India) Jodhpur, p. 118.

European Pharmacopoeia (2005). Senna leaf dry extract. Standardized, 2402-2403.

Gupta DP (2008). The herbs, Habitat, Morphology and Pharmacognosy of Medicinal Plants, pp. 357-358.

Kinjo J, Ikeda T, Watanabe K, Nohara T (1994). An anthraquinone glycoside from Cassia angustifolia leaves. Phytochem., 37(6): 1685-1687.

Martindale (1977) The Extra Pharmacopoeia. 27th edition, The Pharmaceutical House, Delhi, pp. 176-177.

Pareek SK (1983). Investigation in agronomic parameters of Senna (Cassia angustifolia Vahl) as grown in north western India. Int. J. Trop. Agric., 1: 139–144.

Rajpal V (2002). Standarization of Botanicals. Testing Extraction Methods Med. Herbs, 1: 54-65.

Ratnayaka HH, Meurer-Grimes B, Kincaid D (2002). Sennoside yields in Tinnevelly Senna affected by deflowering and leaf maturity. Hortic. Sci., 37(5): 768-772.

Sharma R (2004). Agro-Techiniques of Medicinal Plants. Daya Publishing House, Delhi, pp. 176-177.

Singh M, Chaudhuri PK, Sharma RP (1995). Constituents of the leaves of Cassia angustifolia. Fitoterapia, 66(3): 284.

Singh SP, Sharma JR, Misra HO, Lal RK, Gupta MM, Tajuddin S (1997). Development of new variety Sona of Senna (Cassia angustifolia). J. Med. Aromatic Plant Sci., 19(2).

Srivastava VK, Gupta R, Maheshwari ML (1980) Photocontrol of anthracene compounds formation in Senna (Cassia acutifolia Vahl) leaves. Chem. Abst., 94: 27411n.

Tripathi YC (1999). Cassia angustifolia: A versatile medicinal crop. Int. Tree Crops J., 10(2): 121-129.

7

Standardization of rapid and economical method for neutraceuticals extraction from algae

Prabuthas P.*, Majumdar S., Srivastav P. P. and Mishra H. N.

FST (Financial Services Technology) Laboratory, PHTC, Agricultural and Food Engineering Department, Indian Institute of Technology, Kharagpur, India – 721302.

Spirulina (cyanobacterium) posses a wide range of components useful in food and pharmaceutical industries. Pigments are intracellular compounds whose extraction involves cell disruption. Traditional extraction methods have several drawbacks like slow, time consuming utilize large quantity of solvents and less yield. So a systemic and economical approach was performed for maximum extraction of pigments using ultrasound-assistance. Phycocyanin concentration in the extract was calculated spectrophotometrically by measuring the absorbance at 615 and 652 nm. The purity of phycocyanin is evaluated based on the absorbance ratio of A_{620}/A_{280}. The optimized results obtained from both the solvents were analyzed. The result shows extraction with 1% calcium chloride solution of volume 45.27 ml, time 9.3 min and amplitude 79.72% gave the maximum yield of phycocyanin (0.20244 mg/ml). The maximum phycocyanin purity ratio of 0.62 was obtained in calcium chloride extract. The crude protein content of extracted samples has also been estimated by Kjeldahl method and it was found to vary from 18.23 to 63.63%, the highest being in distilled water extract.

Key words: *Spirulina*, phycocyanin.

INTRODUCTION

Spirulina platensis is a multifilamentous prokaryotic cyanobacterium and can be easily monocultured and harvested. C-phycocyanin (C-PC) is the major phycobiliprotein in *Spirulina*. This blue colored red fluorescing biliprotein of algae was first reported in 1928 by Lemberg (Eocha, 1963). The pigments mainly phycocyanin and chlorophyll has wide range of importance. So their estimation extraction has become the present need. Several factors can influence the phycocyanin extraction. The most important factors affecting phycocyanin yield are cellular disruption method, type of solvent and extraction time (Abalde et al., 1998; Reis et al., 1998).

Traditional methods such as water extraction in which *Spirulina* biomass were suspended and the pigments leached out were estimated spectrophotometrically. In view of the multiple uses of phycocyanin, we have used a new procedure for extraction of the pigments. Ultrasound is being used for disruption of cell and the disrupted cells

are centrifuged and analyzed spectrophotometrically. The wet biomass is immediately utilized by bacteria and starts degradation because of its nutritional composition. Hence to avoid these problems, use of dried biomass is suitable and convenient (Jayant, 2005). In this work, the best solvent for phycocyanin extraction from *S. platensis* was first investigated. Subsequently, the effects of temperature and biomass-solvent ratio on the phycocyanin concentration and extract purity were evaluated to establish the optimum conditions for phycocyanin extraction.

MATERIALS AND METHODS

Spirulina was grown in batch culture in Erlenmeyer flask containing standard Zarrouk's medium (Zarrouk, 1966). At the end of cultivation, the biomass was recovered by filtration, dried at 40°C for 48 h.

Experimental design

A statistical method, central composite rotate design (CCRD) with three independent variables (volume, time and amplitude) and two dependent variables (phycocyanin and chlorophyll concentrations) was used for designing the experiment. The statistical software

*Corresponding author. E-mail: prabuthas@gmail.com.

Table 1. Actual independent variables designed using Design Expert 7.0 software.

Experiment no	Volume (ml)	Time (min)	Amplitude (%)
1	45.27	9.3	36.22
2	104.73	9.3	36.22
3	45.27	20.57	36.22
4	104.73	20.57	36.22
5	45.27	9.3	83.78
6	104.73	9.3	83.78
7	45.27	20.57	83.78
8	104.73	20.57	83.78
9	25.00	15.00	60.00
10	125.00	15.00	60.00
11	75.00	5.00	60.00
12	75.00	25.00	60.00
13	75.00	15.00	20.00
14	75.00	15.00	100.00
15	75.00	15.00	60.00
16	75.00	15.00	60.00
17	75.00	15.00	60.00
18	75.00	15.00	60.00
19	75.00	15.00	60.00
20	75.00	15.00	60.00

Design Expert 7.0 has been used for designing and analyzing the experiment. A total of 20 experiments were performed. The independent variables were added in the design with 5 levels with equal difference (volume in range of 25 to 125 ml, time 5 to 25 min and amplitude 20 to 100%). Table 1 show the experimental design obtained through Design Expert 7.0 software.

Phycocyanin extraction with different solvents

Phycocyanin extraction was evaluated in terms of phycocyanin concentration using distilled water and 1% CaCl₂ solution (Abalde et al., 1998; Bermejo et al., 2003). The extraction was carried out by mixing 0.1 g of dried biomass with amount of the solvent, time and amplitude according to the experimental design. The samples were exposed to ultrasounds using a UP 50 H sonicator (Hielscher), followed by centrifugation for 15 min at 13500 rpm using refrigerated centrifuge model number BL156R (Biolab). The optical density of the supernatant was measured at 615 and 652 nm using an UV-VIS spectrophotometer (Elico). Phycocyanin concentration (PC), according to Bennett and Bogorad (1973), was defined as

$$\text{Phycocyanin} = \frac{(OD_{615} - 0.474(OD_{652})}{5.34}$$

The purity of C-PC is generally evaluated using the absorbance ratio of A_{620}/A_{280} "Optimization experiments was carried out using CCRD with varied levels of independent variables (volume, time and amplitude)"wherein a purity of 0.7 is considered as food grade, 3.9 as reactive grade and greater than 4.0 as analytical grade (Rito-Palomares et al., 2001).

The purity of the optimized experimental sample was monitored spectrophotometrically by the A615/A280 ratio (Abalde et al., 1998). This relationship is indicative of the extract purity of phycocyanin with respect to most forms of contaminating proteins.

RESULTS AND DISCUSSION

Influence of different solvents on extraction of pigments

The phycocyanin content was expressed in terms of mg/ml, which was determined from the equation described by Bennett and Bogorad (1973). The protein content was measured by using Kjeldahl method and expressed as percentage. The phycocyanin and protein content during the extraction with different solvents were calcium chloride and distilled water investigated.

The maximum amount of phycocyanin, 0.3116 mg/ml was obtained in calcium chloride solution followed by 0.299 mg/ml in distilled water. The maximum amount of protein, 63.63% was obtained in distilled water solvent and 54.69% in calcium chloride solution. So CaCl₂ solution was found to the best solvent for extraction of phycocyanin with ultrasound assistance.

Optimization of pigment extraction

Statistical analysis of all the 20 experiments was performed for both the solvent using the same software (Design Expert 7.0). The optimized result for the maximum yield of phycocyanin was derived. The input criteria for software, volume and time of extraction, should be minimum and amplitude of Sonicator should be in range with these criteria. The analysis was made and the result displayed in the Table 2.

Maximum phycocyanin concentration was obtained in

Table 2. Optimized phycocyanin concentration.

Criteria	Distilled water	Calcium chloride
Volume (ml)	45.27	45.27
Time (min)	9.3	9.3
Amplitude (%)	73.865	79.72
Phycocyanin (mg/ml)	0.20168	0.20244
Desirability	1.000	1.000
Purity ratio	0.50	0.62

Table 3. ANOVA for the phycocyanin extraction using distilled water.

Source of variation	Sum of squares	Degree of freedom	Mean square	F-Value
Model	0.057	9	6.366×10^{-3}	119.32(Significant)
Residual	5.335×10^{-4}	10	5.335×10^{-5}	
Lack of fit	4.431×10^{-4}	5	8.861×10^{-5}	4.90(not significant)
Pure error	9.045×10^{-5}	5	1.809×10^{-5}	
Total	0.058	19		

Table 4. ANOVA for the phycocyanin extraction using 1% calcium chloride solution.

Source of variation	Sum of squares	Degree of freedom	Mean square	F-Value
Model	0.062	9	6.907×10^{-3}	18.96 (Significant)
Residual	3.642×10^{-3}	10	3.642×10^{-4}	
Lack of fit	1.402×10^{-3}	5	2.805×10^{-4}	0.63 (not significant)
Pure error	2.240×10^{-3}	5	4.480×10^{-4}	
Total	0.066	19		

calcium chloride solution (0.20244 mg/ml) and followed by distilled water (0.20168 mg/ml). The result obtained with this method was compared with other traditional methods as described by Silveria et al. (2006) for *Spirulina* biomass taken in rotary shaker along with solvent, followed by centrifugation and spectrophoto-metric analysis. The result obtained with ultrasound shows the better yield of phycocyanin than the traditional method.

The analysis of variance made for both water and calcium chloride extracted sample results. The result for water and calcium chloride are presented in Tables 3 and 4 respectively. The results show that the model is best fit with lack of fit not significant.

The quadratic equation for results was made using the statistical software for phycocyanin. Equations (1) and (2) show the quadratic model for distilled water and calcium chloride solution results, respectively.

$$PC = 0.12 - 0.057A + 9.033 \times 10^{-4} B + 1.388 \times 10^{-3} C + 3.431 \times 10^{-4} AB - 1.094 \times 10^{-3} AC + 1.394 \times 10^{-3}BC + 0.029 A^2 - 1.908 \times 10^{-3} B^2 - 3.339 \times 10^{-3} C^2 \qquad (1)$$

$$PC = 0.12 - 0.062A + 2.393 \times 10^{-3} B + 3.547 \times 10^{-3} C + 1.513 \times 10^{-3} AB - 3.087 \times 10^{-3} AC + 7.625 \times 10^{-3}BC + 0.024 A^2 - 5.096 \times 10^{-3} B^2 - 6.209 \times 10^{-3} C^2 \qquad (2)$$

where PC is phycocyanin concentration in mg/ml; A = volume of solvent in ml; B = time of exposure to ultrasound in minutes, and C = amplitude of the sonicator in percent.

The purity of phycocyanin was determined by using the method as described by Liu et al. (2005) both the water and calcium chloride solution extracted sample were analyzed for all the experiments conducted. Of all, the 9th experiment of calcium chloride solvent extract shows the highest purity of phycocyanin of 0.682. Distilled water extracted sample has shown less purity than as prescribed for food grade purpose (0.6)

REFERENCES

Abalde J, Betancourt L, Torres E, Cid A, Barwell C (1998). Purification and characterization of phycocyanin from the marine cyanobacterium *Synechococcus* sp. IO9201, Plant Sci., 136: 109–120.

Bennett A, Bogorad L (1973). Complementary chromatic adaptation in a filamentous blue-green alga, J. Cell Biol., 58: 419–435.

Bermejo A, Acien FG, Ibanez MJ, Fernandez JM, Molina E, Alvarez-Pez JM (2003). Preparative purification of B-phycoerythrin from the microalga *Porphyridium cruentum* by expanded-bed adsorption chromatography, J. Chromatogr. B, 790: 317–325.

Eocha O (1963). Phycpbiliproteins from Spirulina, Biochemistry, 2: 375.

Jayant MD (2005). Drying of Spirulina biomass, Int. J. Food Eng., 1(5): 2.

Liu LN, Chen XL, Zhang XY, Zhang YZ, Zhou BC (2005). One step chromatography method for efficient separation and purification of R-phycoerythrin from *Polysiphonia urceolata*, J. Biotechnol., 116: 91-100.

Reis A, Mendes A, Lobo-Fernandes H, Empis JA, Novais JM (1998). Production, extraction and purification of phycobiliproteins from *Nostoc* sp., Bioresour. Technol., 66: 181–187.

Rito-Palomares M, Nunez L, Amador D (2001). Practical application of aqueous two-phase systems, J. Chem. Technol. Biotechnol., 76: 1273.

Silveria ST, Birkert JFM, Costa JAV, Burkert CAV, Kalil SJ (2006). Optimization of phycocyanin using factorial design, Bioresource Technol., 98: 1629-1634.

8

Competitiveness strategy of Avocado exporting companies from Mexico to USA

Joel Bonales Valencia*, Oscar Hugo Pedraza Rendón and José César Lenin Navarro Chávez

Institute of Economic and Business Research, Universidad Michoacana de San Nicolás de Hidalgo Morelia, Michoacán, Mexico.

This work investigated the competitive factors of 25 avocado exporting companies located in Uruapan, Michoacán (Mexico). This cradle in a census of the twenty-five exporting companies, with its organization, objectives and problematic production could be known. This research paper was focused on the knowledge of the main theories on competitiveness strategy. The article has the objective of identifying the companies of the avocado sector located in Uruapan, Michoacan, Mexico. Such identification is made in order to recognize the variables explaining its competitiveness and especially the higher performance in relation to those that are not part of the cluster, always considering that the competitiveness is here related to international markets; therefore, the export capacity will be a fundamental element to be taken into account. The hypothesis to be demonstrated is the following: The relation that exists between the exporting companies from avocado to the United States of America, located in Uruapan, Michoacán, and its competitiveness depends on the quality of the fruit that takes place for export, of its price, the used technology, the qualifications of its personnel, and of the channels of distribution.

Key words: Competitiveness, cluster, strategy, avocado, export.

INTRODUCTION

The opening of the U.S. market in November 1997 to the commercialization of Mexican avocado production was an excellent opportunity to develop a market in a culture that tended to consume natural foods, of good quality; within that a considerable amount of Mexican customary incorporating the avocado in its diet is included. Mexico's state of Michoacan, in particular the Uruapan municipality, is the largest producer of Hass avocados in the world. Proximity to the large U.S. market of 300 million habitants with high spending power was a unique business opportunity to take advantage of the efficient network of drug dealers with ample experience in the handling of the avocado.

The problem of the industrial sector in the radical process of commercial opening, adopted by Mexico as of the decade of the eighties, generated challenges and opportunities for several Mexican companies, but there was a question of competitive problems in customary companies to work in protected markets. These

distortions with serious social effects are related directly to the competitiveness. One critical issue for the Mexican avocado industry has been U.S. import regulations that have often denounced as green barriers. These regulations concern agricultural pesticide use as well as quality and maturity standards. In spite of the North American Free Trade Agreement (NAFTA), the U.S. has continued to impose six cents per pound tariff avocado imports from Mexico but not on avocado imports from countries such as Chile and the Dominican Republic. With the entry of Mexico into the General Agreement on Tariffs and Commerce (GATT) in 1986, the export of Michoacan's avocados has experienced a number of diverse problems. Noncompetitive intermediaries have assumed greater control over avocado commercialization and distribution. Strong U.S. policies protecting the U.S. California avocado industry have continued. The Mexican avocado sector is underorganized with production automation and commercialization having fallen behind that of other avocado producing countries such as Chile, Israel, the U.S., and Spain. There has been very little research on the competitive success factors of Mexican firms, much less those exporting to the U.S. By

---

*Corresponding author. E-mail: j_bonales@yahoo.com.

Table 1. Organizations in the study.

Nº	Companies	Nº	Companies
1	Agrícola TREDI, S.A. de C.V.	14	Empacadora El Durazno, S.A. de C.V.
2	Aguacates Frutas de Michoacán, S.A. de C.V.	15	Fresch Direction Mexicana, S.A. de C.V.
3	Aguamich, S.A. de C.V.	16	Frutas Finas de Michoacán, S.A. de C.V.
4	AMIMEX, S.A. de C.V.	17	Grupo Purépecha, S.A. de C.V.
5	AVOFRUT, S.A. de C.V.	18	Henry, S.A. de C.V.
6	Avopack, S.A. de C.V.	19	INDEX, S.A. de C.V.
7	AVOPER, S.A. de C.V.	20	Mc Daniel, S.A. de C.V.
8	Best Farmer, S.A. de C.V.	21	Missión de México, S.A. de C.V.
9	Calavo, S.A. de C.V.	22	San Lorenzo, S.A. de C.V.
10	Chiquita, S.A. de C.V.	23	Tropic de México, S.A. de C.V.
11	Del Rey, S.A. de C.V.	24	Vifrut, S.A. de C.V.
12	Dovi, S.A. de C.V.	25	West Pack, S.A. de C.V.
13	ECO, S.A. de C.V.		

Source: Association of Producers and Packers of Avocado from Michoacan, A.C. (2009).

identifying the competitiveness factors for Mexican avocado exporting firms, this study will advance current knowledge about competitive factors for organizations in the Mexican agricultural sector that are dependent on exports to the U.S.

RESEARCH METHODOLOGY

Sample

The total of investigation elements that constitutes the area of analytic interest is all the companies that export avocado to United States of America, located in Uruapan, Michoacan, based on the work plan for the export of Avocado from Mexico to the United States of America from June of 2003, and modernized this way in August of 2009, the population is finite, for what you proceeded to make a census and to apply the questionnaire to each company with base in the list of exporters of Avocado Association of Exporters and Packers of Mexican Avocado obtaining the following thing: the universe is of ninety five companies that take place, they pack, they market and they export the avocado from Michoacan's State.

The sample for this research study were 25 Mexican companies that exported Michoacan's avocados to the U.S.A. The population that is registered by the Agencies of Government of the United States of America (United States Department of Agriculture, USDA) for the export of the Michoacan's are twenty-five companies and they are mentioned in Table 1.

Mexico (2009) participated with 34% of the world production of avocado. Of the 802,000 tons that were harvested in our country in 2009 in an extension of 98,150 hectares, Michoacan collaborated with 81%, with a yield average of 8 tons for hectare. Their exports ascended to 46,616 tons, being the main destinations of France (55%), Canada (16%), Korea (10%), Japan (7%), United States (8%) and Switzerland (4%). The mobilization of this great production volume toward the consumption centers and their distribution, among the population, requires that middlemen should be included in the modification process when running out of time.

General model Ex dice

The general model, in which the relation between the quality, the

price, the technology, the qualification and the channels of distribution presented/displayed like independent variables and the competitiveness like dependent variable is described in Table 2.

Methods

The research conducted is of scientific (hypothetical-deductive), correlational design to test the functional relationship of the variables in competitive models under study. Based on the theoretical framework, using the theory of competitiveness in the export sector, the following hypotheses were made:

General hypothesis

The relation that exists between the exporting companies from avocado to the United States of America, located in Uruapan, Michoacán, and its competitiveness depends on the quality of the fruit that takes place for export, of its price, the used technology, the qualifications of its personnel, and of the channels of distribution.

Working hypothesis

H_1: The application of the quality norms, the improvement of the system of control of quality and an adaptation of the system of inspection in the exporting companies from avocado to the United States of America, will bring like consequence for a greater competitiveness.

H_2: When determining a better price of the avocado, indicated by means of the market that supplies, its production costs and costs of commercialization; a greater competitiveness of the exporting companies from avocado will be maintained to the United States of America.

H_3: For greater use of technology, a greater competitiveness of the exporting companies from avocado to the United States of America is guaranteed.

H_4: The qualification, based by means of the organization and the investment helps to obtain a greater competitiveness of the exporting companies from avocado to the United States of

Table 2. Measures of central tendency and variability.

	Quality	Price	Technology	Training	Distribution channels	Competitiveness
N	25	25	25	25	25	25
Mean	29.40	19.72	20.56	24.68	20.56	114.9
Median	28	19	20	24	21	112
Mode	36	25	24	24	17	133
Standard deviation	4.87	4.34	2.74	4.66	3.34	15.8
Variante	23.75	18.87	7.5	21.72	11.17	248.5
Skewness	0.15	0.038	-0.28	-0.36	-0.05	-0.04
Kurtosis	-1.15	-0.71	-0.99	-0.25	-1.03	-1.4
Range	17	17	9	18	11	45
Minimum	20	11	15	14	15	89
Maximum	37	28	24	32	26	134
Sum	735	493	514	617	514	2833

Source: Obtained information of the field investigation.

Table 3. Descriptive statistics and correlations.

Variable	Mean	σ	1	2	3	4	5
1. Quality	29.40	4.87					
2. Price	19.72	4.34	0.63				
3. Technology	20.56	2.74	0.83	0.64			
4. Training	24.68	4.66	0.55	0.58	0.66		
5. Distribution channels	20.56	3.34	0.28	0.27	0.39	0.38	
6. Competitiveness	114.9	15.8	0.85	0.81	0.89	0.82	0.55

Source: Obtained information of the field investigation. n=25. $p < 0.01$.

America.

H_5: To better selection of the channels of distribution, interpreted by means of the design and administration of the distribution channel and the boarding; a greater competitiveness of the exporting companies from avocado to the United States of America is obtained.

Questionnaire

Following a pre-test with 11 organizations, the interview questionnaire consisted of a total of 38 questions. Information was collected regarding: product quality, market price, production technology, personnel training, distributional channels, and overall competitiveness.

Product quality scale scores were based on 10 questions that asked respondents to define the quality of the product, norms of product quality, quality control systems, and inspection systems. Product price scale scores were based on seven questions regarding export market price and production costs. Production technology scale scores were based on six questions regarding technology used, the presence of technical consultants, the degree of modernization, and technology investment. Training level was determined by eight questions that asked about the provision of operational and administrative personnel training and technical qualification systems. The nature of product distribution channels was measured by seven questions that

asked about the nature of distribution channels and storage facilities used by the company. The overall competitiveness of companies was determined by thirty eight questions that asked respondents all about competitiveness.

In the design of the questionnaire the situation of indecision was not managed because it is not very probable that this happens. Also, they did not think about the individuals reactions to study as of agreement or in disagreement, but in such a way that allows obtaining the answers to the outlined questions. Therefore, the mensuration scale is integrated with the following assigned values 4, 3, 2, 1 that correspond to: totally of agreement, of agreement in general, in disagreement in general and totally in disagreement, respectively.

To evaluate to the mensuration instrument, the pre-test procedure was used, suggested by Bohrnstedt and the reliability coefficients of Cronbach-α.

The volume of data that was obtained when applying the questionnaire to all the companies that export avocado to the United States of America, located in Uruapan, Michoacan, is shown in Table 1. The information of the previous table, which concentrates on the indicators of each of the variables that are studied, is formed in Table 3.

RESULTS AND DISCUSSION

The companies that studied presented good

competitiveness, based on the results. The category was repeated more of 133 points. 50% of the companies are superficially (medium) of the value 112 points. The average of the companies is located in 114.9 points (good competitiveness). Also, 15.8 points are turned aside of the average (standard deviation). No company described like deficient its competitiveness (38 points). Companies 12, 17, 23, 2, 21, 13, 7, 24, 25, 5 and 10 (44%) described excellently their competitiveness. The scores tend to be located in average and elevated values. As far as the amount of dispersion of the data (variance) it was 248.5 points.

With respect to the variable quality, the effect that took place when applying the questionnaires to the study object was good quality. The medium one that was obtained was 28 points. The average that threw was 29.40 points, which indicate that the companies are in relation to the quality of medium one. Also 4.87 points are turned aside of the average. Only 9 companies (36%) described them with excellent quality, without arriving at the maximum value from 40 points. The slant that appeared in the quality of the survey companies was 0.154 points. As far as the amount of dispersion of the data it was of 23.75 points.

The information that was obtained when applying questionnaires to the exporting companies of avocado, with respect to the variable price was good. Since the average observed was 19.72 points, the category that was repeated more (fashion) was 25 points. 50% of the companies are superficially (medium) of the value 19 points. Also, 4.34 points were turned aside of the average. Only 8 companies (32%) described the variable price of their companies excellently, and only a single company obtained the highest level (28 points). The slant that appeared in the price of the interviewed companies was of 0.038 points, representing a positive slant because the average is greater than the medium one. As far as the amount of dispersion of the data (variance) it was 18.87 points.

Test of hypothesis

The general hypothesis that affirms the relation that exists between the exporting companies from avocado to the United States of America, located in Uruapan, Michoacán, and its competitiveness depends on the quality of the fruit that takes place for export, of its price, the used technology, the qualification of its personnel, and of the channels of distribution, was proven (Table 2). The first hypothesis affirmed that with one better quality obtained by means of the application of the quality norms, the improvement of the system of control of quality and the system of inspection in the exporting companies from avocado to the United States of America, will bring a like consequence of greater competitiveness, it presented or displayed an index of

considerable correlation (r) positive of 0.850 (Table 2); whereas its coefficient of determination (r^2) were of 0.723, which means that a narrow entailment between the two variables exists.

The second hypothesis test showed that a positive correlation of 0.811 was obtained considerably; whereas its coefficient of determination were 0.658 to determine a better price of the avocado, indicated by means of the market that supplies, its production costs and costs of commercialization; a greater competitiveness of the exporting companies from avocado will be maintained to the United States of America.

The third hypothesis test, since it turned out to be the highest correlation of all the variables that studied (0.888) what means that there exists a very noticeable association between the two variables because when a high degree of technology, translated in machinery and modern equipment is controlled, use of technical attendance and infrastructure; a greater competitiveness of the exporting companies from Avocado to the United States of America is guaranteed, whereas its coefficient of determination (r^2) were 0.789, which represents that a positive entailment between the two variables exists considerably.

The fourth hypothesis that affirms that through a qualification based by means of the reeducation of the human resource, the integral systems of qualification and the investment; it will mean a greater competitiveness of the exporting companies from avocado to the United States of America is approved. The previous thing which is based on the correlation (r) between the qualification and competitiveness of 0.820 and the determination coefficient of 0.672, symbolizes that a positive entailment between the two variables exists considerably.

The last hypothesis of this investigation test, because of better selection of the channels of distribution, interpreted by means of the design and administration of the distribution channel and the boarding; a greater competitiveness of the exporting companies from avocado to the United States of America is obtained. The previous thing, based on the relation between the channels of distribution and the competitiveness, according to the correlation of Pearson (r), was positively averaged at 0,550, whereas its coefficient of determination was of 0,303. The reason is the narrow entailment that exists between the two variables.

Conclusions

The companies that export avocado of Uruapan, Michoacan to the United States of America, are competitive in 44% of the studied cases. The companies that turned out to be competitive were: 12, 17, 23, 2, 21, 13, 7, 24, 25, 5 and 10. 3, 50% of the companies are superficially (medium) of 112 points. In average, the companies are located in 114.92 (excellent competitive).

However, 15.76 units of the scale are turned aside of the average.

One determined that the independent variables (quality, price, technology and qualification) strongly affect the competitiveness, since a positive entailment between the independent variables and the dependent variable with exception of the distribution channels exists, where an effect was minor. Large quality affects the 0.850, 0.811, 0.888 and 0.820 channel competitiveness and the technology of the price qualification of distribution in 0.550.

As it is observed, those that greater influence has it are the technology, followed of the quality and finalizing with the distribution channels.

The general hypothesis and the five hypotheses of work that were formulated validate in their totality. A code of commerce practices is due to establish fruits to implanter a customs Control of the labeled one, veracity of the information and the certification of the quality that the export avocado shows, with base in an official mark at the time of exporting itself. The precooked one that occurs to the avocado is due to carry in plastic boxes of 400 kg. The truck must happen through a cold water curtain to 6°C, with the purpose of lowering the temperature quickly to 100%; without curing it is introduced to the cold cameras.

It is required to invest in qualification; to carry out a good handling of human, technological and financial resources to establish efficient networks of distribution, with a service of quality, considering the price, punctuality in the service, quality in the product and security in the delivery of the product, that is to say, everything a process of development based on the satisfaction of the consumers.

The channels of distribution are due to select more adapted to export to the United States of America; the tradition has been to sell wholesalers. Once selected the channel, it must be administered properly since an organization common in charge does not exist to distribute the avocado in the United States of America, if she had her exporter would not compete to each other.

REFERENCES

Agustín JA (2009). Frutos del campo michoacano. Cultivos alternativos para el Estado de Michoacá, pp. 133-135.

Bancomext (2005). El Mercado Internacional del Aguacate y sus características. Manual de la Dirección Ejecutiva del Sector Primario e Industria Ligera y la Dirección de Alimentos, 12: 35-37.

Bonales VJ, Chávez FJ (2003). Modelos Competitivos de las agroindustrias. Prospectiva. UMSNH-ININEE, 2: 39-62.

Briones G (2008). Métodos y Técnicas de Investigación para las Ciencias Sociales. Trillas, pp. 178-185.

Campos SM, Naranjo PE, Saucedo CA (2007). La competitividad de los Estados Mexicanos. Escuela de Graduados en Administración Pública y Política Pública. ITESM, pp. 167-175.

CECIC (2009). Programa Regional de Competitividad Sistémica. http://www.contactopyme.gob.mx/estudios/docs/Competitividad_siste mic_ michoacan_CONTENIDO.pdf.

CEPAL (2009). Instrumentos de medición de competitividad. www.eclac.cl/mexico/capacidadescomerciales/Taller%20Honduras/D ocumentosypresentaciones/3.presentacion_Conceptosymedicioncom petitivida_ H.pdf.

FAO (2009). El estado mundial de la agricultura y la alimentación. http://www.rlc.fao.org/prior/desrural/alianza.htm.

FAO (2009). Perspectivas alimentarias. http://www.fao.org/giews/spanish/fo/ index.htm.

Gallegos ER (2003). Algunos aspectos del aguacate y su producción en Michoacán. Universidad Autónoma de Chapingo, pp. 87-93.

Haar J, Ortíz BM (2006). Cómo Exportar a Estados Unidos. Limusa, pp. 137-143.

Kovacic A (2007). Benchmarking the Slovenian Competitiveness by system of indicators. Benchmarking, 5: 553–574.

Rogoff K (2005). Rethinking exchange rate competitiveness. Global Competitiveness Report 2005/2006. W. E. Forum, Ed., pp. 99-105.

Sánchez PJ (2003). La producción del aguacate y su problemática en Michoacán. Manual de la Industria del Aguacate del Banco de México-Fira., 220(12): 182-190.

WEF (2009). Finding from the Global Competitiveness Index 2007-2008. http://www.weforum.org/pdf/Global_Competitiveness_Reports/Mexico .pdf.

9

Effects of temperature and period of blanching on the pasting and functional properties of plantain (*Musa parasidiaca*) flour

Oluwalana IB, Oluwamukomi MO*, Fagbemi TN and Oluwafemi GI

Department of Food Science and Technology, Federal University of Technology, Akure, Ondo State, Nigeria.

The effect of temperature and period of blanching on the pasting and functional properties of the flour obtained from plantain (*Musa AAB*) blanched under three temperature regimes were studied. Plantain fruit fingers were washed, hand-peeled and manually sliced into cylindrical pieces of 2 mm thickness. Blanching was carried out on the sliced samples in hot water at 60, 80 and 100 °C for varied periods of 5, 10 and 15 min and dried in the air oven at 65 °C (24/h), while the un-blanched sample served as the control. Results showed that blanching temperature and time variation had significant effects (P < 0.05) on pasting and functional properties of the plantain flour. The peak viscosity ranged between (355.25 to 527.08 RVU), trough of (235.83 to 335.92 RVU), breakdown of (91.92 to 217.67 RVU), final viscosity of (302.75 to 434.92 RVU), setback of (64.58 to 103.08 RVU), peak time of (4.52 to 4.91 min) and pasting temperature of (81.55 to 83.23 °C) across the blanching temperature and time chosen. The blanched samples were significantly different (P > 0.05) from each other and from the un-blanched control. However, varying the blanching time did not show any significant difference in water and oil absorption capacity for each temperature chosen compared to the control, the bulk density varied between (0.194 to 0.420 g/ml), emulsion capacity (17.86 to 41.12%) while the control was 39.84% which was significantly different from the other samples. Least gelation concentration varied between (2.0 to 6.0%). Samples blanched at 60 °C for 5, 10 and 15 min were not significantly different from each other but were significantly higher than that of the control which was (2%). Results indicated that the blanched plantain flours could be utilized as substitutes for soya and wheat flours in complementary/weaning foods and also useful as an important consideration in the production of pastries. Blanched plantain flour will be useful as a better binder and meat extender than the unblanched plantain flour. Blanching at low blanching temperature produced a flour of low stability against retrogradation than those at high temperature.

Key words: Plantain, flour, blanching, functional and pasting properties.

INTRODUCTION

Plantain (*Musa parasidiaca*) is an important staple food in Central and West Africa (Stover and Simmonds, 1987) which along with bananas provide 60 million people with 25% of the calories (Wilson, 1987). Nigeria produces about 2.11 million metric tonnes annually (FAO, 2004). However, about 35 to 60% post harvest losses had been reported and attributed to lack of storage facilities and inappropriate technologies for food processing (Olorunda and Adelusola, 1997). When processed into flour, it is used traditionally for the preparation of gruel which is made by mixing the flour with appropriate quantities of boiling water to form a thick paste (Mepba et al., 2007). Pere-Sira (1997) had also indicated the use of plantain flour as a component of baby food. However studies have shown that plantain, like other fruits, is susceptible to browning when the pulp is sliced. The browning potential of various fruits and vegetables has been shown to be directly related to the ascorbic acid level, polyphenol

*Corresponding author. E-mail: mukomi2003@yahoo.com.

Treatment	Plant material type	Code	Temperature of conservation (°C)	Time of blanching (mm)
A	Plantain blanched	PB_{60-5}	60	5
		PB_{60-10}	60	10
		PB_{60-15}	60	15
B	Plantain blanched	PB_{80-5}	80	5
		PB_{80-10}	80	10
		PB_{80-15}	80	15
C	Plantain blanched	PB_{100-5}	100	5
		PB_{100-10}	100	10
		PB_{100-15}	100	15
Control	Un-blanched plantain	UPB_{CO}	-	-

Table 1. Experimental design of the effect of blanching at 60, 80 and 100°C for 5, 10 and 15 min on properties of flour from blanched and unblanched plantain (*Musa parasidiaca*).

oxidase activity (Golan et al., 1977). Therefore, dehydration methods such as hot-air blanching and osmo-dehydration are employed to remove water and limit enzymatic and non-enzymatic browning in foods. Processing of plantains into flour is limited as most plantain foods are eaten as boiled, fried or roasted (Tortoe et al., 2008). Essential in determining potential uses for plantain flour is the identification of its functional properties. Functional properties of plantain flour in baked products have being reported by Ogazi (1986) and Gwanfogbe et al. (1988).

Pasting properties is an important index in determining the cooking and baking qualities of flour. Studies have been conducted on the pasting properties of plantain soy flour mixes (Abioye et al., 2006). The important component of pasting properties of starch is associated with a cohesive paste and has been reported (Oduro et al., 2000) to be significantly present in domestic products such as pounded yam, which requires high setback, high viscosity and high paste stability.

Various studies have been carried out on the processing and utilization of plantain. Ukhum and Ukpebor (1991) determined the sensory evaluation and physico-chemical changes during storage of instant plantain flour. Onyejegbu and Olorunda (1995) studied the effects of processing conditions and packaging on the quality of plantain chips. Ogazi (1996) carried a lot of studies ranging from processing and utilization of plantain in various forms including complementary diets. Fagbemi (1999) studied the effect of blanching and ripening on functional properties of plantain (*Musa aab*) flour. He observed that blanching reduced the emulsion capacity and viscosity, while bulk density, water and oil absorption capacities were increased by blanching. Unripe plantain could be used as an emulsifier and thickener in a food system. He also observed that ripening had a negative effect on all the functional properties examined except

the bulk density, and gelation property. However, more studies need to be carried out to determine the effect of different processing conditions on the various quality attributes of the plantain flour. The objectives of this study are therefore to determine the effect of varying the temperature and period of blanching on the pasting and the functional properties of plantain flour.

MATERIALS AND METHODS

Mature green and healthy plantain (*Musa*, AAB group) bunches were obtained from the International Institute of Tropical Agriculture (IITA), Oyo Road, Ibadan, with acceptable appearance for consumption (Dadzie and Orhard, 1997). All reagents used were of analytical grade.

Preparation of plantain flour

The fruits were processed at green (unripe) stages which were subsequently de-fingered. The fingers were washed, hand-peeled and manually sliced into cylindrical pieces of 2 mm thickness. Blanching was carried out on the sliced samples in hot water at 60, 80 and 100°C. For each temperature chosen, the timing was varied for 5,10 and 15 min and dried in the air oven at 65°C (24/h), using the unblanched sample as the control (Baiyeri and Ortiz, 2000). The dehydrated products were milled in a hammer mill to produce flour which passed through a 500 µm screen. The plantain flour obtained was packaged in polyethylene bags labeled and stored at room temperature for further analysis (Table 1).

Analysis

The samples were evaluated for functional and pasting properties. Functional properties such as water and oil absorptions, emulsion capacity were determined as described by Sathe et al. (1982), emulsion stability as described by Beuchat (1977). The bulk density was determined as described by Narayana and Narasinga (1984). The modified procedure of Coffman and Garcia (1977) was used to determine least gelation concentration, swelling power and solubility were determined by the methods of Leach et al. (1957).

The pasting properties of the samples were assessed in a rapid visco analyser (RVA-4) using the RVA general pasting method (Newport Scientific Pty Ltd, Warriewood, Australia). The sample was turned into slurry by mixing 3 g (14% moisture basis) with 25 ml of water inside the RVA test canister which was then lowered into the system (Newport Scientific, 1998). The slurry was heated from 50 to 95 °C and cooled back to 50 °C within 12 min, rotating the can at a speed of 160 rpm with continuous stirring of the content with a plastic paddle. Parameters estimated were peak viscosity, setback viscosity, final viscosity, pasting temperature and time to reach peak viscosity. All the experiments were conducted in triplicates and the mean ± standard deviation were reported. Data were subjected to analysis of variance (ANOVA) and the means separated by Duncan's New Multiple Range test (DMRT) at a significance level of 0.05.

RESULTS AND DISCUSSION

Pasting properties

The pasting properties of starch are used in assessing the suitability of its application as functional ingredient in food and other industrial products. The most important pasting characteristic of granular starch dispersion is its amylographic viscosity (Aviara et al., 2010). Plantain flour forms paste when reconstituted with hot water, hence its amylographic viscosities are important in assessing the suitability of its application as functional ingredient in food and other industrial products (Aviara et al., 2010). When starch or starch-based foods are heated in water beyond a critical temperature, the granules absorb a large amount of water and swell to many times their original size. Over a critical temperature, which is characteristic of a particular starch, the starch undergoes an irreversible process known as gelatinization (Adebowale et al., 2008). When the temperature rises above the gelatinization temperature, the starch granules begin to swell and viscosity increases on shearing. The temperature at the onset of this rise in viscosity is referred to as the gelatinization or pasting temperature (Isikli and Karababa, 2005). The pasting temperature of the plantain flour samples ranged between 79.93 and 83.23 °C for the blanched samples and 81.80 °C for the control (Table 2). The pasting temperature is a measure of the minimum temperature required to cook a given food sample (Sandhu et al., 2005), it can have implications for the stability of other components in a formula and also indicate energy costs (Newport Scientific, 1998).

The peak time is a measure of the cooking time (Adebowale et al., 2005). This ranged between 4.53 to 4.91 min for the blanched samples and 4.52 min for the control sample. Peak viscosity, which is the maximum viscosity, developed during or soon after the heating portion of the pasting test (Newport Scientific, 1998), is lower for the blanched samples (355.25 to 527.08 RVU) and highest for the control sample (543.42 RVU). Peak viscosity is often correlated with the final product quality. It also provides an indication of the viscous load likely to

be encountered during mixing (Maziya-Dixon et al., 2004). Higher swelling index is indicative of higher peak viscosity while higher solubility as a result of starch degradation or dextrinization results in reduced paste viscosity (Shittu et al., 2001). These were corroborated by results of swelling power and solubility reported earlier. The hold period sometimes called shear thinning, holding strength, hot paste viscosity or trough due to the accompanied breakdown in viscosity is a period when the sample was subjected to a period of constant temperature (usually 95 °C) and mechanical shear stress. It is the minimum viscosity value in the constant temperature phase of the RVA profile and measures the ability of paste to withstand breakdown during cooling (Newport Scientific, 1998). This ranged between 248.92 and 325.75 RVU for blanched sample at 100 and 60 °C respectively, while the un-blanched control had a higher value of 325.75 RVU. This period is often associated with a breakdown in viscosity (Ragaee et al., 2006). It is an indication of breakdown or stability of the starch gel during cooking (Zaidhul et al., 2006). The lower the value the more stable is the starch gel. The breakdown is regarded as a measure of the degree of disintegration of granules or paste stability (Dengate, 1984; Newport Scientific, 1998).

The breakdown viscosities recorded by blanched samples were lower than that of the un-blanched control sample. They reduced from 217.67 RVU in the control to a range of 161.08 to 207.25 RVU in the samples blanched at 100 and 60 °C respectively. The viscosity after cooling to 50 °C represents the setback or viscosity of cooked paste. It is a stage where retrogradation or re-ordering of starch molecules occur. It is a tendency to become firmer with increasing resistance to enzymic attack. It also has effect on digestibility. Higher setback values are synonymous to reduced dough digestibility (Shittu et al., 2001) while lower setback during the cooling of the paste indicates lower tendency for retrogradation (Sandhu et al., 2007). The final viscosity was highest for the control gari (439.33 RVU), while it ranged between 302.75 and 414.92 RVU for the blanched samples. The extent of increase in viscosity on cooling to 50 °C reflects the retrogradation tendency (Ragaee et al., 2006; Sandhu et al., 2007). The setback viscosity indicates the tendency of the dough to undergo retrogradation, a phenomenon that causes the dough to become firmer and increasingly resistant to enzyme attack (Ihekoronye and Ngoddy, 1985). It has a serious implication on the digestibility of the dough when consumed. The high setback value for the sample at low blanching temperature indicates that its paste would have a low stability against retrogradation (Mazurs et al., 1957) than those at high temperature.

Functional properties

The water absorption capacity (WAC) of all the treated

Table 2. Pasting properties of plantain flour blanched for 5, 10 and 15 min at 60, 80 and 100°C.

Sample	UPB_CO	PB60			PB80			PB100		
		5	10	15	5	10	15	5	10	15
Peak viscosity(RVU)	543.42±0.20[a]	527.08±0.04[b]	511.92±0.01[b]	445.67±0.04[b]	434.58±0.17[c]	414.08±0.56[c]	403.58±0.30[c]	422.0±0.04[c]	410.0±0.56[c]	355.25±0.17[d]
Trough (RVU)	325.75±0.32[a]	335.92±0.01[a]	304.67±0.01[b]	291.08±0.54[b]	288.92±0.43[b]	276.92±4.60[d]	235.83±2.09[d]	257.33±0.49[d]	248.92±0.04[d]	263.33±0.31[c]
Breakdown (RVU)	217.67±0.51[a]	191.16±0.04[b]	207.25±0.02[b]	154.59±0.51[c]	145.66±0.60[d]	137.16±5.16[d]	167.75±2.39[c]	164.67±0.45[c]	161.08±0.52[c]	91.92±0.14[d]
Final viscosity (RVU)	409.33±0.10[a]	434.92±0.01[a]	407.75±0.03[b]	374.92±0.05[b]	353.50±1.25[c]	366.50±0.21[c]	302.75±0.07[d]	343.75±0.39[c]	320.42±0.10[d]	355.92±0.40[c]
Setback (RVU)	83.58±0.22[b]	99.00±0.00[a]	103.08±0.02[a]	83.84.66±1.04[b]	64.58±0.82[c]	89.58±4.39[a]	66.92±2.13[c]	86.42±0.10[b]	71.50±0.06[c]	92.59±0.09[a]
Peak time (min)	4.52±0.01[b]	4.73±0.01[a]	4.53±0.01[b]	4.64±0.04[b]	4.75±0.05[a]	4.67±0.00[a]	4.59±0.02[b]	4.64±0.04[b]	4.56±0.05[b]	4.91±0.02[a]
Pasting temperature(°C)	81.80±0.10[ab]	82.65±0.00[ab]	79.93±0.03[c]	81.55±0.01[b]	82.77±0.58[a]	81.20±0.4[b]	82.55±0.01[b]	80.90±0.70[b]	81.32±0.34[ab]	83.23±0.03[a]

Table 3. Functional properties of plantain flour blanched for 5, 10 and 15 min at 60, 80 and 100°C.

Sample	UPB_CO	PB60			PB80			PB100		
		5	10	15	5	10	15	5	10	15
Water absorption capacity (%)	124.50±4.50[a]	120.50±0.50[b]	118.50±0.50[b]	115.50±0.00[b]	123.50±1.50[a]	123.50±0.50[a]	123.50±0.50[a]	125.50±0.50[a]	125.50±0.50[a]	125.50±0.50[a]
Oil absorption capacity (%)	108.50±4.50[a]	115.50±0.50[a]	111.50±0.50[a]	109.00±1.0[a]	112.50±0.50[a]	112.50±0.50[a]	111.50±0.50[a]	110.50±0.50[a]	112.50±0.50[a]	111.50±0.50[a]
Bulk density (g/ml)	0.420±0.07[a]	0.249±0.02[b]	0.219±0.08[b]	0.290±0.02[b]	0.182±0.05[c]	0.164±0.08[c]	0.140±0.01[c]	0.172±0.03[c]	0.159±0.07[c]	0.140±0.03[c]
Emulsion stability (%)	41.27±0.51[a]	29.83±0.27[b]	26.82±0.50[b]	31.07±0.00[b]	24.58±1.41[c]	20.04±0.29[c]	30.23±0.35[b]	18.48±0.09[d]	12.03±0.07[d]	20.25±0.10[c]
Emulsion capacity (%)	39.84±0.19[a]	41.12±0.28[a]	30.45±0.72[c]	36.68±0.37[b]	30.46±0.62[c]	25.00±0.11[c]	36.19±0.32[c]	27.03±0.12[c]	17.86±0.17[d]	24.14±0.18[c]
Swelling power (%)	48.09±0.23[a]	36.04±0.18[c]	39.98±0.12[b]	40.03±0.08[b]	30.95±0.07[d]	28.47±2.61[d]	35.98±0.07[c]	39.80±1.24[b]	38.19±0.33[b]	35.16±0.21[c]
Solubility (%)	5.89±0.33[a]	4.24±0.03[b]	3.77±0.04[d]	3.47±0.05[d]	4.69±0.09[b]	4.59±0.07[b]	4.38±0.09[b]	4.02±0.07[c]	4.14±0.07[c]	4.31±0.06[a]
Least gelation concentration (%)	2.00±0.00[c]	6.00±0.00[a]	6.00±0.00[a]	6.00±0.00[a]	4.00±0.00[a]	4.00±0.00[a]	6.00±0.00[a]	4.00±0.00[a]	6.00±0.00[a]	6.00±0.00[a]

samples ranged between (118.50 to 125.50%) across the blanching temperature and time (Table 3). There were no significant differences in the control and blanched samples (P > 0.05) except for samples blanched with 60 °C which recorded lower values of 115.5 to 120.50 °C. This is contrary to the finding of Fagbemi (1999) who observed that water absorption capacity could be enhanced by blanching. However, the WAC values are higher than 36 and 85% reported for fluted pumpkin seed flour, but lower than the values of 100% for gourd seed flours (Olaofe et al., 1994) many of oil seeds defatted flours (100 to 260%) (Ige et al., 1984), 120% for lupin seed flour, 130% for soy flour, 138% for pigeon pea (Oshodi and Ekperigin, 1989), and 157% for calabash seed flour (Olaofe et al., 2009) and 610 and 670% for full fat and defatted Dioclea reflexa seed flour respectively (Akinyede et al., 2005). The comparatively higher water absorption capacity of 125% recorded for the samples blanched at 100 °C is an indication of its use in composite flours for bread making. WAC is considered critical in viscous foods such as soups and gravies; thus, the flours may find use as functional ingredients in soups, gravies and baked products (Akinyede et al., 2005). The oil absorption capacity (OAC) of the samples ranged between (108.50 to 115.50%) while the treated samples were not significantly (P > 0.05) different from the unblanched/control (108.50%). The OAC were however higher than 89.7% reported for pigeon pea (Oshodi and Ekperigin, 1989), jack bean (105.6%), gourd seed flour (96%) (Olaofe et al., 1994) but lower than the values of 140, 193, 142, and 142%, reported for chickpea, soy, yam bean flours, fluted pumpkin seed flour respectively (Ige et al., 1984; Oshodi and Ekperingin, 1989; Oshodi et al., 1999; According to Fagbemi (1999), good oil absorption capacity of flour samples suggest that they may be useful in food preparations that involve oil mixing like in bakery products, where oil is an important ingredient. The water/fat binding capacity of proteins is an index of its ability to absorb and retain oil, which in turn influences the texture and mouth feel of food products like ground meat formulations, doughnuts, pancakes, baked goods and soups.

The seed flours may thus be used to replace some legumes and oil seeds as thickeners used in some liquid and semi liquid foods (Akinyede et al., 2005). This is an indication that the flours can act as flavour retainer and be used to improve the mouth feels of food. The bulk-density of the samples reduced with blanching from 0.420 g/ml for the control sample to a range of values between 0.140 and 0.420 g/ml for the blanched samples. The bulk densities of the blanched samples at 60 °C for 5, 10 and 15 min were significantly higher than other samples blanched at 80 and 100 °C. This is contrary to a report made by Tagogoe (1994) and Fagbemi (1999) which showed that bulk density increased as a result of blanching/heat treatment prior to drying. The reduction in the bulk densities in samples blanched at 80 and 100 °C will be an advantage in the bulk storage and transportation of the flour. The emulsion capacity (EC) and stability of the blanched samples were generally lower than the unblanched/control (39.84 and 41.27%) respectively. As blanching temperature and time increased, emulsion capacity and stability of the samples reduced. This is consistent with the observation of Fagbemi (1999) which could be attributed to the heat of blanching. The EC of the blanched and unblanched/control flours ranging from 17.86 to 41.12%, are however higher than those of soya bean (15.0%), breadnut flour (18%), 8.1 and wheat flours (7 to 11%) (Lin et al., 1974) and calabash seed flour (23.2%) (Olaofe et al., 2009) suggesting that the blanched and unblanched/control plantain flours could be utilized as substitutes for soya and wheat flours in complementary/ weaning foods. The values are comparable with those of *Pachira glabra* (35.5%) and *Afzelia africana* seeds (41.5%) (Ogunlade et al., 2011), pigeon pea (49.1%) and sunflower flours (75.1%) (Adeyeye et al., 1994; Oshodi and Ekperijin, 1989; Oshodi et al., 1999). EC is an important considera-tion in the production of pastries, coffee whiteners and frozen desserts. Emulsion capacity denotes the maximum amount of oil that can be emulsified by protein dispersion, whereas emulsion stability denotes the ability of an emulsion with a certain composition to remain unchanged (Enujuigha et al., 2003).

Blanched and unblanched/control flours may thus be useful in such food formulations (Akinyede et al., 2005).

The emulsion stability values (50.0 ± 0.0% and 51.00 ± 0.0%) were higher than 11.5 and 29.5% for gourd seed and yellow melon respectively (Ogungbenle, 2006). The swelling power (SP) of the blanched samples ranging between 28.47 and 40.03% were generally lower than that of the unblanched/control (48.09%) while the solubility ranging from 3.47 to 4.69% were also lower than that of the control (5.89%). Increase in blanching temperature and time had a significant (P < 0.05) decrease on the swelling power on the samples. The least gelation concentration (LGC) of the unblanched/ control (2.00%) was significantly lower (P < 0.05) than those of the blanched samples which ranged from 4.0 to 6.0%. Blanching increased the least gelation concentra-tion implying that the flour of the blanched plantain will act as a better binder. The values were lower than those reported for cowpea (10 to 14%), great northern bean (10%) and lupine seed flours (14%) The ability of seed flours to form gel is desirable in the preparation of extended meat products (Akinyede et al., 2005).

Conclusion

This study has shown that blanching caused a reduction in emulsion capacity. However, their higher values than soy and wheat flours suggests that the plantain flours could be utilized as substitutes for soya and wheat flours in complementary/weaning foods and also useful as an important consideration in the production of pastries, coffee whiteners and frozen desserts. Blanching increased the least gelation concentration, implying that the flour of the blanched plantain is a better binder and more useful as a meat extender than the unblanched plantain flour. The high setback value for the sample at low blanching temperature indicates that its paste would have a low stability against retrogradation than those at high temperature. The lower breakdown viscosity recorded by blanched samples over the un-blanched control sample was an indication of paste stability of the starch gel during cooking.

REFERENCES

Abioye VF, Ade-Omowaye BIO, Adesigbin MK (2006). Effect of Soy Flour Fractions on some Functional and Physicochemical Properties of Plantain Flour In: (O.A. T. Ebuehi, ed), pp. 142-143.

Adebowale AA, Sanni LO, Onitilo MO (2008). Chemical composition and pasting properties of tapioca grits from different cassava varieties and roasting methods. Afr. J. Food Sci., 2: 077-082

Adebowale AA, Sanni LO, Awonorin SO (2005). Effect of texture modifiers on the physicochemical and sensory properties of dried fufu. Food Sci. Technol. Int., 11(5): 373-382.

Adeyeye EI, Oshodi AA, Ipinmoroti KO (1994). Functional properties of some varieties of African yam bean (*Sphenostylis sternocarpa*) flour II. Int. J. Food Sci., Nutr., 45: 1-12.

Akinyede AI, Amoo IA, Eleyinmi AF (2005). Chemical and functional properties of full fat and defatted *Dioclea reflexa* seed flours J. Food, Agricul. Environ., 3 (2): 112-115. www.world-food.net.

Aviara NA, Igbeka JC, Nwokocha LM (2010). Physicochemical properties of sorghum (sorghum bicolor I. Moench) starch as affected by drying temperature. Agric. Eng. Int. CIGR J. Open access at http://www.cigrjournal.org 12:2 85

Baiyeri KP, Ortiz R (2000). Effect of Nitrogen Fertilization on Mineral Concentration in Plantain (Musa species AAB) Fruit Peel and Pulp at Unripe and Ripe Stages. Plant Product Res. J., 5: 38-43.

Beuchat LR (1997) Functional and Electrophoretic Characteristics of Succinylated Peanut Flour. J. Agric. Food Chem., 25: 258-261.

Coffman C, Garcia VV (1977). Functional Properties of Protein Isolate from Mung Bean Flour. J. Food Tech., 12: 473-484.

Dadzie BK, Orchard JE (1997): Routine Post-Harvest Screening of Banana / Plantain Hybrids: Criteria and Methods. INIBAP Technical Guidelines.

Dengate HN (1984). Swelling, pasting, and gelling of wheat starch. In: Pomeranz Y (Ed) Adv. Cereal Sci. Technol. AACC, USA, pp. 49-82.

Enujuigha VN, Badejo AA, Iyiola SO, Oluwamukomi MO (2003). Effect of germination on the nutritional and functional properties of African oil bean (Penthaclethra macrophylla Benth) seed flour. J. Food, Agricul. Environ., 1 (3&4): 72-75.

Fagbemi TN (1999). Effect of Blanching and Ripening on Functional Properties of Plantain (Musa aab) Flour. Foods Hum. Nutr., 54: 261-269.

FAO (Food and Agriculture Organization) (2004). Statistics Series No 95 FAO, Rome.

Golan A, Kahn V, Saduski AY (1977). Relationship between Polyphenols and Browning in Avocado Mesocarp-Comparison between the Fuerte and Lerman Cultivars. J. Agric. Food Chem., 25: 1253-1259.

Gwanfogbe PN, Cherry JP, Simmons JG, James C (1988) Functionality and Nutritive Value of Composite Plantain (Musa paradisiaca) Flours. Trop. Sci., 28: 51-66.

Ige MM, Ogunsua AO, Oke OL (1984). Functional properties of the proteins of some Nigerian oil seed, conophor seeds and three varieties of melon seeds. J. Agric. Food Chem., 32: 822-825

Ihekoronye AI, Ngoddy PO (1985). Browning reaction. In: Integrated Food Science and Technology for the Tropics. Macmillan Education Ltd, London, pp. 224-229.

Isikli ND, Karababa EA (2005). Rheological characterization of fenugreek paste (cemen). J. Food Eng., 69:185-190.

Leach HW, McCowan LD, Schoch TJ (1957). Structure of the Starch Granule: Swelling Power and Solubility Patterns of Different Starches. Cereal Chem., 36: 534-544.

Lin MJY, Humbert ES, Sosulski FW (1974). Certain functional properties of sunflower meal products. J. Food Sci., 39: 368-370.

Maziya-Dixon B, Dixon AGO, Adebowale AA (2004). Targeting different end uses of cassava: genotypic variations for cyanogenic potentials and pasting properties. A paper presented at ISTRC-AB Symposium, 31 October – 5 November 2004, Whitesands Hotel, Mombasa, Kenya.

Mazurs EG, Schoch TJ, Kite FE (1957). Graphical analysis of the brabender viscosity curves of various starches. Cereal Chem., 34(3): 141-153

Mepba HD, Eboh L, Nwaojigwa SU (2007). Chemical Composition, Functional and Baking Properties of Wheat – Plantain Composite Flours. Afr. J. Food Agric, Nutr. Dev., 7(1): 4-5.

Narayana K, Narasinga Rao MS (1984). Effects of partial proteolysis on the functional properties of winged bean (Phosphorcapus tetragonolobus) flour. J. Food Sci., 49: 944-947.

Newport Scientific (1998). Applications manual for the Rapid Visco Analyzer using thermocline for windows. Newport Scientific Pty Ltd., 1/2 Apollo Street, Warriewood NSW 2102, Australia, pp. 2-26.

Oduro I, Ellis W, Dziedzoave N and Nimako-Yeboah K (2000). Quality of gari from selected processing zones in Ghana. Food Control, 11: 297-303.

Ogazi PO (1986). Quality Assessment of Plantain Fruits for Dehydration. Nig. Food J., 4(1): 125-130.

Ogazi PO (1996). Plantain: production, processing and utilization. Paman and Associates Limited, Uku-Okigwe, p. 305.

Ogungbenle HN (2006). Chemical composition, functional properties and amino acid composition of some edible seeds. La Rivista Italiana Delle Sostanze Marzo, 73: 81-86.

Ogunlade I, Ilugbiyin A, Osasona AI (2011). A comparative study of proximate composition, antinutrient composition and functional properties of Pachira glabra and Afzelia africana seed flours. Afr. J. Food Sci., 5(1): 32-35

Olaofe O, Adeyemi FO and Adediran GO (1994). Amino acid, chemical composition and functional properties of some oil seeds. J. Agric. Food Chem., 42(4): 879-881.

Olaofe O, Ekuagbere AO, Ogunlade I (2009). Chemical, amino acid composition and functional properties of calabash seed's kernel. Bull. Pure Appl. Sci., 28(1- 2): 13-24.

Olorunda AO, Adelusola MA (1997). Screening of Plantain / Banana Cultivars for Import, Storage and Processing Characteristics. Paper Presented at the International Symposium on Genetic Improvement of Bananas for Resistance to Disease and Pests 7 – 9th September, (IRAD, Montpellier, France).

Onyejegbu CA, Olorunda AO (1995). Effects of raw materials, processing conditions and packaging on the quality of plantain chips. J. Sci. Food Agric., 68: 279-283.

Oshodi AA, Ekperijin NM (1989). Functional properties of pigeon pea flour (Cajanus cajan) Food Chem., 34: 187-191.

Oshodi AA, Ogungbenle HN, Oladimeji MC (1999). Chemical composition, nutritionally valuable minerals and functional properties of benniseed (Sesamum radiation), pearl millet (Pennisetum typhoides) and quinoa (Chenopodium quinoa) flours. Int. J. Food Sci. Nutr., 50: 325-331.

Pere-Sira E (1997). Characterization of Starch Isolated from Plantain (Musa paradisiacal normalis) Starch, 49: 45-49.

Ragaee S, Abdel-Aal EM (2006). Pasting properties of starch and protein in selected cereals and quality of their food products. Food Chem., 95: 9-18.

Ramteke RS, Eipeson WE (2000). Predehydration Steaming Changes Physicochemical Properties of Unripe Banana Flour. Dept. of Food Sci. and Tech., Makerere University Kampala Uganda.

Sandhu KS, Singh N, Malhi NS (2005). Physicochemical and thermal properties of starches separated from corn produced from crosses of two germ pools. Food Chem., 99: 541-548.

Sandhu KS, Singh N, Malhi NS (2007). Some properties of corn grains and their flours I: Physicochemical, functional and chapati-making properties of flours. Food Chem., 101: 938-946.

Sathe SK, Desphande SS, Salunhkhe DK (1982). Functional properties of lupin seed (Lupinus mutabilis) proteins and protein concentrates. J. Food Sci., 47: 491-497.

Shittu TA, Lasekan OO, Sanni LO, Oladosu MO (2001). The effect of drying methods on the functional and sensory characteristics of pukuru-a fermented cassava product. ASSET-An Intern. J., 1(2): 9-16.

Stover RH, Simmonds NW (1987). Bananas: Tropical Agriculture Series (3rd Edition) Inc. New York

Tagodoe A (1994). Functional Properties of Raw and Precooked Taro Colocasia Esculenta Flour. Int. J. Food Sci. Tech., 29: 457-482.

Tortoe CT, Paa – Nii TJ, Apollonius IN (2008). Effects of Osmo-Dehydration, Blanching and Semi – Ripening on the Viscoelastic, Water Activity and Colorimetry Properties of Flour from Three Cultivars of Plantain (Musa AAB). CSIR – Food Research Institute, Accra, Ghana.

Ukhum ME, Ukpebor IE (1991). Production of instant plantain flour, sensory evaluation and physico-chemical changes during storage. Food Chem., 42: 287-299.

Wilson CF (1987). Status of Bananas and Plantains in West Africa in Bananas and Plantain Breeding Strategies Proceedings of International Workshop on Plantain and Banana. Australian Council for International Research.

Zaidhul ISM, Hiroaki Y, Sun-Ju K, Naoto H and Takahiro N (2006). RVA study of mixtures of wheat flour and potato starches with different phosphorus contents. Food Chemistrydoi:10.1016/j.foodchem.2006.06.056.

10

Quality assessment of vacuum packaged chicken snacks stored at room temperature

V. P. Singh*, M. K. Sanyal, P. C. Dubey and S. K. Mendirtta

Division of Livestock Products Technology, Indian Veterinary Research Institute, Izatnagar-243122, India.

Chicken snacks were prepared by utilizing spent hen meat, sodium caseinate and rice flour, spice mix, condiments, common salt, phosphate and baking powder. The control was prepared in a similar manner except that, spent hen meat was substituted by equal quantity of rice flour. Chicken snack and control were packaged under vacuum in laminated (polyethylene/aluminium foil) pouches (size 25 × 20 cm), stored at 30 ± 2°C. The changes in physico-chemical characteristics, sensory attributes and microbiological profile of vacuum packaged chicken snacks, as well as control were analyzed during storage at room temperature (30 ± 2°C) for 30 days with regular intervals of six days. Both chicken snacks and control indicated non-significant effect of treatment on days of storage with respect to the contents of fat, protein, ash, pH, total plate count (TPC), yeast and mould counts (YMC). However, shear force value in treated products were significantly (P<0.05) different on day 0 and 6 from the rest of the storage days. The TBA values for control on day 0, 6, 12 were found significantly different from the rest of the storage days. Sensory attributes for both control and treated products were found to be less affected by the days of storage in the whole of the storage period. Overall comparison of physico-chemical, microbiological and sensory profiles of control and treated products found highly significant (P<0.01) difference except for some values of moisture, shear force and pH. The study revealed that both products can be stored under vacuum in very good condition up to 30 days at room temperature.

Key words: Chicken snack, spent hen meat, rice flour, sodium caseinate, vacuum.

INTRODUCTION

Snacks are convenient fast food and their consumption is increasing day by day due to rapid urbanization and sociological changes. It is a food of choice for school going children, adolescent girls and high mobility groups. The market of snack food industry including semi-processed/cooked and ready to eat foods was around Rs 82.9 billion in 2004 to 2005 and is rising rapidly with a growth rate of 20%. Most of the snacks available in the market are mainly based of cereals which are high in calorie and low in protein contents. So, the incorporation of meat in these snacks is a good alteration in its nutritional value particularly high value animal protein.

By incorporation of spent hen meat, we can enhance nutritive value, palatability and can help in utilizing this poultry industry by-product. The spent hens are old and

culled chickens, which have completed their productive and reproductive phase of life (Mahapatra, 1992). The meat of such birds is tougher, less juicy due to high collagen contents (Abe et al., 1996) and high degree of cross linkages (Wenham et al., 1973; Bailey, 1984) as compared to broiler meat. These shortcomings of using spent hens meat in different products can be overcome with suitable food additives or extenders like flours, starch and milk proteins (Chung et al., 1989; Tarte et al., 1989). Non-meat proteins from a variety of plant sources can be utilized in different meat products in various ways (Gujral et al., 2002; Dzudie et al., 2002; Bhat and Pathak, 2009; Serdarouglu and Degirmencioglu, 2004).

The broiler production in our country is increasing in a faster pace. Similarly, the number of broiler spent hens in poultry industry is also increasing. These birds are heavier in weight and their meat is high in fat contents (10 to 15%) (Kondaiah, 1990). The meat of such birds is poor in quality similar to the spent hen's meat. Very few workers have attempted the still inconclusive study of

*Corresponding author. E-mail: vetvpsingh@rediff.com.

Table 1. Formulations for chicken snack preparations.

Ingredients (w/w)	Chicken snack	Control snack
Broiler spent hen meat	50.0	0
Rice flour	41.0	91
Sodium caseinate	2.5	2.5
Common salt	2.0	2.0
Condiments	2.5	2.5
Spice mix	1.5	1.5
Baking powder	0.5	0.5

Phosphate: 0.3% of meat used (on weight basis), ice water: 100% of flour used (on weight basis).

Table 2. Composition of spice mixture.

Ingredients	Percent (%)
Coriander powder	15.0
Cumin seeds	15 .0
Red chilli powder	20.0
Black pepper	15.0
Cloves	5.0
Cardamom	5.0
Turmeric	10.0
Cinnamon	5.0
Aniseed	10.0

chicken snacks from spent hens, particularly from broiler spent hens meat. Thus, the study was conducted and chicken snacks prepared were envisaged to evaluate the effect of vacuum packaging to know the suitability of its storage at ambient temperature.

MATERIALS AND METHODS

Sources of chicken meat

Fifty weeks old broiler spent hens were procured from Central Avian Research Institute, Izatnagar. The birds were slaughtered and dressed in the abattoir of the Institute by humane method of slaughter. The body fat was removed and deboning of dressed chicken was done manually removing all tendons and separable connective tissues. The lean meat was packed in low density polyethylene bags and frozen at -20 ℃ until use.

Condiments and rice flour

Onion, garlic and ginger in the ratio of 3:1:1 were ground in a mixture to the consistency of fine paste. Rice flour used in the study was procured from the standard flour mill of Izatnagar, Bareilly.

Spice mixture

The spice mix formula shown in Table 2 was formulated on the

basis of the trials conducted among the scientists and students of the Livestock Products Technology Division of the Institute. The ingredients used in this formulation were purchased from the local market. After removal of extraneous matter, all spices were dried in an oven at 80 ℃ for 3 h and then ground in a grinder to powder. The course particles were removed using a sieve of 100 mesh and fine powdered spices were mixed in required proportion to obtain spice mixture for chicken and control snacks preparation.

The spice mix was stored in plastic airtight container for subsequent use.

Sodium caseinate, common salt, baking powder and phosphate

Sodium caseinate was procured from Central Drug House (P) ltd., Mumbai, India. Common salt of the brand Tata and baking powder of the brand Rex were purchased from the local market. Sodium phosphate of food grade was procured from the local market.

Packaging materials

Two layered laminated pouches (aluminium foil/polyethylene) of food grade quality (size 25 x 20 cm) were procured from Sadar Bazaar, Delhi for packaging of chicken snack as well as control snack.

Preparation of chicken and control snacks

For the preparation of chicken snack and control, standardized formulation (Table 1) on the basis of several trials was used. Dressed and deboned meat was cut into small cubes and minced twice through the mincer (Electrolux, Sweden) after microwave thawing of the stored chicken meat. Minced chicken meat was blended with ice water (5% of calculated amount of water), common salt and sodium hexametaphosphate and chopped in a bowl chopper (Seydelmann, Germany) for 1 min. Condiment mixture was added to the emulsion and chopped again for 30 s followed by mixing of sodium caseinate and rechopped for 1 min. Spice mix powder, rice flour and the rest, 95% of the water was added to the mixture and chopped again for 1 min. Thus, the emulsion was prepared for chicken snacks. The emulsion was extruded through a manually operated stainless steel extruder into the shape of chips (size 20 x 2.5 x 0.3 cm) which were cooked in a microwave oven (Kelvinator, India) for about 8 to 10 min to prepare crisp snacks. Control snacks were prepared following the procedure mentioned earlier, except that no spoilt hen meat was used in its preparation.

Analytical techniques for physico-chemical characteristics

Moisture, fat, protein and ash of treated as well as control samples were analyzed as per the method described in AOAC (1995). The pH was determined following the method of Strange et al. (1977), whereas, thiobarbituric acid (TBA) value by the procedure of Witte et al. (1970). The procedure of Smith et al. (1991) was followed with suitable modifications for determining the shear force value of chicken and control snacks using Bratzler shear press.

Sensory evaluation

Chicken snacks as well as control snacks were subjected to sensory evaluation by a panel of seven judges comprising of scientists of the institute by using 8-point Hedonic scale (Keeton, 1983).

Table 3. Physico-chemical properties of chicken snacks as affected by the packaging under aerobic and vacuum packaging during storage at 30 ± 2°C (Mean*± SE).

Particulars	Days of storage					
	0	6	12	18	24	30
Moisture %						
Control	8.27±0.28	8.20±0.29	8.18±0.12	8.16± 0.16	8.13± 0.07	8.10± 0.13
Treated	8.80±0.14	8.73±0.21	8.70±0.29	8.68± 0.30	8.64± 0.27	8.61± 0.24
Fat %						
Control	0.65±0.08	0.63±0.06	0.62±0.07	0.60± 0.13	0.56± 0.04	0.52± 0.09
Treated	3.54±0.25	3.50±0.25	3.48±0.11	3.44± 0.12	3.41±0.26	3.39±0.09
Protein %						
Control	9.08±0.71	9.03±0.12	8.98±0.36	8.92±0.09	8.87±0.07	8.82±0.17
Treated	22.10±1.13	22.05±0.20	22.01±0.52	21.96±0.62	21.89±0.09	21.86±0.14
Ash (%)						
Control	1.50±0.14	1.48±0.26	1.45±0.26	1.42±0.08	1.41±0.22	1.39±0.10
Treated	2.60±0.18	2.58±0.23	2.54±0.25	2.51±0.10	2.48±0.11	2.47±0.22
Thiobarbituric acid value (mg malonaldehyde/kg)						
Control	0.25[abc]±0.03	0.23[bcd]±0.02	0.24[cd]±0.01	0.21± 0.02	0.26±0.01	0.27±0.02
Treated	0.89[ab]±0.02	0.87[abc]±0.03	0.84[abc]±0.03	0.87[abc]±0.04	0.89[ab]±0.02	0.90[a]±0.02
Shear force value (kg/cm²)						
Control	5.30±0.21	5.38±0.08	5.40±0.20	5.43± 0.16	5.48±0.08	5.51±0.06
Treated	4.40[b] ±0.31	4.43[b] ±0.09	5.46[a] ±0.15	5.52[a] ±0.24	5.58[a]± 0.28	5.61[a]±0.29
pH						
Control	6.22±0.14	6.47±0.22	6.57±0.19	6.59±0.14	6.40±0.15	6.35±0.15
Treated	5.50 ±0.20	6.13±0.27	6.38 ±0.13	6.53±0.18	6.36±0.21	6.30±0.20

*Means with different superscript in a row differ significantly (P<0.05).

Micro-biological quality assessment

The total plate count (TPC), Enterobacteriaceae count (EC), yeast and mould count (YMC) in chicken snack, as well as control were determined following the methods of APHA (1984). The experiment was repeated three times for each and every parameter.

Statistical analysis

Data collected in study were analyzed statistically, following the procedure of Snedecor and Cochran (1980) in the computer center of the Institute. Mean and standard errors were calculated for different parameters. The data were subjected to analysis of variance and paired comparison test. In significant effects, least significant differences were calculated at an appropriate level of significance.

RESULTS

The mean values and degree of significance for various

parameters such as moisture, fat, protein, ash, thiobarbituric acid (TBA) value, shear force value and pH for the chicken snacks, as well as for control snack at regular interval of 0, 6, 12, 18, 24 and 30 is presented in Table 3. Same values for microbiological profile and different sensory attributes are depicted in Tables 4 and 5 respectively.

Physico-chemical characteristics of chicken and control snacks

The physico-chemical characteristics of chicken snacks and control for storage period of 30 days showed none significant differences (P>0.05) in the contents of moisture, fat, protein, ash and pH in both of the treatments. The contents of shear force value (kg/cm²) and pH were found in increasing order with advancement of the days of storage while moisture, fat, protein and ash

Table 4. Changes in microbiological profile of chicken snacks packaged under vacuum in laminated pouches during storage at 30±2°C (Mean± SE)*.

Particulars	Days of storage					
	0	6	12	18	24	30
Total plate count (cfu/g)						
Control	NDS	$14.7 \times 10^{1a} \pm 0.09$	$28 \times 10^{1b} \pm 1.53$	$60 \times 10^{1c} \pm 1.73$	$9.8 \times 10^{2d} \pm 0.14$	$31 \times 10^{2e} \pm 2.08$
Treated	NDS	$22 \times 10^{1a} \pm 1.15$	$34 \times 10^{1b} \pm 3.18$	$58 \times 10^{1c} \pm 2.08$	$13 \times 10^{2d} \pm 1.73$	$42 \times 10^{2e} \pm 1.53$
Enterobacteriaceae count (cfu/g)						
Control	NDS	NDS	NDS	NDS	$10.9 \times 10^{1a} \pm 0.15$	$32.4 \times 10^{1b} \pm 0.87$
Treated	NDS	NDS	NDS	NDS	$14.6 \times 10^{1a} \pm 1.03$	$47.5 \times 10^{1b} \pm 1.61$
Yeast and Mould count (cfu/g)						
Control	NDS	NDS	NDS	NDS	$12.7 \times 10^{1a} \pm 3.60$	$25.3 \times 10^{1b} \pm 17.57$
Treated	NDS	NDS	NDS	NDS	$15.0 \times 10^{1a} \pm 14.42$	$29.7 \times 10^{1b} \pm 22.11$

*Means with different superscript row-wise differ significantly (P<0.05); NDS- Not detected significantly.

Table 5. Sensory attributes of chicken snacks as affected by aerobic and vacuum packaging during storage at 30±2°C (Mean*± SE).

Particulars	Days of storage					
	0	6	12	18	24	30
Colour and appearance						
Control	6.47±0.07	6.45± 0.07	6.42± 0.07	6.34± 0.08	6.30± 0.08	6.30± 0.08
Treated	7.29[a]±0.07	7.26[ab]±0.07	7.21[ab]±0.07	7.20[ab]±0.07	7.16[ab]±0.07	7.09[ab]±0.06
Flavour						
Control	6.20[ab]±0.06	6.15[ab]±0.07	6.21[ab]± 0.07	6.16[ab]±0.07	6.10[ab]±0.07	6.23[b] ±0.06
Treated	7.17[a]±0.06	7.09[a]±0.06	7.02[abc]±0.06	7.00[abcd]±0.07	6.95[abcd]±0.07	7.15[bcd]±0.06
Texture						
Control	6.03[bc]±0.05	6.00[a]±0.05	5.97[ab] ±0.05	5.92[abc] ±0.05	5.88[ab] ±0.05	5.87[bcd]±0.05
Treated	7.34±0.08	7.31±0.08	7.29±0.08	7.25±0.07	7.21±0.07	7.19±0.07
Crispness						
Control	6.00[ab]±0.06	5.98[a]±0.06	5.95[ab] ±0.06	5.97[ab] ±0.06	5.92[b] ±0.06	5.88[b] ±0.06
Treated	7.03[a]±0.08	7.00[a]±0.07	6.96[ab]±0.07	6.92[ab]±0.07	6.90[ab]±0.07	6.84[ab]±0.07
Aftertaste						
Control	6.33[a]±0.07	6.31[abc]±0.07	6.28[abc]±0.07	6.30[abc]±0.07	6.28[abc]±0.07	6.22[c]±0.06
Treated	7.13[a]±0.05	7.10[ab]±0.05	7.06[abc]±0.05	7.02[abc]±0.05	7.00[abc]±0.05	6.94[bc]±0.05
Meat flavour intensity						
Treated	6.27±0.06	6.25±0.06	6.23±0.06	6.20±0.06	6.18±0.06	6.20±0.05
Overall acceptability						
Control	6.13±0.07	6.11±0.06	6.10±0.06	6.07±0.06	6.05±0.06	6.00±0.06
Treated	7.06[a]±0.15	7.17[ab]±0.06	7.15[abc]±0.06	7.10[abc]±0.06	7.09[abc]±0.06	7.03[abc]±0.06

*Means with different superscript in a row differ significantly (P<0.05).

showed decreasing trend in the whole of the storage period of 30 days. TBA value of chicken snacks initially decreased up to the 12th and 18th day in control snacks and thereafter increased. TBA (mg malonaldehyde/kg) values of chicken snacks were none significantly different in the entire storage period while control snacks of 0, 6 and 12th day was significantly (P<0.05) different from the product of the 18th, 24th and 30th days. The values of shear force on day 0 and 6 were significantly (P<0.05) different from the rest of the storage values in chicken snacks while a non significant difference was observed on control snacks during the whole of the storage period.

On comparative assessment of chicken snacks and control snacks, we found highly significant difference (P<0.01) in the contents of fat, proteins, ash and TBA value during the whole of the storage period and in moisture contents and shear force value on day 6 and in pH on both day 0 and 6. However, a significant difference (P<0.05) in the contents of moisture was observed in the rest of the storage time and shear force value on day 0, in pH on day 12, among the treated and control snacks. There were non-significant differences also observed in shear force value between the 12th to 30th day, in pH on the last three studied days.

Microbiological profile

In general, TPC, EC and YMC profiles of chicken snacks and control snacks at different intervals during storage were in increasing trend. The TPC (cfu/g) of the products, irrespective of its product type indicated an increasing trend during storage after the 6th day of storage and increased significantly (P<0.05) after every 6 days until the 30th day of storage. EC (cfu/g) of the products in both the treatments was not detected significantly until the 18th day, after that it indicated an increasing trend. EC in the products during storage differed significantly (P<0.05) to each other from the 24th to 30th day. YMC (cfu/g) in chicken snacks was also not significantly detected until the 18th day; after that it showed increasing trend during the entire period of storage. Like EC, YMC of both the products increased significantly (P<0.05) after the 24th day. Higher count for TPC, EC and YMC were noticed in chicken snacks, as compared to control snacks which might be due to presence of meat and higher moisture content. Comparative study of chicken and control snacks revealed significant differences during the entire period of the microbiological profile.

Sensory attributes

In general, all the sensory attributes that is colour and appearance, flavour, texture, crispness, aftertaste, meat flavour intensity and overall acceptability indicated decreasing trend during the entire storage period at ambient temperature in both chicken and control snacks. However, this statement is reversed on day 12 and 30 for flavour score of control snack, day 30 for flavour and meat flavour intensity of chicken snacks and on 0 day of chicken snack for overall acceptability. The scores for colour and appearance of the product did not change significantly during the whole of the storage period for both treatments.

This statement is also true for flavour scores of control snacks, but flavour scores of chicken snack on day 0 and

6 were significantly (P<0.05) different from other scores during storage. Texture scores of both products were non-significantly different during whole storage except the scores of day 6 as compared to 0 and 30th day scores for control.

The scores for crispness on day 6 for control were significantly (P<0.05) different from the scores of control on days 24 and 30, while the rest of the scores for crispness in all days were non-significantly different. The scores for aftertaste, meat flavour intensity and overall acceptability were none significantly different during the whole of the storage, except the aftertaste scores on day 0 which was significantly (P<0.05) different from day 30 in both of the treatment. Comparative study between control and treated snacks revealed overall highly significant difference, irrespective of the days of storage with some exceptions in the scores of flavour and overall acceptability. Flavour score of day 0 and overall scores for whole storage period except 0 day were found non-significantly different. Though, meat flavour intensity scores was not observed in control snacks, so the comparative study was not conducted for meat flavour intensity score.

DISCUSSION

Physico-chemical characteristics of chicken and control snacks

A non-significant difference in contents of moisture, fat, protein and ash was noticed in the products but quantitative trend decreased in order during the entire period of storage. This trend was very well in the range of the findings of Kalra et al. (1987) but the qualitative trend for moisture in his study was in increasing order rather than in decreasing. Values for TBA were also similar to the findings of Park et al. (1993) for beef snacks. The trend of gradual increase in pH with the advancement of the storage time is very well agreed according to the findings of Huang et al. (1996), Prabhakara and Janardhana (2000), Kumar and Sharma (2006), and Bhat and Pathak (2009) for different meat products. The increasing trend of shear force value during the entire storage period in both of the products, might be due to gradual decline in moisture content with the advancement of the storage period.

Microbiological profile

TPC (cfu/g) was not detected significantly on day 0, as the total colony count was less than 30, so we did not consider it as significant. Thereafter, it showed increasing trend from day 6 to 30 of storage. Enterobacteriaceae count, yeast and mould count were also not detected significantly, until day 18 of the storage and then showed increasing trend. The counts for all three parameters were greater than the values obtained by Hobbs and Greene (1976) for beef snacks stored at 37°C for 5 months which might be due to post processing contaminations. However, the values obtained in this study were very well in the standards microbiological limits for meat products. Higher TPC, enterobacteriaceae count, yeast and mould count were noticed in treated products, as compared to control during the entire storage period. It could be due to the incorporation of meat which is a good medium for the growth of micro-organisms.

Sensory attributes

In general, sensory attributes showed insignificant decreasing trend during the whole of the storage period irrespective of the product type. Kalra et al. (1987) also observed slight decrease in the scores for colour and texture of snacks packaged in low density polyethylene (LDPE) bags of 100 and 150 gauge thickness, as well as in friction top tins during storage at room temperature up to 6 months. Decline in colour and appearance scores during storage could be due to dilution of meat pigments.

These findings are also supported by Zyl and Zayas (1996), Kumar and Sharma (2006), and Bhat and Pathak (2009). The decrease in flavour and meat flavour scores with the advancement of the storage period might be due to dilution in meaty flavour. Similar reports were published by Padda et al. (1989), Kumar and Sharma (2005, 2006), and Bhat and Pathak (2009) for various meat products. The decline in overall acceptability scores, could be reflective of changes in scores of flavour, colour, texture and other sensory attributes. Similar findings were reported by Nag et al. (1998) and other workers.

Conclusion

Chicken snacks prepared by utilizing 50% broiler spent hen meat, sodium caseinate and rice starch, as well as control snacks kept well for 30 days at ambient temperature (30±2°C) under vacuum in laminated pouches. During entire storage, chicken as well as control snacks did not show much change in their physico-chemical characteristics, microbiologlu profile and sensory attributes. Although, they all were in decreasing trend but their values were very well under the acceptable limit. So we can say that, vacuum packaging for such type of self sustained meat snacks may be a good alternative of packaging. Though, the study was only for 30 days storage, we cannot definitely comment on the shelf life of the product.

REFERENCES

Abe HA, Kimura T, Yamuchi K (1996). Effect of collagen on the toughness of meat from spent laying hens. J. Jpn. Soc. Food Sci. Technol., 43(7): 831-834.

APHA (1984). In: Compendium of methods for the microbiological examination of foods. Speck, M.L (Ed), American. Public Health Association, Washington, DC.

AOAC (1995). Official methods of analysis. 16th Edn. Association of Official Analytical Chemists, Washington, DC.

Bailey AJ (1984). The chemistry of intra molecular collagen. The Royal Society of Chemistry, Burlington House. Recent Adv. Chem. Meat., pp. 22-47.

Bhat ZF, Pathak V (2009). Effect of mung bean (Vigna radiate) on quality characteristics of oven roasted chicken seekh kababs. Fleischwirtschaft Int., 6: 58-60.

Chung S, Bechtel P, Villota R (1989). Production of meat based intermediate moisture snack foods by twin-screw extrusion. 49th Annual Meeting, Institute of Food Technologists, Chicago, IL. June 25-29.

Dzudie T, Joel Scher J, Hardy J (2002). Common bean flour as an extender in beef sausages. J.Food Eng., 52 (2): 143-147.

Gujral HS, Kaur A, Singh N, Sodhi SN (2002). Effect of liquid whole egg, fat and textured soy protein on the textural and cooking properties of raw and baked patties from goat meat. J. Food Eng., 53: 377-385.

Hobbs WE, Greene VW (1976). Cereal and cereal products. In: Compendium of methods for the microbiological examination of foods. Speck, M.L (Ed), American. Public Health Association, Washington, DC. pp. 559 (cf Park et al., 1993).

Huang JC, Zayas JF, Bowers J (1996). Functional properties of Sorghum flour as an extender in ground beef patties. IFT Annual Meeting: Book of abstracts, ISSN 1082-1236, pp. 63-64.

Kalra CL, Kaur S, Sharma TC, Kulkarni SC, Berry SK (1987). Studies on the preparation, packaging and storage of potato snacks from cold stored potatoes. Indian Food Packer, 45: 30-39.

Keeton JT (1983). Organoleptic quality assessment methods for foods of animal origin. J. Food Sci., 48: 878.

Kondaiah N (1990). Poultry meat and its place in market. Poult. Guide, 27: 41-45.

Kumar RR, Sharma BD (2005). Evaluation of the efficacy of pressed rice flour as extender in chicken patties. Indian J. Poult. Sci., 40(2): 165-168.

Kumar RR, Sharma BD (2006). Efficacy of barley flour as extender in chicken patties from spent hen meat. J. Appl. Anim. Res., (30): 53-55.

Mahapatra CM (1992). Poultry Products Technology-Prospects and Problems. Poult. Guide, 29: 69-70.

Nag S, Sharma BD, Kumar S (1998). Quality attributes and shelf life of chicken nuggets extended with rice flour. Indian J. Poult. Sci., 33(2): 182-186.

Padda GS, Sharma N, Bisht GS (1989). Effect of some vegetative extenders on organoleptic and physico-chemical properties of goat meat balls. Indian J. Meat Sci. Technol., 2: 116-122.

Park J, Rhee KS, Kim BK, Rhee KC (1993). High protein texturized products of defatted soy flour, corn starch and beef: Shelf life, physical and sensory properties. J. Food Sci., 58: 21-27.

Prabhakara RK, Janardhana RB (2000). Effect of binders and precooking meat on quality of chicken loaves. J. Food Sci. Technol., 37 (5): 551-553.

Serdarouglu M, Degirmencioglu O (2004). Effect of fat level (5, 10 and 20%) and corn flour (0,2 and 4%) on some properties of Turkish type meat balls (koefte). Meat Sci., 68(2): 291-296.

Smith GL, Stadler JW, Keeton JT, Papadopoulos LS (1991). Evaluation of partially defatted chopped beef in fermented beef snack sausage. J. Food Sci., 56: 348-351.

Snedecor GW, Cochran WG (1980). In: Statistical Methods. 7th Edn. Oxford and IBH Publishing Co., Calcutta.

Strange ED, Benedit RC, Smith JL, Swift CE (1977). Evaluation of rapid test for monitoring alterations in meat quality during storage. J. Food Protect., 40: 843-847.

Tarte R, Molin RA, Koymazadeh M (1989). Development of beef/corn extruded snack model product. Proc. 50th Annual Meeting, Institute of Food Technologists, Chicago, IL. June 25-29.

Wenham LN, Fairbain K, McLeod W (1973). Eating quality of mutton compared with lamb and its relationship to freezing practices. J. Anim. Sci., 36: 1081.

Witte VC, Krauze S, Bailey ME (1970). A new extraction method for determining 2-thiobarbituric acid values of pork and beef during storage. J. Food Sci., 35: 582-585.

Zyl HV, Zayas JF (1996). Effect of three levels of sorghum flour on the quality characteristics of frankfurters. IFT Annual Meeting: Book of abstracts, ISSN 1082-1236, p. 64.

11

Sensory profile and chemical composition of *Opuntia joconostle* from Hidalgo, Mexico

Contreras L. E.[1]*, Jaimez O. J.[1], Castañeda O. A.[1], Añorve M. J.[1] and Villanueva R. S.[2]

[1]Chemical Research Center, Universidad Autonoma del Estado de Hidalgo. Carretera Pachuca-Tulancingo km 4.5, C. P. 42076, Pachuca, Hidalgo, Mexico.
[2]Research and Assistance in Technology and Design of the State of Jalisco A. C. Normalista Avenue No. 800, Normal Hill, C. P. 44270, Guadalajara, Jalisco. Mexico.

The objective of this study was to compare the sensory profile and the chemical composition of four samples of *Opuntia joconostle* cultivated in Sahagun and Cuautepec, Hidalgo (Mexico). A group of ten trained panelists built the flavor profile of the samples according to ISO 13299:2003. The chemical composition was determined according to AOAC. The differences observed between sensory profiles and chemical composition of the four *O. joconostle* studied were significant even though all fruits belonged to the same specie and variety. The maturity stage, the chemical composition of soil and the geographic region could have influence on the results.

Key words: *Opuntia joconostle*, flavor profile, sensory analysis, chemical composition.

INTRODUCTION

Cacti have traditionally been an important part of Mexican culture. The Cactaceae family is a botanical group of the new world and Mexico is the country with the largest center of diversity of this family (Ortega-Nieblas et al., 2001). This natural resource has been and is being used for multiple purposes since pre-Columbian times. Some of the current uses include: Food for humans as vegetable and fruit, forage for animals, source for alcoholic beverages, sweetener, live fences, industrial products such as cosmetics and dye, and as a medical source against diabetes and other diseases (Saenz et al., 1998; Badii and Flores, 2001; Basurto et al., 2006).

In Mexico, the genus *Opuntia* is represented mainly by the following species: *Opuntia ficus-indica, Opuntia albicarpa, Opuntia robusta, Opuntia streptacanta* and *Opuntia joconostle*. The prickly pear from the last specie is known as "Xoconostle" and is characterized by a smaller size, a weak pink color and an acid flavor. This fruit is widely used for the preparation of jam, marmalades, beverages, sauces and has traditionally

been used for alternative treatments of diabetes mellitus.

In despite of the potential of prickly pears in the food, cosmetic and pharmaceutical industries (Ruiz-Feria et al., 1998; Moreno-Alvarez et al., 2003; Garcia-Pantaleon et al., 2009) there are few information about their composition and medicinal properties. The physicochemical and sensory properties (odor, aroma, taste and texture) of some fruits of *O. joconostle* have been determined. The objective of this study was to compare the sensory profile and the chemical composition of four samples of *O. joconostle* cultivated in Sahagun and Cuautepec, Hidalgo (Mexico).

MATERIALS AND METHODS

Sample collection and preparation

Eighty fruits of four samples of *O. joconostle* were collected from Cuautepec (samples 1, 2 and 3) and Sahagun (sample 4), Hidalgo (Mexico). Fruit samples were taken from ten different plants pursuant to the criteria set forth by Viloria-Matos and Moreno-Álvarez (2001) for fruits of the same species. The samples were transported in thermally insulated containers at a temperature of 7±1ºC. Fruits were washed and thorns were removed. Later, the pulp was separated manually and kept in plastic containers under

*Corresponding author. E-mail: eliclopez@yahoo.com.mx.

Table 1. Chemical composition of *O. joconostle* fruits analyzed.

Parameter (g/ 100 g)	Sample 1	Sample 2	Sample 3	Sample 4
Protein	0.71	1.16	1.56	1.11
Carbohydrates	7.98	6.33	5.81	6.95
Crude fat	<0.10	<0.10	<0.10	<0.10
Moisture	87.30	87.69	87.70	89.05
Ash	0.49	0.54	0.65	0.54
Crude fiber	3.52	4.28	4.28	2.35
Energetic content (kcal)	34.76	29.96	29.48	32.24
pH	3.00	3.20	3.00	2.80

Samples 1, 2 and 3 were collected in Cuautepec, Hidalgo, Mexico. Sample 4 was collected in Sahagun, Hidalgo, Mexico.

refrigeration before analyses.

Chemical composition

Moisture content, protein, ethereal extract, ash and crude fiber were analyzed according to AOAC (1990) methods.

The pH of the "Xoconostle" pulp samples was determined by taking 10 g of homogenized pulp sample in 50 ml clean beaker, using a digital pH meter (Jenway 3510-UK) at 25ºC.

Sensory evaluation

The sensory evaluation was performed in the "Centro de Investigación y Asistencia en Tecnología y Diseño del Estado de Jalisco" - Research and Assistance in Technology and Design of the State of Jalisco (CIATEJ), Mexico. For the determination of the flavor profile of four samples of "Xoconostles" ten judges were selected and trained according to the standards of the ISO 3972:1991, ISO 8586:1993, ISO 5496:1992 and the ISO 6564:1985. The sensory profile was developed by using descriptive quantitative and qualitative analysis (QDA). A list of descriptive terms was generated for each sample. A sensory score sheet with a 150 mm non-structured scale was used to rate the perceived intensity of each descriptive term. The intensity of each descriptor was the average of the intensity attributed by the 10 panelists and two repetitions. Only the descriptive terms that presented a variation coefficient less than 10% were selected.

RESULTS AND DISCUSSION

Chemical composition

The chemical composition parameters of four samples of *O. jononostle* analized are listed in Table 1. Later analysis revealed low amounts of protein (0.71 to 1.56%), crude fat (<10%), crude fiber (2.35 to 4.28%), ash (0.49 to 0.65%) and carbohydrates (5.81 to 7.98%). These results were consistent with the low energetic content observed (29.48 to 34.76 kcal/100 g). Moisture content was the higher parameter found in all samples analyzed (>80%). All the fruits presented low and homogeneous pH values (2.8 to 3.2).

Excepting pH, chemical composition values found were in agreement with those reported for some prickly pears varieties (*O. ficus-indica*, *O. robusta, O. albicarpa*). pH was lower to those reported for most prickly pear fruits (4.27 to 5.75) (Sawaya et al., 1983; Chavez-Santoscoy et al., 2009).

Sensory profiles

At the beginning of the experiment, a list of 42 descriptor terms for appearance, odor, aroma, taste, texture, and trigeminal sensations was proposed. However, the descriptors that represented similar characteristics or those that were not detected by all the judges were eliminated. A final vocabulary list of six odor, five aroma, one taste and two texture descriptors was selected and used to build the QDA of each sample. "Xoconostles" collected in Sahagun showed a larger number of descriptors compared with those from Cuautepec (Figure 1). The odor terms found for Sahagun samples were green fruit, peach, green oxidized fruit, melon, cucumber and wet straw, aroma descriptor for these samples included citric, wet straw, green fruit, dry grass and dirty cloth. Regarding Cuautepec samples, they showed a similar flavor profile among them but markedly different from "Xoconostles" from Sahagun. Their common odor descriptors were green fruit, peach and cucumber while melon was detected only in samples 1 and 3. Sample 3 showed green oxidized fruit odor descriptor perceived in sample 4. Green fruit was the only aroma descriptor detected in samples 1, 2 and 3. Bittersweet taste, firmness and fiberousity were detected in all prickly pear fruits studied.

Conclusion

Despites the analyzed samples belonged to the same genus and specie (*O. joconostle*), there were notorious differences in their flavor profile attributed mainly to the

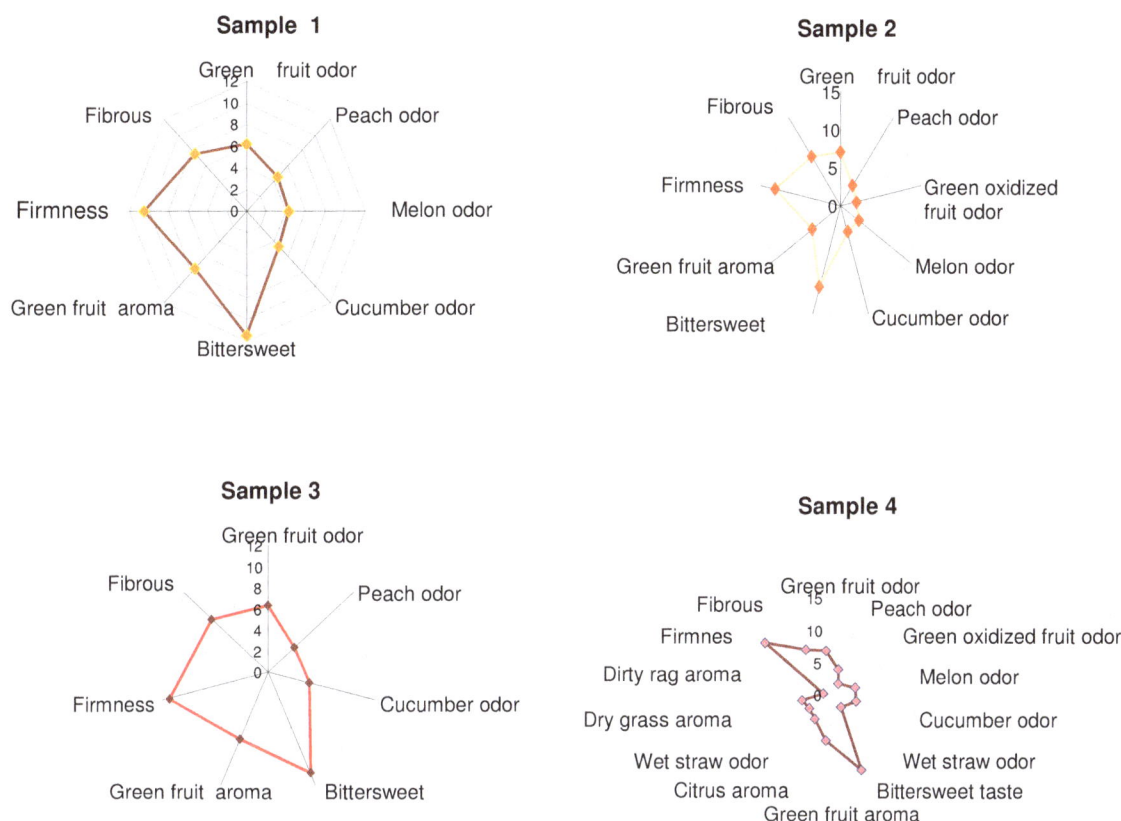

Figure 1. Sensory flavor profiles of samples 1, 2, 3 and 4 collected in Cuautepec and Sahagun, Hidalgo, Mexico.

geographical conditions of the regions where they were collected. However their chemical composition was similar. Further research is needed in order to find the factors affecting the sensory properties that determine the flavor profile of each fruit.

REFERENCES

Badii MH, Flores AE (2001). Prickly pear cacti pests and their control in Mexico. Fla. Entomol., 84(4): 503-505.

Basurto DS, Lorenzana-Jiménez M, Magos G (2006). Cactus utility glucose control in diabetes mellitus type 2. Rev. UNAM School Med., 49: 157-161.

Chavez-Santoscoy RA, Gutierrez-Uribe JA, Serna-Saldívar SO (2009). Phenolic Composition, Antioxidant Capacity and *in vitro* Cancer Cell Cytotoxicity of Nine Prickly Pear (*Opuntia* spp.) Plant Foods Hum. Nutr., 64: 146–152.

Garcia-Pantaleon DM, Flores-Ortiz M, Moreno-Álvarez MJ, Belén-Camacho DR, Medina-Martínez CA, Ojeda-Escalona CE, Padrón-Pereira CA (2009). Chemical, biochemical, and fatty acids composition of seeds of *Opuntia boldinghii* Britton *et* Rose. J. Prof. Assoc. Cactus., 11: 45-52.

Moreno-Álvarez MJ, Medina C, Antón L, García D, Belén-Camacho DR (2003). Using the pulp of prickly pear (*Opuntia boldinguii* Britt. et Rose) in the pigmented citrus beverage. Interciencia, 28: 539-543.

Ortega-Nieblas M, Molina-Freaner F, Robles-Burgueño M, Vázquez-Moreno L (2001). Proximate composition, protein quality and oil composition in seeds of *Columnar* cacti from the Sonoran Desert. J. Food Comp. Anal., 14: 575-584.

AOAC (1990). Official Methods of Analysis of the Association of Official Analytical Chemists, Published by AOAC, Inc. Helrich K (editor), 15th edition, Arlington, I&II: 17-18, 40-62, 69-83, 1012.

Ruiz-Feria CA, Lukefahr DS, Felker P (1998). Evaluation of *Leucaena leucocephala* and cactus (*Opuntia* sp.) as forages for growing rabbits. Livestock Res. Rural Dev., 10: 1-13.

Saenz C, Estévez M, Sepúlveda E, Mecklenburg P (1998). Cactus pear fruit: A new source for natural sweetener. Plant Foods Hum. Nutr., 52: 141-149.

Sawaya WN, Khatchadourian HA, Safi WM, Al-Muhammad HM (1983). Chemical characterization of prickly pear pulp, *Opuntia ficus-indica*, and the manufacturing of prickly pear jam. Int. J. Food Sci. Technol., 18: 183–193.

Viloria-Matos AJ, Moreno-Álvarez MJ (2001). Betalainas una síntesis de su proceso. Biotamns. 12: 7-18.

12

Phytochemical and mineral content of the leaves of four Sudanese *Acacia* species

Hassan Abdalla Almahy[1]* and Omaima Dahab Nasir[2]

[1]Department of Chemistry, Taif University, Faculty of Science and Education, Alkhurma, Kingdom of Saudi Arabia.
[2]Department of Medical Laboratory, Taif University, Faculty of Applied Medical Sciences, Turaba, Kingdom of Saudi Arabia.

The phytochemical and mineral composition of *Acacia nilotica, Acacia mellifera, Acacia albida* and *Acacia radiana* were investigated. Both freshly expended and older leaves of the plant species were used for the analysis. Averagely, the phytochemical constituent of the plants are as follows: Tannins (0.18 to 0.21 mg ml^{-1}), phenols (0.03 to 1.01 mg rnl^{-1}), saponins (0.11 to 0.23 mg ml^{-1}) and flavonoids (0.02 to 0.24 mg ml^{-1}). Alkaloids were found only in *A. albida* and *A. radiana* (0.04 to 0.06 mg ml^{-1}). The result also indicates that the leaves of the plant species contained in percentage, the following elements; phosphorous (0.50 to 1.04%), calcium (0.80 to 2.00%), magnesium (0.24 to 0.49%), potassium (1.05 to 1.54%), sodium (0.51 to 1.03%) and nitrogen (2.38 to 4.06%). The result obtained indicate that the leaves of the plant are good services of phytochemicals and minerals needed for maintaince of good health and can also be exploited in the manufacture of drugs.

Key words: *Acacia* species, leaves, phytochemical, mineral elements.

INTRODUCTION

The roles of plants in maintaining human health is well documented (Moerman, 2004). In Sudan, many of these indigenous spices and their extracts are used in traditional medicine (Okwu and Ekeke, 2003), many of these plants possess bioactive compounds that exhibit physiological activity against bacteria and other microorganisms. These species, hence are used in treatment or many diseases such as rheumatism, diarrhoea, dysentery, cough asthma, diabetes, malaria, elephantiasis, cold and others (Bartram, 1998; Burkill, 2000; Gill, 1999; Nnimh, 1998; Oliver, 2007).

The general assumption is that the active dietary constituents contributing to the protective effects of these plant materials are phytochemicals, vitamins and minerals. Phytochemicals are present in a variety of plants utilized as important components of both human and animal diets. These include spices, fruits, herbs and vegetables (Urquiaga, 2000). Thus, diets containing an abundance of fruits and vegetables give protection against a variety of diseases, particularly cardiovascular diseases (Urquiaga, 2000). Many rare and useful herbs occur in Sudan, from which important drugs could be prepared or agents which may serve as starting materials for the partial synthesis of some useful drugs (Sofowora, 2005). The usefulness of these plant materials medicinally is due to the presence of bioactive constituents such as alkaloids, tannins, flavonoids and phenolic compounds (Hill, 1998). These chemical are known to carry out important medicinal roles, in human body.

Alkaloids play some metabolic role and control development in living systems (Edeoga and Eriata, 2001). They are also involved in protective function in animals and are used as medicine especially the setriodal alkaloids (Stevens et al., 1997). Tannins are known to inhibit pathogenic fungi (Burkill, 2005). Saponin prevent disease invasion of plants by parasitic fungi (Bidwell, 1999), hence have some antifungal properties, they have anti-fertility effects on rats (Zhang et al., 1996). Studies revealed that flavonoids apart from their antioxidant protective effects, inhibits the initiation, promotion and progression of tumors (Kim et al., 2002;

Okwu, 2004). *Acacia* species belong to the family Mimosaceae and they occur through out the tropics, the Sudanese species are trees with bright yellow flower and usually alternate leaflets (Keay et al., 1999). *Acacia nilotica* has distinct pickles on twigs and young branches.

The trees are about 3 m tall, with spreading crown, the fruits are dark brown and velvety when young, with short soft and often sticky pickles mostly surrounding the centre.

Acacia mellifera, trees are about 2.5 m tall. The fruit is flat and papery 6 to 9 cm across, densely velvet with short hairs. It becomes finely hairy, when ripe (Keay et al., 1999). *Acacia albida* is a water side tree very conspicuous during the flowering season. The trees are usually about 2.4 to 3.8 m tall, branching low down with staggling branches. The fruits are light brown, glabrous and 3.75 to 6.25 cm across, including the soft fleshy narrow wing which extends about three quarters way round the body with a corky and knobby flesh surrounding a hard woody shell (Keay et al., 1999). *Acacia radiana* trees are about 2 m tall having a compact rounded crown and drooping branches (Keay et al., 1999). The leaves of *A. nilotica* and *A. albida* aides have medicinal properties; they are used in treatment of skin disease such as eczema, candidiasis and acnes. The concoction made from their barks is also used in treating asthmatic patients (Gill, 1999). The leaves of *A. mellifera*, *A. albida* and *A. radiana* are used as vegetables in food preparation. *A. albida* leaves are used as fodder for feeding live stock, while the fruits of *A. mellifera* are edible (FAO, 2000). This study investigates the presence and the amount of some phytochemicals in the leaves of the *Acacia* species used in the investigation, thus determing their medicinal values.

MATERIALS AND METHODS

Sample collection

The leaves of *A. nilotica*, *A. mellifera*, *A. albida* and *A. radiana* were collected from (Central of Sudan) Elgazira State. They were identified by the taxonomy unit of the Forestry Department College of Natural Resources and Environmental Management University of Khartoum. The leaves were separated into freshly expanded and older and air dried in the laboratory for one week. The leaves were separated due to the fact that the freshly expanded leaves are the only ones used for phytochemical screening.

Preparation of sample for analysis

The dried leaves were ground to a fine powder using Thomas-Willey Milling Machine. The milled sample bottles at room temperature in the laboratory of the Biological Sciences Department, University of Khartoum.

Phytochemical analysis (qualitative analysis for presence of phytochemicals)

Alkaloid determination

The presence of alkaloid was determined using the Mayer and Wagner's test as descried by the Harbone, 1998. 2 g of each portion of the powdered sample were put into a conical flask and 20 ml of dilute sulphuric acid in ethanol was added into it and then heated in water bath to boil for 5 min. The mixture was filtered and the filtrates were separated and treated with 2 drops of Mayer and Wagner reagents in test-tubes. Development of an orange colouration indicated positive result.

Saponins determination

The froth test and emulsion test as described by Harbone 2001 were used to determine the presence of saponins. 20 ml of water was added to 0.25 g of the powdered sample in 100 ml beaker and boiled was filtered and then used for the test.

Froth test: 5 ml of the filtrate was diluted with 20 ml of water and shaken vigorously. A stable froth (foam) up on standing indicates the presence of saponins.

Emulsion test: 2 drops of olive oil was added to the frothing solution and shaken vigorously the formation of emulsion indicates the presences of saponins.

Tannin determination

The presence of tannins was carried out using the Harbone method (Harbone, 2001). 1.0 g of the powdered sample was boiled with 50 ml of water, filtered and the filtrate used to carryout the ferric chloride test. Few drops of ferric chloride was added to 3 ml of the filtrate in a test tube. A greenish black precipitate indicates the presence of tannins.

Flavonoids determination

The presence of flavonoids in the samples was determined using the Harborne and Sofowora methods (Sofowora, 2005; Harbone, 2001). 10 ml of ethyl acetate was added to 0.2 g of the powdered sample and heated in a water bath for 5 min. The mixture was cooled filtered and the filtrates used for the test.

Ammonium test: About 4 ml of filtrate was shaken with 1 ml of dilute ammonia solution. The layers were allowed to separate and the yellow colour in the ammoniacal layer indicates the presence of flavonoids.

Aluminum chloride solution test: 1 ml of 1% aluminum chloride solution was added to 4 ml of the filtrate and shaken. A yellow coloration indicates the presence of flavonoids.

Quantitative analysis of phytochemical

Alkaloid determination

The quantity of alkaloids in the sample was determined using Harbone method (Harbone, 2001). 5 g of the powdered sample was extracted with 10 ml of petroleum ether. The petroleum ether was removed using aspirator. 1.0 g of the extract was suspended in 10 ml of double distilled water and the pH adjusted to 7.6, after shaking for one hour, the suspension was centrifuged. 1 ml of the supernatant was diluted to 50 ml with phosphate buffer. The absorbance was measured spectrophotometrically at 580 nm wavelength.

Saponins determination

Saponins determination was carried out using Harborne (2001) method. 0.1 g of the sample boiled with 5 ml of double distilled water for 5 min decanted filtered while still hot. 2 ml of olive oil was added to it and shaken for 30 s. The absorbance was measure at 620 nm wavelength.

Tannin determination

Okeke and Elekwa (2003) method was used for tannin determination 0.5 g of the sample was shaken with 10 ml of 2 M HCl in a test tube for 5 min. The contents were then transferred into a volumetric flask made up to 50 ml and then filtered. 5 ml of the filtrates, was introduced into a test tube and 3 ml of 0.1 M $FeCl_3$ in 0.1 HCl and 3 ml of 0.008 M of potassium ferro cyanide $(KFe(CN)_3$ were added. The absorbance was read at 720 nm within 10 min.

Flavonoids determination

Flavonoids determination was done using Boham and Kocipai method (Boham and Kocipai, 1997). 10 g of the plant material was extracted repeatedly with 100 ml of 80% aqueous methanol at room temperature. The whole solution was filtered through Whatman filter paper number 42 (125 mm). The filtrate was then transfered into a crucble and evaporated into dryness over in a water bath and weighed to a constant weight.

Total phenol determination

The total phenol determination was done using Harborne (2001) method. The fat free sample was boiled with 50 ml of ether for the extraction of phenolic component for 15 min. 5 ml of the extract was pipetted into a 50 ml flask and 10 ml of distilled water was added 2 ml of ammonium hydroxide solution and 5 ml of concentrated amyl alcohol were added and made up to mark and left to react for 30 min for color development. The absorbance of the solution was read at 505 nm wavelength using a spectrophotometer.

Mineral element determination

The levels of the mineral elements calcium, phosphorus, sodium, magnesium, potassium and nitrogen were determined using the wet digestion extraction methods as described by Ojuwale (1998), Andrew (1999) and Nivozamsky et al. (2007). 0.2 g of the samples were weighed into a 15 ml flask. 5 ml of the extraction mixture (H_2SO_4-Selenium salicylic acid was added to the sample and allowed to stand over night. The mixture was heated initially at 20°C for 3 h and 5 ml of concentrated perchloric acid ($HClO_4$) added. This was then heated vigorously until digestion was completed. The solution was allowed to cool and filtered using an acid washed filtered paper into 50 ml volumetric flask and finally made up to mark with distilled water. The potassium and sodium content were determined using the flame photometer method, phosphorus by the vanado-molybdate yellow method using the spectrophotometer method. Calcium and magnesium determined by the versanate EDTA complexometric titration method and nitrogen by the semi micro distillation method using the Markham apparatus.

RESULTS

The phytochemical composition of the Acacia species determined is summarized in Table 1. The freshly expanded and older leaves of the Acacia species contain tannins, phenols, saponins and flavonoids. Alkaloids were present in the leaves of A. albida and A. radiana and not in A. nilotica and A. mellifera. The quantitative estimation of the phytochemicals content of the leaves of the Acacia species is summarized in Table 2. A. radiana had more alkaloid content (0.24 and 0.06 mg ml^{-1}) when compared with that of A. albida (0.04 and 0.06 m gm^{-1}). Tannins were relatively fairly distributed in all the Acacia species used. However, older leaves of A. nilotica had the highest amount (0.22 mg ml^{-1}) followed by those of A. radiana (0.21 mg ml^{-1}). A. mellifera contained more phenols and saponins than other species, its phenols content is 0.06 mg ml^{-1} and 1.01 mg ml^{-1}, respectively for fresh and older leaves and saponins is 0.23 mg ml^{-1} and 0.23 mg ml^{-1} for fresh and older leaves. A. radiana had the least amount of phenol and saponins (0.06 and 0.08 mg ml^{-1}) respectively. The freshly expanded leaves of all the Acacia species had more flavonoids content when compared with the older leaves.

The mineral element content of the Acacia species determined is summarized in Table 3. The leaves of all the Acacia species investigated contained phosphorus, calcium, maganesium, potassium, sodium and nitrogen in varying quantities. Generally, the freshly expanded leaves of the Acacia species have more phosphorus when compared with the older leaves. The phosphorus content of freshly expanded leaves of A. nilotica, A. mellifera, A. albida and A. radiana are 0.84, 0.74, 1.04 and 1.02% while that of the older leaves are 0.38, 0.50, 0.10 and 0.75% in that order. A. radiana and A. albida had more phosphorus when compared to those of other species. The calcium content of the species ranged from (0.80 to 2.00%), magnesium (0.24 to 0.49%), potassium

Table 1. Qualitative analysis of the phytochemical in the leaves of the four *Acacia* species investigated.

Acacia species	Alkaloid	Tannins	Phenols	Saponins	Flavonoids
Acacia nilotica (fresh)	-	+	+	+	+
Acacia nilotica (old)	-	+	+	+	+
Acacia mellifera (fresh)	-	+	+	+	+
Acacia mellifera (old)	-	+	+	+	+
Acacia albida (fresh)	+	+	+	+	+
Acacia albida (old)	+	+	+	+	+
Acacia radiana (fresh)	+	+	+	+	+
Acacia radiana (old)	+	+	+	+	+

+ = Presence; - = absence.

Table 2. The alkaloids, tannins, phenol, saponins and flavonoids contents mg ml^{-1} of the *Acacia* species investigated.

Acacia species	Alkaloid	Tannins	Phenols	Saponins	Flavonoids
Acacia nilotica (fresh)	0.000	0.210	0.960	0.220	0.240
Acacia nilotica (old)	0.000	0.215	0.025	0.150	0.020
Acacia mellifera (fresh)	0.000	0.210	0.645	0.230	0.230
Acacia mellifera (old)	0.000	0.208	0.010	0.228	0.020
Acacia albida (fresh)	0.060	0.180	0.470	0.120	0.200
Acacia albida (old)	0.050	0.200	0.470	0.160	0.150
Acacia radiana (fresh)	0.024	0.210	0.047	0.140	0.180
Acacia radiana (old)	0.055	0.212	0.080	0.110	0.040

Table 3. The percentage of some mineral elements in the leaves of the *Acacia* species investigated.

Acacia species	P	Ca	Mg	K	Na	N
Acacia nilotica (fresh)	0.84	1.50	0.37	1.45	0.70	3.36
Acacia nilotica (old)	0.38	2.00	0.30	1.54	0.85	3.85
Acacia mellifera (Fresh)	0.74	1.00	0.49	1.18	0.70	3.92
Acacia mellifera (old)	0.50	0.80	0.24	1.52	0.86	2.38
Acacia albida (fresh)	1.04	0.80	0.49	1.05	1.03	3.15
Acacia albida (old)	0.10	0.37	0.37	1.15	0.82	3.36
Acacia radiana (fresh)	1.02	1.32	0.36	1.24	0.51	3.68
Acacia radiana (old)	0.75	1.60	0.31	1.43	0.53	4.06

(1.05 to 1.54%), sodium (0.51 to 0.86%) and nitrogen (2.38 to 4.06%). *A. nilotica* and *A. radiana* had the highest calcium content while *A. albida* the least calcium content. The nitrogen content of the leaves of the *Acacia* species was higher than that of the other minerals element.

DISCUSSION

The fully expanded and older leaves of the *Acacia* species were found to have saponins, tannins, phenols and flavonoids. Alkaloids were present only in *A. albida* and *A. radiana*. These phytochemicals are known to have antimicrobial activity (Ebana et al., 2009). The presence of phenolic compounds in the leaves of these plants indicates that they may act as antimicrobial agents. Phenols and phenolic compounds are extensively used in disinfections and remain the standard with which other bacteriacides are compared. Thus, the presence of phenoilc compounds in the *Acacia* species may be the reason for the therapeutic, antiseptic, antifungal or bactericidal properties of the plants (Gill, 1999). The presence of flavonoids in the leaves of all the *Acacia* species indicates their medicinal value. Flavonoids are antioxidants and free radical scavengers which prevent oxidative cell damage, have strong anticancer activity and protect the cell against all stages of carcinogenesis (Salah et al., 2002; Okeke and Elekwa, 2003). Flavonoids in intestinal-tract lower the risk of heart disease (Okwu,

2005). *Acacia* species have been used in the treatment of arthritis in herbal medicine (Stray, 1998). The leaves of the *Acacia* species are found to contain saponins. The freshly expanded leaves of *A. mellifera* are of high economic value due to their high saponins content. Saponins is useful in medicine and pharmaceutical industry due to its foaming ability that produces frothy effects in the food industry (George, 2002). Saponins are also used in the manufacture of shampoos, insecticides, various drug preparation and synthesis of steroidal hormones (Sodipo and Akiniyi, 2000).

A. albida and *A. radiana* leaves had very little amount of alkaloids others lacked alkaloids. Alkaloids are known to exhibit marked physiological activity when administered to animals (Okwu, 2004). Pure isolated plant alkaloids and their synthetic derivatives are used as basic medicinal agents for analgesic antispasmodic and bactericidal effects (Stray, 1998). The high concentration of tannins detected in the older leaves of *A. nilotica* and *A. radiana* makes them of high demand in the world market. Tannins have been found to posses astringent properties, hasten the healing of wounds and inflamed mucous membranes (Morton, 2001; Kozioc and Marcia, 2004). The *Acacia* species studied were known to contain calcium, potassium, magnesium, nitrogen and sodium. These elements are very important in human nutrition. They are required for repair of worn out cells, strong bone and teeth, building of red blood cells, maintaining osmotic balance and for body mechanisms (WHO, 2008). The *Acacia* species, contained high quantity of nitrogen which is an essential constituent of protein.

In general, there are indications that *Acacia* leaves constitute rich sources of mineral elements. The low percentage of sodium in the samples might be an added nutritive advantage due to the direct relationship of sodium intake and hypertension in human (Dahl, 2006). This investigation has revealed that the *Acacia* species studied are of high medicinal value due to their phytochemical and mineral contents which can be utilized in the treatment of many diseases and also be exploited for use in pharmaceutical and cosmetic industries. As a result of their high phytochemical and mineral content, the *Acacia* species are potential sources of useful food and drugs. Studies are on going to ascertain the nutritive and vitamin potential of *Acacia* species in order to know comprehensively their food and medicinal values, so as to be fully exploited for enhancement of life of mankind.

REFERENCES

Moerman DE (2004). An Analysis of the food plants of native North. Am. J. Ethnopharmacol., 52: 1-22.

Okwu DE, Ekeke OM (2003). Phytochemical screening and mineral composition of chewing sticks in South Eastern Nigeria. Global J. Pure Appl. Sci., 9, 235-238.

Bartram T (1998). Encyclopedia of Herbal Medicinal. Robinson. London, pp. 26-30.

Burkill IH (2000). The useful plants of West Tropical Africa. Royal Bot. Gardens, 3: 522.

Gill LS (1999). Medicinal uses of trees and plants in Africa. University of Benin Press: Benin. Nigeria, p. 276.

Nnimh AC (1998). The encyclopedia of medicinal plants. Dorling Kindersley press: London, pp. 24-48.

Oliver B (2007). Constituents of the African Plants. Planta Med., 24: 13-34.

Urquiaga I (2000). Leighton, F., Plant polyphenol antioxidants and oxidative stress. Biol. Res., 33: 159-165.

Sofowora A (2005). Medicinal plants and traditional medicine in Africa. Spectrum Books Ltd: Ibadan. Nigeria, p. 289.

Hill AF (1998). Economic Botany. A text book of useful pants and plant products. Mccraw-Hill Book Company Inc: New York, 2nd edn., p. 248.

Edeoga HO, Eriata DO (2001). Alkaloid, tannin and saponins contents of some medicinal plants. J. Med. Aromat. Plant Sci., 23: 344-349.

Stevens AH, Hendriks JF, Malingre TM (1997). Alkaloids of some European and macaronesian sediodege and sempervivodeae (crassulaceae). Phytochemistry, 31: 3917-3924.

Burkill IH (2005). The useful plants of West Tropical Africa families A-D Royal Bot. Garden, 1: 691.

Bidwell RS (1999). Plant physiology. Macmillian: London, 2nd edn., pp. 225-227.

Zhang YE, Raun JL, Ding WP (1996). Studies on the antifertility constituent of *Marsdenia oreophila*. Acta Pham. Sin., 29, 281-184.

Kim SY, Ohandy MJ, Jung MY (2002). Antioxidant activities of selected oriental herb extracts. J. Am. Oil Chem. Soc., 71: 633-640.

Keay RJ, Onochie CA, Stanfield DP (1999). Nigerian trees, Nigerian National Press Ltd: Apapa Lagos, 11: 495.

FAO (2000). Woody species in compounds farms in South Eastern Nigeria and their functions, 94.

Harbone JB (1998). Introduction of ecological biochemistry. Academic press: London, 3rd edn., 42.

Harbone JB (2001). Phytochernical methods. Chapman and Hall Ltd: London, 111-113.

Boham AB, Kocipai AC (1997). Flavonoids and condensed tannins from leaves of Hawaiian *Vaccinium vaticulatuin* and *Vaccinium calyciniurn*. *Pacific Science* , 48, 458-463.

Ojuwale JA (1998). Experimental procedure of analysis in the laboratory. Oxford University Press: Ibadan, pp. 46-48.

Andrew OC (1999). Official method of analysis. Association of Official Analytical Chemists Washington D. C.: Washington, 13th edn., pp. 248-254.

Nivozamsky I, Reck VI, Houba VN, Van VW (2007). Novel digestion techniques for multiple element plant analysis. Common Soil Sci. Plant Anal., 14, 239-248.

Ebana RU, Essien AI, Ekpa OD (2009). Nutritional and potential medicinal value of the leaves of *Lasianthera africana* (Beauv). Glob. J. Pure Appl. Sci., 1: 1-7.

Salah W, Miller NJ, Pagauga G, Tijburg AP, Bolwel F, Evans C (2002). Polyphenolic flavones as scavenger of aqueous phase radicals as chain breaking oxidant. Res. J. Biochem., 2: 339-346.

Okeke CU, Elekwa IZ (2003). Phytochemical study of the extracts of *Gongronema latifoliuin* (Asclepiaclaceae). J. Health Visual Stud., 5: 47-55.

Okwu DE (2005). Phytochemical, vitamins and mineral contents of two Nigerian medical plants. J. Mol. Med. Adv. Sci., 1: 378-381.

Stray F (1998). The National guide to medicinal herbs and plants. Tiger Books International: London, pp. 12-16.

George AG (2002). Legal status and toxicity of saponins in food and cosmetics. Toxicology, 3, 85-91.

Sodipo OA, Akiniyi JA (2000). Studies on certain characteristics of extracts from bark of *Pansinystalia macruceras*. Glob. J. Pure Appl. Sci., 6, 83-87.

Morton J (2001). Purple mombin fruits of warm climates. Miami Publishers: New York, p. 245.

Kozioc MJ, Marcia MJ (2004). Chemical composition nutritional evaluation and economic prospects of *Spondias purpurea* (Anarcardiaceae). Econ. Bot., 52: 373-380.

WHO (2008). World Health Organization Technical Series. Trace elements in human nutrition and health. World Health Organization: Geneva, pp. 199-205.

Dahl LK (2006). Salt and Hypertension. Am. J. Clin. Nutr., 25: 231-238.

13

Moisture contents, mouldiness, insect infestation and acceptability of market samples of dried 'tatase' pepper and tomato in Kano, Nigeria

Oyebanji, A. O.*, Ibrahim M. H., Okanlawon S. O., Okunade S. O. and Awagu, E. F.

Nigerian Stored Products Research Institute, Onireke-Dugbe, P. M. B. 5044, Dugbe, Ibadan.

Triplicate samples of dried 'tatase' pepper and dried tomato in early and late storage periods were randomly obtained monthly from 2 major markets in Kano and analyzed for moisture contents, mould loads, insect infestation and consumer acceptability and relationships of parameters. Respective mean±SD values of moisture for new and old stocks; 11.05±1.04 and 9.93±2.85% and 9.69±0.98 and 9.74±1.31% were not significantly different at P = 0.05. Moisture contents varied widely, with low correlation coefficients, r = -0.47 and r = -0.14 for moisture contents/mould loads relationships in dried 'tatase' pepper and dried tomato respectively. However, correlation coefficients for log.mould load / acceptability by brightness of colour relationships, r = -0.71 and r = -0.8; and by discoloration, r = +0.71 and r = +0.81 were high and related to mouldiness. Mean±SD of mould loads in new stocks were 2.7±1.9 $\times$ 10^3 and 4.3 $\times$ 10^3 ± 4.6 $\times$ 10^2 cfu/g for pepper and tomato respectively, and showed increases to 3.0±2.3 $\times$ 10^4 and 3.4±2.6 $\times$ 10^4 cfu/g respectively in old stocks. Insect infestations were common on tomato but rare even on old 'tatase' pepper.

Key words: Dried 'tatase' pepper, dried tomato, new stock, old stock, moisture contents, mould loads, insect infestation, consumer acceptability, correlation coefficient.

INTRODUCTION

'Tatase' pepper (*Capsicum annum* L.) and tomato (*Lycopasicum esculentum* Mill) were some of the most abundantly produced crops in Northern Nigeria annually, especially in Kano. However, Akanbi et al. (2006) described fresh tomato fruits as highly perishable and were often lost to deterioration and wasted during the peak harvesting period. Oyebanji et al. (2008) reported that *Geotrichum candidum, Alternaria altanata, Aspergillus niger, Fusarium chlamydosporium, Mucor* and *Botrytis* species grew in succession on ripe tomato as initiated by damage to fruits within 2 to 9 days as influenced by different storage conditions of ambient ventilated shed, refrigerator and ambient moist jute covered cane box, with 70, 65 and 40% spoilage losses respectively in 20 days of storage. The high perishability of fresh 'tatase' pepper and tomato under ordinary

condition in a tropical developing country like Nigeria compelled drying of the crops to extend shelf life. Doymaz (2007) and Sobukola et al. (2007) had also reported that drying was the most common form of food preservation to extend shelf life under ambient storage in packs.

Similarly, Hell et al. (2009) reported that vegetables including okra, hot chilli pepper, tomato, melon seeds, onion and baobab leaves were dried to preserve them for lean periods and decrease their perishability. Sun drying of crops has been particularly relevant in Nigeria, especially in the drier northern states for conservation and reduction of post harvest losses. 'Tatase' pepper and tomato were dried annually in Northern Nigeria including Kano area spreading 'tatase' pepper and tomato on bare floor or ground, perhaps to meet the large scale or commercial drying in the abundance of harvest. Idah and Aderibigbe (2007) reported that such dried crops were exposed to contamination with dirt and risked wetting during rains or flood and needed proper packaging for

*Corresponding author. E-mail: femcovc@yahoo.com.

storage. Contaminations of crops with dirt during open air sun drying of tomato reported by Adesina et al. (2010) and were imagined in the drying of 'tatase' pepper and tomato in Kano. Meanwhile, no official monitoring of the dried products (for standard in terms dryness for packaging and storage or mould quality and regulation against risk of mycotoxin contamination) existed for 'unpackaged-unlabelled' products such as dried 'tatase' in Nigeria.

However, Atanda and Akano (1997) reported that dried 'tatase' pepper was widely acceptable to consumers, being affordable and available in the off season. The early period of storage of dried 'tatase' pepper and dried tomato was from February to April annually. This period was when the products were abundant in the markets in good visual quality. They were then bright coloured and seemingly wholesome. The late period of storage was from November to January each year and the products were often discoloured and unappealing, but buyers were then left with no choice in the market. Dried 'tatase' pepper and tomato were used in proportions as condiments in preparation of soup like the fresh forms by soaking in water to rehydrate. Preliminary visits to markets and stores in Kano showed varied visual qualities of 'tatase' pepper and tomato as bright colour or discoloration at different periods of a year. Buyers of dried 'tatase' pepper and dried tomato from markets in Kano were local consumers, traders and dealers from other parts of Nigeria, especially from the south.

Ordinarily, these consumers and traders demanded good quality dried products that were bright looking, not mouldy and not insect infested. Desire for quality in the dried pepper and tomato manifested in the buyers' inspection, selection and rejection tendencies for the discoloured. As well, traders sorted the products to improve the grading for acceptability according to Idah and Aderibigbe (2007). Kano remained a major commercial centre for dried crops including 'tatase' pepper and tomato in Nigeria, with the yearly abundance of production. The drying, packaging and storage had to be adequately done to preserve, conserve and grant consumers satisfaction for their expenses. Ogunkoya et al. (2011) had since reported that demand for high quality dried crops manifest more in international markets, due to existence of monitoring, appreciation of standards, crops' abundance and availability of choice in the market. Due reward for all (traders and consumers of dried 'tatase' pepper and tomatoes) was expected. Parameters of quality and preference of dried pepper and dried tomato were known to include brightness of colour, moisture content, mouldiness and insect infestation according to Idah and Aderibige (2007).

There was the need to document quality status, relevance and relationships of parameters of mould growth and customer appeal concerns of dried 'tatase' pepper and dried tomato in Kano. This study was therefore a market survey, of dried 'tatase' pepper and dried tomato during the early and late storage periods in Kano, which determined the moisture contents, mouldiness, insect infestation quality parameters and consumers' acceptability of samples as well as their relationships.

MATERIALS AND METHODS

Dried 'tatase' pepper and dried tomato samples from Kano markets

Market samples of dried 'tatase' pepper and dried tomato from the City and Yankaba markets in Kano, consisting of respective triplicate bowl measures of the products per month during early storage period of February to April for new stocks and during the late storage periods of November to January for old stocks, were obtained into sterile black polyethylene sampling bags. The market samples of dried 'tatase' pepper and dried tomato were respectively sub-sampled for analysis of moisture contents, mouldiness, insect infestation and consumers' acceptability.

Moisture content determination

Moisture content was determined by the oven drying method according to Atanda and Akano (1997) and AOAC (2000) for 10 g sub-sample of dried 'tatase' pepper or dried tomato, dried to a constant weight at 80°C in 4.5 h. Moisture content was expressed as percentage of moisture loss by weight during drying of a 10 g subsample.

Mouldiness

Mouldiness as mould load of bulked triplicate samples of dried 'tatase' pepper or tomato, each month from a market, was determined by serial dilution pour plate technique as according to Akani and Madumere (2008) for a 10 g sub-sample rinsed in 90 ml Tween 80 incorporated sterile water for 1 ml suspension dispensing in series into 9 ml sterile water in test tubes. Pour in sterile Petri-dishes consisted of 1 ml suspension with sterile cool molten Malt Extract Agar incorporated with 50 ppm of Tetracycline and Chloramphenicol. Plates were incubated at 28°C for 3 days and mould colonies were enumerated in plate with less than 100 colonies. Mould load was expressed as colony forming units per gram of sub-sample (cfu/g). The isolated moulds were identified by macroscopic examination of growth of pure cultures for colour, texture and surface appearance. Microscopic (Yashima Tokyo OS.K No. 800038) examination of mount of pure culture for sporangia, conidial heads, conidia and vegetative mycelium morphology was according to Samson and Reenen-Hoekstra (1988) and Harrigan and McCance (1990).

Percentage acceptability

Acceptability of market samples of dried 'tatase' pepper and dried tomato were determined by consumers' sorting of respective samples into Undiscoloured-Uninfected or Discoloured-Infected-Infested lots and percentages of acceptable component of samples computed by weight of sorted lots. These values were respectively matched with predetermined weight acceptability percentage groups of 0 to 10, 10 to 20, 20 to 30, 30 to 40, 40 to 50, 50 to 60, 60 to 70, 70 to 80, 80 to 100% that corresponds with hedonic scores on a 1 to 9 scale of 1= disliked extremely, 2 = disliked, 3 = disliked

Table 1. Moisture contents of new and old samples of dried 'tatase' pepper and dried tomato from markets in Kano.

Market sample		Moisture content (%)	
Stock	Period and market	Pepper	Tomato
New	February at Yankaba market	10.94	8.62
	February at City Market	11.34	8.71
	March at Yankaba Market	9.20	9.82
	March at City Market	12.10	9.40
	April at Yankaba Market	10.80	11.08
	April at City Market	11.90	10.50
	Mean ± SD	(11.05±1.0)	(9.60±0.98)
Old	November at Yankaba market	11.07	7.73
	November at City market	10.64	11.20
	December at Yankaba Market	8.41	11.11
	December at City Market	13.26	9.80
	January at Yankaba market	5.05	9.02
	January at City market	11.15	9.60
	Mean± SD	(9.93±2.85)	(9.74±1.31)

Values in parenthesis were grand mean ± SD for new and old stocks of dried pepper and dried tomato respectively.

moderately, 4 = disliked slightly, 5.= neither disliked nor liked, 6 = liked slightly, 7 = liked moderately. 8 = liked, 9 = liked extremely; according to Munoz and King (2007) and Olayemi et al. (2011).

Insect infestation

Presence of insect infestation in samples was determined by stereomicroscope (Reichert Austria Nr 259314) aided visual detection of adult, dead or life stages or their absence as in Williams et al. (2002).

Statistics

Mean values of moisture contents of triplicate samples of old and new stocks of dried 'tatase' pepper and tomato from City and Yankaba markets in Kano were respectively computed and grand means for new and old stocks were determined and compared for differences for dried 'tatase' pepper and tomato respectively at P=0.05. Also, mould loads of bulked monthly samples of dried 'tatase' pepper and tomatoes from markets were respectively computed for mean values for new and old stocks and analysed for difference by t-test for null hypothesis. Correlation coefficients with critical correlations of mean moisture contents and logarithms of mould loads; and of logarithms of mould loads and percent acceptability of market samples of dried 'tatase' pepper and tomato were computed respectively from scatter diagrams at P=0.05 and 10 degrees of freedom.

RESULTS AND DISCUSSION

Moisture content

Table 1 shows moisture contents of market samples of dried 'tatase' pepper and tomato in Kano. Moisture contents ranged from 9.2 to 12.10% (new), from 5.05 to

13.26% (old); and from 8.62 to 11.08 (new) and from 7.73 to 11.20% (old) for samples of dried 'tataste' pepper and dried tomato respectively. The widely varied moisture contents of market samples of new dried 'tataste' pepper or dried tomato, showed the varied extents of drying achieved for packaging and storage of the samples, usually in exposure and often in torn polyethylene lined sacks (hessian or jute), all of which determined storability against mouldiness. Safe moisture content of crop was put as the moisture content in equilibrium with relative humidity of 70% at 27°C. While the safe moisture content of dried 'tatase' pepper and that of dried tomato need to be established, Idah and Aderibigbe (2007) reported that 4.2% moisture content, was adequate drying of tomato for storage in polyethylene bags, or hermetic packaging which should not be prone to damage or exposure of dried crop.

Hence, market samples with higher levels of moisture therefore suggest risk of mouldiness and possibly mycotoxin contamination in storage. Findings in this study showed the varied adequacy of drying and storage of 'tatase' pepper and tomato in Kano. Even wider moisture content variation in the range of 24.80% in the wet season and 5.3% in the dry season in Ibadan, as reported by Atanda and Akano (1997) in open storage. Therefore, sensitization needs to continue for the awareness of the need for conscious monitoring during drying and determination of safe moisture content of 'tatase' pepper and tomato for specification of necessary dryness for storage. Mean ± SD of moisture contents of 11.05±1.0% for new and 9.93±2.85% for old; as well as 9.6±0.98% for new and 9.74±1.31% for old stock of dried

Table 2. Mould loads of new and old samples of dried 'tatase' pepper and dried tomato from markets in Kano.

Sample		Mould load (cfu/g)	
Stock	Period and market	Pepper	Tomato
New	February at Yankaba market	3.2×10^3	4.1×10^3
	February at City market	1.8×10^3	5.2×10^3
	March at Yankaba market	5.3×10^3	4.0×10^3
	March at City market	5.5×10^2	4.1×10^3
	April at Yankaba market	4.3×10^3	4.1×10^3
	April at City market	7.6×10^2	4.5×10^3
	Mean±SD	$(2.7 \times 10^3 \pm 1.9 \times 10^3)$	$(4.3 \times 10^3 \pm 4.6 \times 10^2)$
Old	November at Yankaba market	1.3×10^4	4.4×10^4
	November at City market	7.4×10^4	1.1×10^4
	December at Yankaba market	2.0×10^4	2.4×10^4
	December at City market	2.0×10^4	8.1×10^4
	January at Yankaba market	3.6×10^4	1.4×10^4
	January at City market	1.4×10^4	2.9×10^4
	Mean±SD	$(3.0 \times 10^4 \pm 2.3 \times 10^4)$	$(3.4 \times 10^4 \pm 2.6 \times 10^4)$

Open values were for 3 bulked samples from a market in a month. Values in parenthesis were means for new or old stocks of dried pepper or dried tomato respectively.

'tataste' pepper and dried tomato respectively were respectively not significantly different at P = 0.05 and these moisture contents were considered high.

High moisture content of samples called for caution against mouldiness. That is, care should be taken to ensure adequate drying for hermetic storage against ingress of moisture during rains and flood. The idea of packaging crops in high or low density polythene bags for the hermetic character assumes adequate drying of crop to safe moisture content and the maintenance of the integrity of packaging material was reported by Williams (1981). Whereas polythene lined jute sacks were in use for packaging dried 'tatase' pepper and tomato in Kano, many bags were found torn on market displays and in stores, risking ingress of moisture during rains or flooding. Hence, there was the need for care in the handling of packs against damage.

Mouldiness

Table 2 shows mould loads of new and old dried pepper (tatase) and dried tomato samples from Kano markets.

Mean ± SD mould loads of new dried 'tatase' pepper and tomato were 2.7±1.9 × 10^3 and 4.3 × 10^3 ± 4.6 × 10^2 cfu/g were lower than those of respective old stocks, 3.0±2.3 × 10^4 and 3.4±2.6 × 10^4 cfu/g, which were not significantly different but showed increases in mould loads or mould growth in the prolonged storage under the inadequate packaging and storage condition practiced in Kano. Low mould loads in the order of 10^2 cfu/g was reported by Atada and Akano (1997) on dried 'tatase' pepper samples from markets in Ibadan between November and February that mostly corresponded with

the time of old stocks in Kano have shown possibility of prolonged storage in good condition attributed to dry season in Ibadan.

The higher mould load in old samples in this study contrasted the report of decreased microbial load reported for adequately dried and packaged stored dried tomato by Idah and Aderibigbe (2007), as should be expected for well stored dried crops even under tropical condition. The lower mould loads of the new stock samples showed relatively of goodness in the microbial quality of dried 'tatase' pepper and tomato in the short duration of holding (1 to 3 months) when samples were drawn for this study. However, the goodness of drying needed to be confirmed by determination of the moisture content of sample for safe storage against mouldiness. The seeming retention of high moisture of old stocks in the long storage duration (8 to 12 months) before samples were drawn may explain the higher mould loads or mould growth that must have occurred on them. Therefore, the need for monitoring, regulation or penalty for carelessness during drying, packaging and storage cannot be overemphasized in Kano. While laboratory determination of moisture content may not be readily accessible to processors in Kano area, there should be conscious monitoring of sample during drying and storage at least by sensory approach of feeling products with hand to detect moisture, flexed feel when not dry enough or hard, light sometimes brittle feel of pepper or tomato if dryness was considerable.

The findings of high mould loads in old stocks of dried pepper and tomato confirms the earlier suggestion from the observed high moisture contents of samples which were not adequately dried and were maintained at the

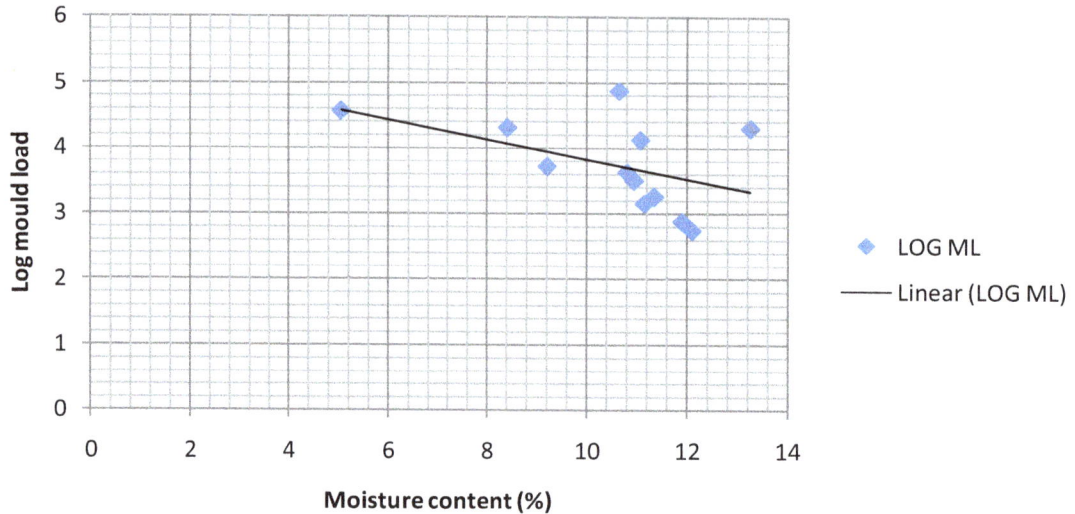

Figure 1. Moisture content / mould load relationship on dried 'tatase' pepper in Kano.

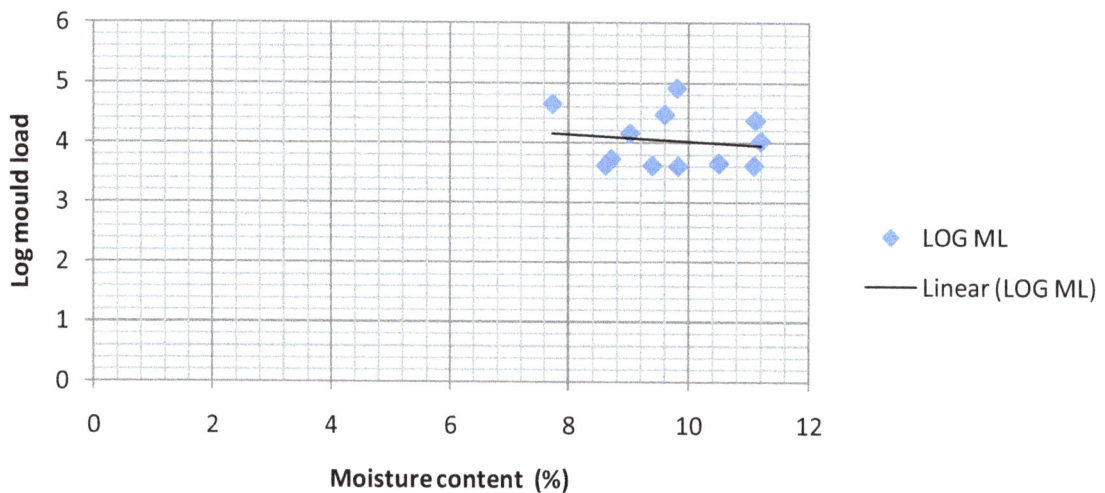

Figure 2. Moisture content / mould load relationship on dried tomato in Kano.

high level during storage under warm ambient temperature condition in Kano that permitted mouldiness. The non significant difference in mould loads of new and old stocks at P=0.05 may be due to post storage sifting of product where by much of the dirt were sorted off in the market prior to display and sampling for analyses. The moulds: *A. niger, A. falvus, Penicillium, Mucor hemalis* and *G. candidum* were associated with samples of dried 'tatase' pepper and dried tomato in this study. Of these, *A. niger* was the most commonly isolated while *A. flavus, Penicillium* and *Mucor* species were common on the dried pepper and dried tomato irrespective of whether the sample was new or old stock, but the yeast (*G. candidum*) was only occasionally isolated on new stocks, perhaps dying off during storage. The isolation of these

moulds supported the report of Atanda et al. (1990) that *A. niger, A. falvus* and *G. candidum* were commonly isolated from dried 'tatase' pepper. The mouldiness of old stocks may considerably explain the discoloration of the samples rather than much of light because of the opaque packaging practice in Kano markets, where polythene lined sack was in practice for storage in market stores, though often found torn.

Moisture content / mould load relationships on dried 'tatase' pepper and tomato

Figures 1 and 2 showed that moisture contents/mould loads relationships on dried tatase pepper and dried

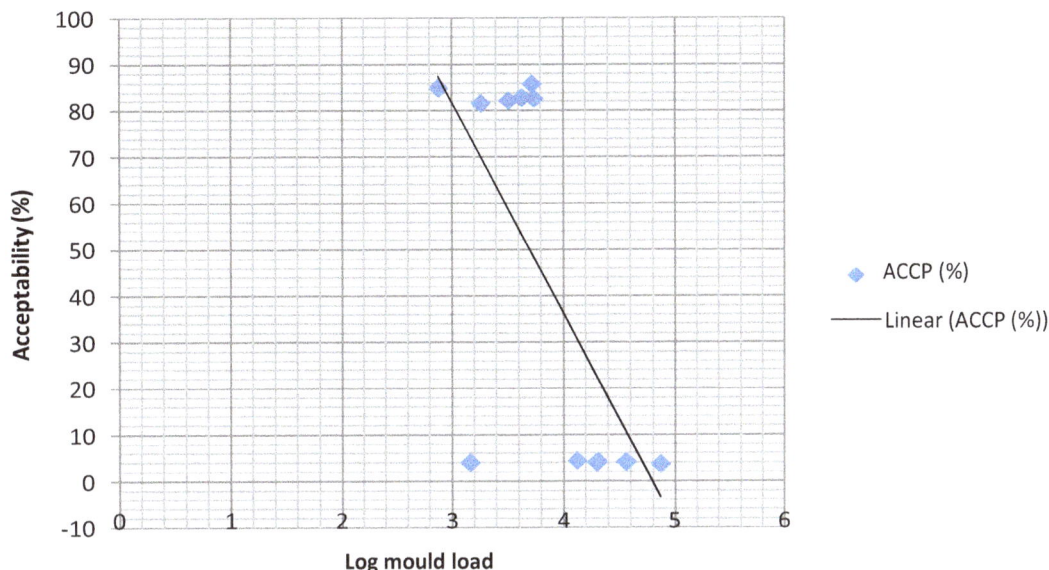

Figure 3. Log. mould loud / acceptability relationship of samples of dried pepper from markets in Kano.

tomato were of insignificant negative correlation coefficients of r = -0.47 and r = -0.14 respectively. These moderate and low negative correlation coefficients were not significant at P = 0.05, with the critical coefficient of correlation of 0.22 and 0.02 respectively, each at 10 degrees of freedom. Therefore, the mould quality of market samples of dried pepper or dried tomato may not be ascertained on the basis of the moisture content whether high or low. These findings were indications of the importance of information on storage time and moisture content in the interpretation of mouldiness of samples.

Insect infestation

Insect were rarely detected on dried 'tatase' pepper samples whether new or old. Only the cast of unknown insect was found in one sample from a market in December. On the other hand, insects including Tribolium, Ephestia, Ants and Corcyra species were found as living adults or dead in old stock of tomato. The non or occasional detection of insects in samples of dried 'tatase' pepper, suggested the unsuitability as food for insect pests of the stored products or was related to repulsion effect, perhaps due to some allergic or toxicity effect on insect pests that did not encourage infestation. As such, 'tatase' pepper may be considered for possible insecticidal uses, but dried tomato would not be so considered.

Mould load / acceptability relationship

Figures 3 and 4 showed significantly high negative

correlation coefficients of r = -0.71 and r = -0.85 between log of mould load and acceptability at P = 0.05 for critical correlations of 0.51 and 0.71 with 10 degrees of freedom for dried 'tatase' pepper and dried tomato samples from markets respectively. Thus, the higher the acceptability of a sample, the lower is the mould load and vice versa. On the other hand, unacceptability significantly correlated positively, r = +0.71 and r = +0.81 for dried 'tatase pepper and for dried tomato respectively, with critical correlations of 0.51 and 0.66 respectively at 10 degrees of freedom as shown in Figures 5 and 6. This meant that the more unacceptable a sample was the higher was the mould load and vice versa.

The acceptability sample was a good and easy indicator of low mouldiness associated with bright coloured samples while discoloured samples were high in mould load. High mouldy samples of dried 'tatase' pepper and dried tomato had the possibility of contamination with mycotoxins perhaps including Aflatoxins. These should ordinarily limit marketability of the product, especially in elite super markets and international markets where more gains might be expected for sales of the dried products. Similarly, Delcourt et al. (1994) reported the need for determination of Aflatoxin content in determination of quality of dried pepper.

Conclusion

Results of the analyses of dried 'tatase' pepper and dried tomato samples from City and Yankabar markets in Kano for moisture contents suggested that drying might have been inadequate for prolonged storage under the local circumstance of packaging and storage as mould load increases. Mould loads or mouldiness is related

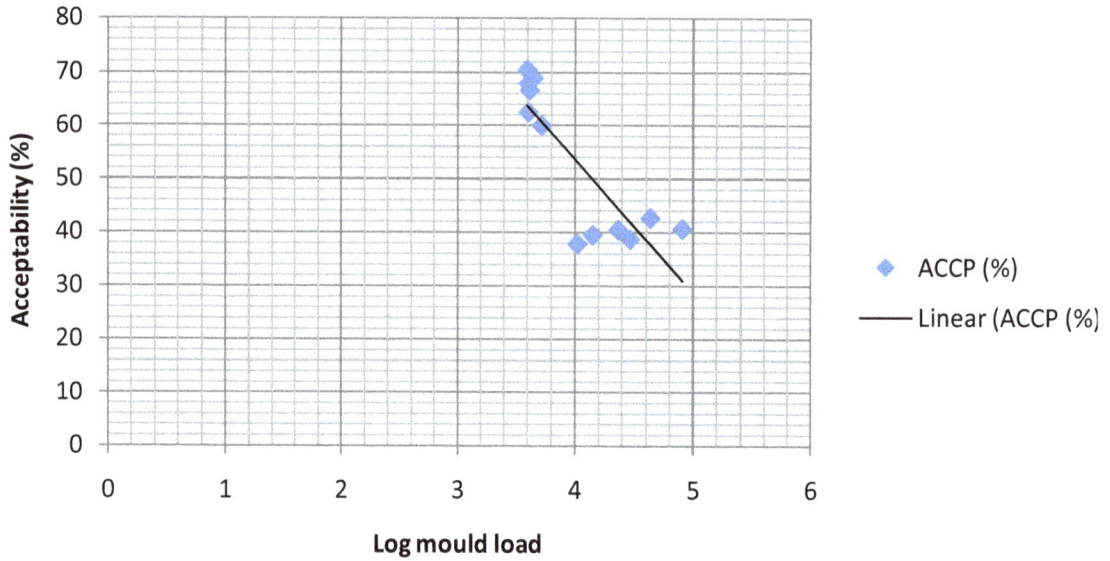

Figure 4. Log mould load/acceptability relationship of samples of dried tomato from markets in Kano.

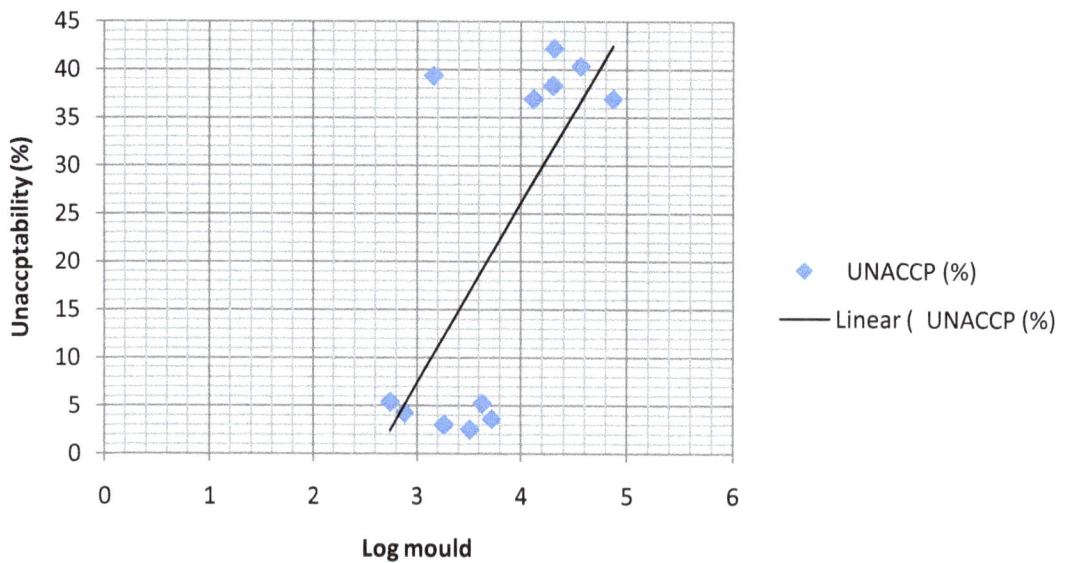

Figure 5. Log mould load/unacceptability relationship of samples of dried pepper from markets in Kano.

negatively to quantity and brightness of colour, and positively to the quantity of discolouration of dried 'tatase' pepper and dried tomato. However, moisture contents of samples which varied widely did not correlate significantly with mouldiness, which was observable in the market samples that were randomly drawn, and this indicated the importance of handling the stock history of state in interpreting results of moisture and mould quality analysis.

Further, discolouration which suggested mouldiness resulted in loss of appeal as indicated by the associated acceptability decrease, which meant risk of rejection and loss of sales ordinarily. Hence, the need for adequate

and safely dried, packaged and stored dried 'tatase' pepper and dried tomato was globally due to the fact that the acceptable protocol cannot be overemphasized. Also, while this study suggested a minor insect infestation problem for dried 'tatase' pepper, it did not for dried tomato, which must necessarily be handled against insect infestation.

ACKNOWLEDGEMENTS

We acknowledge the sponsorship of Nigerian Stored Products Research Institute for this work and thank

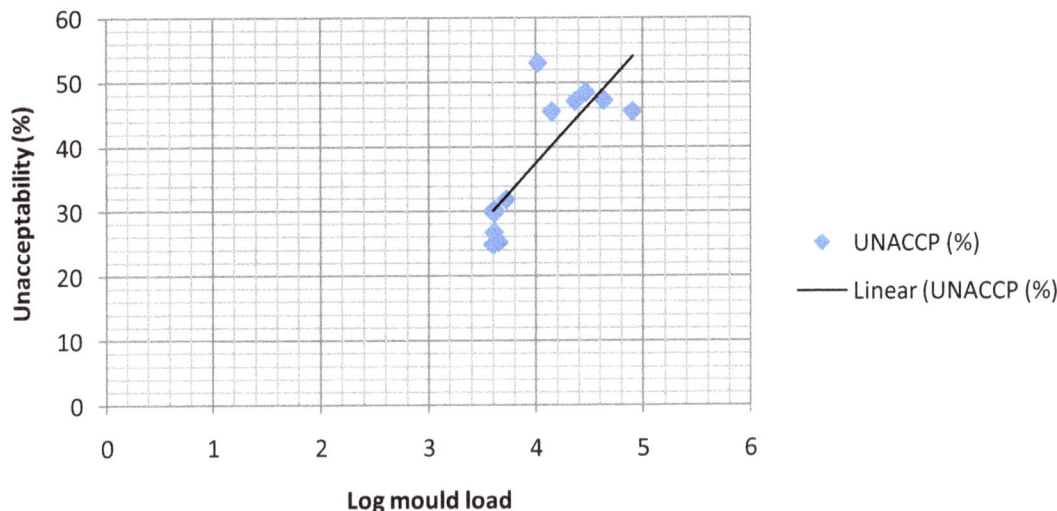

Figure 6. Log mould load / unacceptability relationship of samples of dried tomato from markets in Kano.

technicians in Kano station of the institute for assistance in sample collection and analysis.

REFERENCES

Adesina KA, Adade L, Ayodele RO (2010). Design, fabrication and performance evaluation of a parabolic concentrator dryer. J. Res. Nat. Dev., 8(1): 1-4.

Akanbi CT, Adeyemi RS, Ojo A (2006). Drying characteristics and sorption isotherm of tomato slices J. Food Engr., 73: 157-163.

Akani NP, Madumere B (2008). Mycoflora of partially processed Periwinkles (*Tympanotonus fascatus*) from local markets in Port Harcourt, Nigeria. Nig. J. Mycol., 1(1): 138-142.

Association of Official Analytical Chemists (2000). Official Methods of Analysis of the Association of Analytical Chemists, Arlington, 2000 17th Edition.

Atanda OO, Akano DA, Afolabi JF (1990). Mycoflora of dry 'tatase' pepper (*Capsicum annum* L.) stored for sale in Ibadan markets. Letter Appl. Micro, 10(1): 35-37.

Atanda OO, Akano DA (1997). The incidence of different types of Microorganisms and Aflatoxin content of Dried "Tatase pepper" (*Capsicum annuum* L.) stored for sale in the dry season. Nig. J. Microbio. 11: 36 -38.

Delcourt A, Rousset A, Lemaitre JP (1994) Microbial and mycotoxin contamination of peppers and food safety. Boll. Chim. Farm, 133(4): 235-238.

Doymaz I (2007) Air drying characteristics of tomato. J. Food Engr., 78: 1291-1297.

Harrigan WF, McCance ME (1990). Laboratory Methods in Food and Dairy Microbiology, 8th Edition. Williams and Wilkins W. Baltimore, USA.

Hell K, Gnonlonfin BGJ, Kodjogbe G, Lamboni Y, Abdourhamane IK (2009) Mycoflora and occurrence of aflatoxin in dried vegetables in Benin, Mali and Togo, West Africa. International J. Food Microbiol., 135(2): 99-104.

Ogunkoya AK, Ukoba KO, Olunlade A (2011). Development of low cost solar dryer. Pac. J. Sci. Technol.,12 (1):98 - 101.

Olayemi FF, Adedayo MR, Bamishaiye EI, Awagu EF (2011) Proximate composition of catfish (*Clarias gariepinus*) smoked in Nigerian Stored Products Research Institute (NSPRI): Developed kiln. Int. J. Fish. Aquacult., 3 (5):96-98.

Oyebanji AO, Robert SI, Pessu P (2008). Fungal growth, sequence and spoilage of ripe tomato fruits as influenced by different storage conditions. Nigerian J. Mycol., 1(1): 18-24.

Idah PA, Aderibigbe BA (2007) Quality changes in dried tomato stored in sealed polythene and open storage. Leonardo Electron. J. Pract. Technol., 10: 123 -136.

Munoz AM, King SC (2007) International consumer product testing across cultures and countries. ASTM International, MNL 55.

Samson RA, Reenen–Hoekstra ES (1988). Introduction to food Borne Fungi 3rd. ed. C.B.S., Netherland, 299 pp.

Sobukola OP, Dairo OU, Sanni LO, Odunewu AV, Fafiolu BO (2007) Thin layer drying process of some leafy vegetables under open sun. Food Sci. Tech. Int., 13(1): 34-35.

Williams AA (1981). Quality of stored and processed vegetables and fruits. Academic Press Inc. Ltd London, pp. 112-150.

Williams JO, Otitodun OG and Okunade SO (2002). Insect attack on Stored Dried Tomato. Trop. Sci., 42 (1):20-21.

14

Development of trifoliate yam: Cocoyam composite flour for *fufu* production

Ezeocha VC*, Omodamiro RM, Oti E and Chukwu GO

National Root Crops Research Institute, Umudike, Abia State, Nigeria.

Trifoliate yam (*Dioscorea dumetorum*) tubers and two varieties of cocoyam (Nxs001 and Nxs003) were processed separately into flour and mixed at the following proportions: 100% trifoliate yam flour; 100% cocoyam flour; 85:15, 75:25, 25:75, and 50:50 cocoyam- trifoliate yam flour. The functional and pasting properties of the composite flours were determined. The trifoliate yam-cocoyam flours were then reconstituted, made into *fufu* and subjected to sensory evaluation. The study showed that there were significant differences (p>0.05) in the functional and pasting properties of the composite flours. The ratings of the sensory panelists showed that NXS003-trifoliate yam composites were preferred for *fufu* production than the NXs001-trifoliate yam composites. 85:15, 75:25 and 50:50 NXs003: trifoliate yam flours were specifically preferred for *fufu* production.

Key words: Trifoliate yam, composite flour, *fufu*, sensory evaluation and functional properties.

INTRODUCTION

Yam is one of the staple foods in Nigeria and other tropical African countries. Yam is grown and cultivated for its energy-rich tuber. Only a few species of yams are cultivated as food crops. *Dioscorea dumetorum* has not been widely studied as other yam species, notwithstanding that it grows readily on various soils, the yield being 3 to 7 times that of other widely grown yam species (Treche and Guion, 1980). This is because in some landraces, tubers with bruises cannot be cooked to softness few days after harvest due to a severe hardening which develops after harvest (Sefa-dedeh and Afoakwa, 2001). Some works have been done on ways of minimizing the post-harvest problem associated with trifoliate yam but no solution has been suggested yet. Processing the yam tuber into a shelf –stable product offers an alternative to fresh storage.

The production of cocoyam otherwise called taro is low compared to other roots and tubers (Aderolu et al., 2009) but its superiority in terms of digestibility of starch (98.8%), the size of starch grain (1/10th of potato) and the sulphur amino acid, make it a better choice than other root crops in developing a composite *fufu* flour with

trifoliate yam (Ezedinma, 1987). Fufu is a thick paste usually made by boiling starchy root vegetables in water and pounding with a mortar and pestle until the desired consistency is reached. Cocoyam and *D. dumetorum* have been neglected in attempts to process roots and tubers into more durable forms. A greater part of these tubers are consumed fresh with oil, the rest if not boiled either harden or spoil. The objectives of this work are therefore to encourage the utilization of trifoliate yam and cocoyam and to determine which ratio of trifoliate yam and cocoyam that has the best rheological properties and is most acceptable.

MATERIALS AND METHODS

D. dumetorum setts and two varieties of cocoyam cormels (NXs 001 and NXs 003) were supplied by the yam and the cocoyam programmes respectively of the National Root Crops Research Institute, Umudike. The method of Martins et al. (1983) was used in the development of the *D. dumetorum* flour. The yams were peeled, cut into 0.3 to 0.4 cm slices, boiled for 45 min and spread thinly on perforated trays to dry. The dried slices were ground into flour. The cocoyam flours were developed with the method of Sanful and Darko (2010). The two varieties of cocoyam were washed, peeled, washed again, cut into 0.3 to 0.4 cm thick discs and then blanched at a temperature of 60°C. They were spread thinly on a tray and sun dried. The dry samples were then milled into flour with a double disc attrition milling machine.

*Corresponding author. E-mail: avezeocha@yahoo.com.

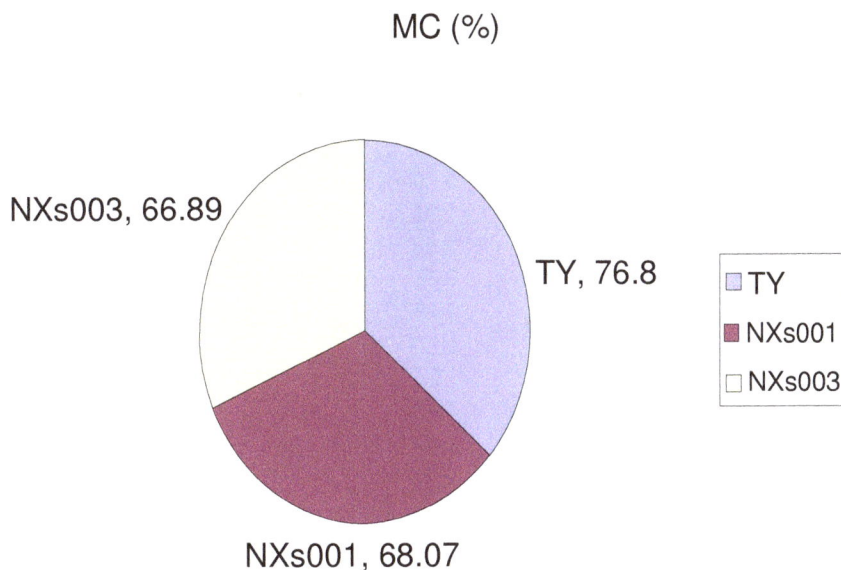

Figure 1. Moisture content of the fresh trifoliate yam and cocoyam (TY stands for Trifoliate yam).

Preparation of composite flours

The trifoliate yam flour and the cocoyam flours were combined at the following proportions: 85:15, 75:25, 25:75, 50:50 NXs001: Trifoliate yam flour and the same proportion for NXs003: Trifoliate yam flour. The moisture content was determined by the standard AOAC (1980) method.

Functional properties

Functional properties such as bulk density was determined using the method of Okezie and Bello (1998), water absorption capacity, swelling capacity and Gelatinization tem perature were evaluated using the method of Okaka et al (1997)

Pasting characteristics

The pasting characteristics of the composite *fufu* flours were determined using Brabender Amylograph (Brabender OHG Duisburg, kulturstrasse 51-55, type-800145010). Gelatinization temperature, peak viscosity, parameters analyzed were peak viscosity, set back viscosity, pasting time and pasting temperature.

Sensory evaluation

Fufu was prepared from the composite flour samples and the commercial yam flour (control). The colour, texture, mouldability and general acceptability of each *fufu* was evaluated by a 15 member panel using a 7-point hedonic scale, with 1 corresponding to dislike extremely and 7 corresponding to like extremely (Iwe, 2003).

Statistical analysis

Data obtained were subjected to statistical analysis of variance (ANOVA) using SAS (2009 version). Means were separated using Duncan's Multiple Range Test.

RESULTS AND DISCUSSION

The moisture content and percentage dry matter of the fresh trifoliate yam and cocoyam samples are shown in Figure 1.

Functional properties

The functional properties determine the use of food material for various food products. The results for functional properties of the composite flours are shown in Table 1. Bulk density of the flours ranged between 0.687 to 0.87g/cm^3. There were significant differences between the bulk density of 100% NXs001, 100% NXs003 and 100% trifoliate yam (p<0.05). 100% trifoliate yam flour had the highest bulk density which means that trifoliate yam flour was denser than the cocoyam flours. The bulk density is influenced by particle size and the density of the flour and is important in determining the packaging requirement and material handling (Karuna et al., 1996).

Water absorption capacity is the ability of flour particles to entrap large amounts of water, such that exudation is prevented (Chen and Lin, 2002). There were significant differences in water absorption capacity of the flours. 100% NXs003 had the highest value of 5.50 g/ml while 100% NXs001 had the lowest value of 2.83 g/ml hence the NXs003: trifoliate yam composite flours have higher water binding capacity than the NXs001: trifoliate yam composite flours. This means that the NXs003-trifoliate

Table 1. Functional properties of the composite trifoliate yam: cocoyam flour.

Samples (%)	Bulk density (g/cm^3)	Water absorption capacity (%)	Swelling capacity	Gelation temperature (°C)
100 trifoliate yam	0.86^{ab}	3.33^{cde}	1283.33^e	89.17^a
100 NXs001	0.76^{ef}	2.83^{ef}	6500.00^a	89.33^a
100 NXs003	0.69^h	5.50^a	806.67^f	75.17^e
NXs001:trifol(85:15)	0.77^d	2.50^f	233.33^j	76.17^d
NXs003:trifol (85:15)	0.72^f	3.83^c	683.30^{fg}	71.83^f
NXs001:trifol*(75:25)	0.77^{de}	2.83^{ef}	700.00^f	70.50^{gh}
NXs003:trifol (75:25)	0.71^g	4.83^b	1733.33^c	75.67^{de}
NXs001:trifol (50:50)	0.85^b	3.17^{de}	516.67^h	80.83^b
NXs003:trifol (50:50)	0.76^{ef}	3.67^{de}	316.67^i	71.00^g
NXs001:trifol (25:75)	0.87^a	2.83^{ef}	1533.33^d	70.17^h
NXs003:trifol (25:75)	0.83^c	3.17^{de}	2666.67^b	76.00^c
Control^	0.62^i	2.33^f	533.33^{gh}	61.67^i

Means with the same superscript in the same column are not significantly different (p>0.05). *trifol stands for trifoliate yam flour, # Ratio of cocoyam flour to yam flour is in parenthesis, ^ control is the commercial yam flour.

Table 2. Pasting properties of the composite trifoliate yam: cocoyam flours.

Sample	Gelation start (°C)	Peak gelation (°C)	Peak viscosity (Bu)	Set back (Bu)	Stability (Bu)	Pasting time (min)
100 trifoliate yam	62.47^f	65.37^j	2500.00^a	626.67^d	1873.33^a	2.90^i
100 NXs001	61.40^h	85.70^a	2033.33^b	620.00^d	1484.00^b	24.30^a
100 NXs003	69.87^d	74.73^f	1933.33^c	883.33^a	1050.00^g	4.87^h
NXs001:trifol*(85:15)	72.63^b	81.53^c	2000.00^b	786.67^b	1213.33^a	2.90^i
NXs003:trifol(85:15)	74.87^a	76.17^e	1483.33^f	730.00^c	753.33^h	1.30^j
NXs001:trifol(75:25)	63.33^f	82.87^b	1550.00^e	416.67^f	1133.33^f	19.53^b
NXs003:trifol(75:25)	52.37^k	66.13^i	1000.00^g	716.67^c	283.33^i	13.77^d
NXs001:trifol(50:50)	60.93^i	67.53^h	2000.00^b	516.67^e	1483.33^b	6.60^g
NXs003:trifol(50:50)	70.37^c	78.03^d	1930.00^c	626.67^d	1303.33^d	7.67^f
NXs001:trifol(25:75)	64.43^e	67.37^h	2016.67^b	516.67^e	1500.00^b	2.93^i
NXs003:trifol(25:75)	62.03^g	67.50^h	1826.67^d	760.00^{bc}	1066.67^{fg}	5.47^h
Control^	56.17^j	72.73^g	2000.00^b	616.67^d	1383.33^c	16.57^c

Means with the same superscript in the same column are not significantly different (p>0.05). *trifol stands for trifoliate yam flour. ^ control is the commercial yam flour.

yam composite flours have better reconstitution ability (Adebowale et al., 2008) than the NXs001: trifoliate yam composite flours. Water absorption capacity varies with size, shape, presence of proteins, carbohydrates and lipids, pH and salts. Previous processing, such as heating, alkali processing, disulfide linking, etc may also influence it (Iwe, 2003).

Pasting properties

The pasting properties of the composite trifoliate yam: cocoyam flour is shown in Table 2. Visser and Thomas (1987) reported that heat causes thickening and then gelation in concentrations above 7% by weight, with a temperature threshold of 65°C. Rate of gelling and gel

firmness are reported to depend on temperature, time of heating and protein concentration. Gelatinization and pasting affect the quality and aesthetic considerations in the food industry, since they affect texture and digestibility, as well as the end use of starchy foods (Adebowale et al., 2005). Viscosity is an important functional property of foods that affects mouth feel, the textural quality of foods and the design of processing lines. Peak viscosity is the ability of starch to swell freely before their physical breakdown (Sanni et al., 2004). 100% trifoliate yam had the highest peak viscosity of 2500 Bu which is even higher than that of the commercial yam flour (control), 2000 Bu. There were significant differences in the peak viscosity of the composite flours with the peak viscosity increasing with increased proportion of trifoliate yam flour. Stability is related to

Table 3. Sensory evaluation of the composite trifoliate yam: cocoyam *fufu*.

Samples	Texture	Colour	Mouldability	General acceptance
100% trifoliate yam	4.20[cd]	4.73[bc]	4.20[abcd]	4.60[abc]
100% NXs001	4.27[bcd]	4.73[bc]	4.20[abcd]	4.27[abcde]
100% NXs003	5.53[a]	4.33[cde]	5.40[a]	5.07[ab]
Nxs001:trifoliate yam (85:15)	4.07[cd]	6.20[a]	3.53[cd]	3.73[cde]
NXs003:trifoliate yam (85:15)	5.40[abc]	4.73[bc]	5.40[a]	5.40[a]
NXs001:trifoliate yam (75:25)	3.33[d]	4.60[bcd]	3.53[cd]	3.33[e]
NXs003:trifoliate yam (75:25)	4.87[abc]	5.67[ab]	4.87[ab]	4.73[abc]
NXs001:trifoliate yam (50:50)	4.80[bcd]	3.33[e]	4.60[abc]	3.87[cde]
NXs003:trifoliate yam (50:50)	5.00[abc]	4.53[bcd]	5.07[ab]	5.20[ab]
NXs001:trifoliate yam (25:75)	4.47[abcd]	4.73[bc]	3.87[bcd]	4.20[bcde]
NXs003:trifoliate yam (25:75)	3.93[cd]	3.53[de]	3.27[d]	3.40[de]
control	4.6[abc]	3.80[cde]	4.40[abcd]	4.53[abcd]

Means with the same superscript in the same column are not significantly different (p>0.05).

setback which is the rate at which the gel formed loses its water (retrogradation). The higher the setback value, the lower the retrogradation during cooling. 100% NXs003 had the highest setback of 883.33 Bu while 75:25 NXs001: trifoliate yam flour had the lowest setback value of 416.67 Bu. Stability is highest in trifoliate yam flour and lowest in NXs003: trifoliate yam flour (75:25).

SENSORY EVALUATION

Table 3 shows the mean scores of the sensory evaluation of the composite flours. There were significant differences on the level of preference of the different samples. The texture of 100% NXs003 was more preferred than the other samples while the texture of NXs001: trifoliate yam *fufu* (75:25) was least preferred. Colour is a quality attribute that plays an important role in food acceptability. NXs001: trifoliate yam *fufu* (85:15) had the highest mean score for colour, 6.20 while the colour of NXs001: trifoliate yam *fufu* (50:50) was least preferred. 100% NXs003 and NXs003: trifoliate yam *fufu* (85:15) had the highest score for mouldability (5.40) with score even higher than that of the control. NXs003: trifoliate yam *fufu* (85:15) was the most generally accepted of the composite flour samples, followed by NXs003: trifoliate yam flour (50:50) while NXs001: trifoliate yam flour (75:25) had the least general acceptability.

Conclusion

A study of the results obtained from the functional, pasting and sensory evaluation show that Nxs003 forms a better composite flour with trifoliate yam for *fufu* than Nxs001. Sensory evaluation result showed that there was no significant difference in the texture, mouldability and general acceptance of the NXs003: trifoliate yam

composite flours and the control (commercial yam flour). 85:15, 75:25 and 50:50 NXs003: trifoliate yam flour are highly recommended for *fufu* production.

REFERENCES

Adebowale AA, Sanni LO, Awonorin SO (2005). Effect of texture modifiers on the physicochemical and sensory properties of dried fufu. Food Sci. Technol. Intern., 11(5): 373-382.

Adebowale AA, Sanni SA, Oladapo FO (2008). Chemical, Functional and Sensory Properties of Instant Yam-Breadfruit Flour. Nig. Food J., 26(1): 2-12.

Aderolu AZ, Lawal MO, Oladipupo MO (2009). Processed cocoyam tuber as carbohydrate source in the diet of Juvenile African Catfish (*Clarias gariepinus*). Euro. J. Sci. Res., ISSN 1450-216X 35(3): 453-460.

AOAC (1980). Association of Official Analytical Chemists. Official Methods of Analysis. Washington D.C.

Chen MJ, Lin CW (2002). Factors affecting the water holding capacity of fibrinogen/ plasma protein gels optimized by response surface methodology. J. Food Sci., 67(7): 2579-2582.

Ezedinma FOC (1987). Response of Taro (Colocasia esculenta) to water management, plot preparation and population. 3rd Intl. Symp. Trop. Root Crops. Ibadan, Nigeria.

Iwe MO (2003). The science and technology of Soyabean. Rojoint communication services LTD, Nigeria, pp. 123-128.

Karuna D, Noel G, Dilip K (1996). Food and Nutrition Bulletin, 17: 2, United Nation University.

Martin G, Treche S, Noubi L, Agbor ET, Gwangwa S (1983). Introduction of Flour from Dioscorea dumetorum in a Rural Area. Proceedings of the Second Triennial Symposium of the International Society for Tropical Root Crops-African Branch held in Douala Cameroon.

Okaka JC, Anosike GN, Okaka ANC (11997). Effect of particle size profile of sun-dried and oven dried cowpea flours on their physical and functional characteristics in model system. J. Food Sci. Technol.

Okezie BO, Bello AB (1988). Physicochemical and functional properties of winged Bean flour and isolate compared with soy isolate. J. Food Sci., 53(2): 450-454.

Sefa-Dedeh S, Afoakwa EO (2001). Changes in cell wall constituents and mechanical properties during post-harvesthardening of trifoliate yam Dioscorea dumetorum (kunth) pax tubers. Food Res. Int., 35(2002): 429-434.

Sanful ER, Darko S (2010). Production of Cocoyam, Cassava and Wheat flour composite rock cake. Pakistan J. Nutr., 9(8): 810-814.

Sanni LO, Kosoko SB, Adebowale AA, Adeoye RJ (2004). The influence of Palm oil and Chemical Modification on the Pasting and Sensory Properties of Fufu flour. Intern. J. Food Properties, 7(2):229237.

Treche S, Guion P (1979). Etude des potentialities nutritionnelles de quell gues tubercules tropicaux au Cameroun L'agronomie Tropicale 84(2): 127.

Visser A, Thomas A (1987) Review: Soya protein products-their processing, functionality, and application aspects. Food Rev. Int., 31(1&2): 1-32.

15

The contribution of the genetic factors of potato lines to the northern area of Western Romania

Ioan GONTARIU[1]*, Ioan-Catalin ENEA[2] and Danela MURARIU[3]

[1]Faculty of Food Engineering, Stefan cel Mare University, Street. Universitatii no. 13, 720229, Suceava, Romania.
[2]Agricultural Research and Development Station of Suceava, B-dul 1 Decembrie 1918, no. 15, 720246, Romania.
[3]Suceava Genebank, B-dul 1 Mai no. 17, 720246, Suceava , Romania.

The creation of some potato genotypes with a high genetic homeostasis could confer a greater ecological plasticity with an increased resistance to the attack of the pathogen agents, such that notable performances concerning the main quality features represent the objectives reached in the potato breeding laboratory from ARDS of Suceava. Taking into consideration the area offered by the institution, the climatic conditions are favorable to many diseases, knowing the contribution of the genotype to the attenuation of its limitative effects represented by a great preoccupation. Besides the mentioned aspects, some quality features have been taken into consideration, such as: the form of the tubers, the depth of the eyes, the skin suberization level, the color of the pulp which confers a commercial quality and of course the economic increase of the production. This work is part of a more vast thematic research, initiated with the purpose of creating new potato advanced cultivars through an increased production capacity. With good resistance to the biotic and abiotic stress factors, advanced cultivars with superior culinary qualities are suitable for industrialization. The quantity and quality features depend to a great extent on the genetic income of the partners, in such a way that these characters have been transmitted in pedigree and the recombination specific capacity.

Key words: Advanced cultivars, potato lines, vegetation period.

INTRODUCTION

Potato has probably more related wild species than any other crops, since the genus Solanum comprises around 2000 species (Hawkes, 1990). Potato breeders usualy use wild *Solanum* species as sources of disease resistance genes (Hijmans and Spooner, 2001). The cultivated potato in Europe, having its origin in the South and Central America, cannot have the possibility to improve the genetic material with the other species of *Solanum* genus, and as such, it is evaluated as having low number of genotypes (Hawkes, 1990). The variety is a main resource for increasing the yield without supplementary cost and energy (Bodea, 1994). Improvement of the cultivar sets with this new genotype represents one of the most efficient ways for increasing the yield productivity, quality and stability, since it less undergoes stress as a result of the less favorable biotic factors (Grădinaru et al., 1986).

Backcrosses with varieties which have different levels of resistance to late blight and valuable cultural characters, such as shape, size and tuber quality, were made to improve the cultural characters of the plants (Blundy et al., 1991). The correlation between the resistant genotype frequency in each hybrid generation and the late blight resistance of the parents used was taken into consideration (Ceapoiu and Floare, 1983). Revealing of the genetic variability of local populations of *Pytophthora infestans* is a crucial step for an efficient potato with late blight control. However, at the moment, it is difficult to perform an accurate identification of intraspecific variation of *P. infestans* by morphological feature (Tooly et al., 1997).

MATERIALS AND METHODS

The biologic material consisted of nine potato lines created at the Agricultural Research and Development Station of Suceava through an intraspecific hybridization which have proved to be valorous with regards to the resistance to diseases (viruses and late blight of

---

*Corresponding author. E-mail: ioang@fia.usv.ro.

Table 1. The main traits of the tubers of the analysed potato lines.

Line	Main traits of the tubers
Sv 01-884-4	Light yellow flesh, round shape, shallow eyes, thin skin and large number of tubers from nest
Sv 01-884-8	Yellow flesh, round shape, shallow eyes and thin skin. The size of the tubers is medium
Sv 01-884-2	Light yellow flesh, oval shape, shallow eyes and thin skin
Sv 00-847-5	Light yellow flesh, round-oval shape, rough skin, shallow eyes and big yields.
Sv 00-847-22	Yellow flesh, round-oval shape and resistant to common scab
Sv 00-847-25	Yellow flesh, round-oval shape, shallow eyes and big yields.
Sv 99-789-11	Red flesh, round-oval shape, rough skin, quasi profound eyes, large number of tubers from nest
Sv 99-789-12	Yellow flesh, round-oval shape and quasi profound eyes. In the humid conditions, the tubers crack due to high accumulation rate.
Sv 99-789-10	Yellow flesh, round shape and quasi profound eyes. The size of the tubers is medium

Table 2. The vegetation period length.

Line	The sprouting time	The vegetation period length (days)
Sv 01-884-4	2.06	94
Sv 01-884-8	1.06	87
Sv 01-884-2	5.06	92
Sv 00-847-5	1.06	91
Sv 00-847-22	30.05	87
Sv 00-847-25	4.06	86
Sv 99-789-11	4.06	91
Sv 99-789-12	1.06	95
Sv 99-789-10	4.06	88
Astral	3.06	89

potato) and with a greater production capacity. The main traits of the potato lines are presented in Table 1. The experiment has been located in blocks with randomized structure plots, in three replications, and the plot surface was 16.8 m². The tubers have been planted manually at a distance of 30 cm on a row and 70 cm between rows, assuring a density of 47.6 thousands of plants on a hectare.

In the nurturing stage of the experiment, it should be noted that annual weeds were destroyed by chemicals using Dancor 1.2 l/ha. Other treatments have been carried out for the late blight of potato infection attenuation using Antracol 2 kg/ha, Bravo 2 kg/ha, Ranman 200 g/ha as contact fungicides and two treatments with systemic fungicides, using Secure 2 kg/ha and Tatoo C 21/ha. In order to combat the carrying aphides of viruses and the larva of the Colorado potato beetle, treatments were carrried out with the following insecticides: Actara (0.06 kg/ha) and Calypso (0.09 kg/ha). However, fertilization was carried out during the spring with 200 kg ammonium nitrate /ha (16:16:16) and 500 kg/ha, reaching 150 kg of Nitrogen, 80 kg of P_2O_5 and 80 kg of K_2O/ha. The soil type for the experiment is leached chernozem. Nevertheless, the interpretation of the results was done through the variance analyses (Ceapoiu, 1968). Concerning the thermo, the following aspects were observed:

1. The monthly averages have been close to the multiannual averages only in the months of April and August. In July, a supplementary caloric contribution was registered due to the temperatures from the third decade when its average was 22.4°C.

As a consequence, the monthly average was 1.1°C bigger than the multiannual value;
2. In comparison with the multiannual average, the coldest month was June (0.6°C), and this was due to the coldness of the air from the first decade, of which the average had been only 14.0°C.

RESULTS

Concerning the influence of the meteorological conditions on the main levels of the potato vegetation, some extracts were made. The big wet soil in the month of May and its heating in the third decade of May (17.7°C) have favored the rapid increase of the tillers, making the flowering of the plants faster in the first days of May (17.7°C). Thus, it favored quick breeding of the tillers, thereby making the breeding of the plants faster in the first days of June. As a consequence, the breeding of plants takes place after 26 to 32 days from the plantation. From the tested lines, the earliest have been noticed for lines 00-847-22 (3005), 01-884-8, 00-847-5 and 99-789-12 (01.06) and the latest for lines 00-847-25, 99-789-11, 99-789-10 and 01-884-2 (04-05.06) (Table 2).

The notes done during vegetation, concerning the frequency of the infested bushes with easy viruses of the

Table 3. The observations concerning the resistance to late blight of potato and viroses.

Lines	The bushes frequency infected with viroses (total %)	The foliar tolerance at the late blight of potato attack (notes)[*]	
		18.07	4.08
Sv 01-884-4	6.0	8	8
Sv 01-884-8	12.9	7	7
Sv 01-884-2	2.5	6	6
Sv 00-847-5	5.6	8	8
Sv 00-847-22	4.2	7	7
Sv 00-847-25	0.5	9	9
Sv 99-789-11	27.3	8	8
Sv 99-789-12	8.0	9	8
Sv 99-789-10	3.2	8	7
Astral	13.4	9	9

[*] 1- the very small tolerance; 9- very tolerant.

mosaic type and the hard viruses (leaf rolling and streak mosaic), permit the outlining of the fact that the majority of the lines own a real resistance to the virotic degeneration. Among these, only line 99-789-11 would represent (after the data subscribed in Table 3) a notable sensitivity. Regarding the total production, beside the 01-884-4 line that has been attached with the maximum production, there are also lines 01-884-8 and 00-847-25 in the next levels that represent a great interest (Table 4).

In order to reach the synthetic valuation (Table 5) of the lines that are closely similar to the "agronomic index" used for maize, they have been given marks from one to five as follows:

1. Very unsatisfactory,
2. Unsatisfactory,
3. Average,
4. Good,
5. Very good.

DISCUSSION

The significant heating that began in the second decade of June and the pluviometrical regime that was favorable to plants but not interrupted in June have favored the acceleration of the developing rhythm of the plants. In this way, the partial recuperation of the negative influence was generated by the lasting plantation. When this was compared with that of June, the meteorological conditions from July and August have been very unfavorable for the reserve substance accumulation.

Concerning the vegetation period of the lines, it is observed that these lines frame themselves in the semi earlier genotypes group, and are similar to one of the Astral cultivar of 89 days. Among these lines, lines 00-847-25, 01-884-8 and 99-789-10 have been outlined through a shorter period (86 to 88 days) of the vegetation

period in 7 to 8 days when compared with lines 01-8844 and 99-789-12 through its vegetation period of 94 to 95 days (Table 2). Taking into consideration the meteorological conditions of plants, it was observed that these conditions were more favorable for the presence of the pathogen pressure which has drastically increased in no time, and were also favorable for the attenuation of the late blight of potato attack against genotypes for the modification of the tolerance grade of the leaves in comparison with the attack produced by the fungus *P. infestans*.

Marking the forms with very high tolerance as 9 and those with a very low tolerance as 1, it can be observed that line 00-847-25 was presented in 2009 as the line that has the biggest tolerance conditions of the leaves for the asexual form of the late blight potato followed by line 00-789-11. However, the lowest tolerance has been presented by line 01-884-2 (Table 3).

Concerning the total production of tubers shown in Table 4, some observations were made:

1. In comparison with the decreasing average (241.1 t/ha), it was reported that only line 01-884-4 has been attached to the total production of 28 t/ha, and since line 01-884-8 registered an increase of 2.1 t/ha, it deserves to be retained at least a year for edification;
2. If the comparison is reached out by evaluating it with the less productive line (99-789-10), lines 01-884-4, 01-884-8 and 00-847-25 can be outlined through significant increases.

The four criteria (among which two are identical in this case) used in evaluating the decreases observed in the lines can contribute to the increase of the selection pressure. Consequently, through this process, the amelioration activity can be increased randomly. At the same time, they can contribute to the diminution of the elimination risk of a very valorous potential material.

Table 4. The tubers yield.

No.	Lines	The total yield (t/ha)	The comparative differences — With average	The comparative differences — With standard	$\frac{X \cdot 100}{X_{max}}$	$\frac{X \cdot 100}{min}$
1.	Sv 01-884-4	28.0	3.9^x	st	100 (st)	138^{xx}
2.	Sv 01-884-8	26.2	2.1	-1.8	93	129^x
3.	Sv 01-884-2	24.6	0.5	-3.4^o	88°	121
4.	Sv 00-847-5	20.7	-3.4^o	-7.3^{oo}	74^{oo}	102
5.	Sv 00-847-22	23.2	-0.9	-4.8^o	83°	114
6.	Sv 00-847-25	25.5	1.4	-2.5	91	126^x
7.	Sv 99-789-11	24.0	-0.1	-4.0^o	86°	118
8.	Sv 99-789-12	24.9	0.8	-3.1	89	122
9.	Sv 99-789-10	20.3	-3.8^o	-7.7^{oo}	72^{oo}	100 (st)
10.	Average	24.1				
	DI-5%		3.3		20%	27%
	DI-1%		5.5		28%	38%
	DI-0.5%		10.3		38%	50%
	Sv 01-884-4	14.2	1.5	mt	90	143
	Sv 01-884-8	15.7	3,0^x	1.5	100 mt	158^x
	Sv 01-884-2	14.4	1.7	0.2	92	145
	Sv 00-847-5	9.9	-2.8^o	-4.3^o	63°	100 mt
	Sv 00-847-22	12.5	-0.2	-1.7	80	126
	Sv 00-847-25	12.7	-	-1.5	81	128
	Sv 99-789-11	14.0	1.3	-0.2	89	141
	Sv 99-789-12	10.8	-1.9	-3.4^o	69°	109
	Sv 99-789-10	10.7	-2.0	-3.5^o	68°	108
.	Average	12.7				
	DI-5%		2.2		30%	48%
	DI-1%		4.8		42%	67%
	DI-0.5%		6.7		59%	92%

Table 5. The syntetic valuation of the descendants.

No.	Lines	The yield — Total	The yield — Commercial	Tolerance — To the late blight of potato	Tolerance — To viroses
1	Sv 01-884-4	5	4	4	4
2	Sv 01-884-8	4	5	4	1
3	Sv 01-884-2	3	4	3	5
4	Sv 00-847-5	1	1	4	4
5	Sv 00-847-22	3	3	4	5
6	Sv 00-847-25	4	3	5	5
7	Sv 99-789-11	3	4	4	1
8	Sv 99-789-12	3	2	4	4
9	Sv 99-789-10	1	2	4	5

Concerning the commercial production dimension (for alimentary use), a remark was made, in the first place, on line 01-884-8 that has been attached with the biggest production of big tubers (over 100 g). If it is taken into consideration that although line 01-884-8 had been over-reached with 1.8 t for the total production of line 01-884-4, whose commercial production had been overreached with 1.5 t/ha, it can be considered that line 01-884-8 can produce big tubers and can outline itself in other vegetation conditions.

Conclusions

The obtained results in year 2009 showed that line 01-884-4 had a very good random activity of tubers and a good tolerance for late blight of potato and viroses. Line 01-884-8 represents some notable qualities and can represent an interest for the areal with a virotic infestation that reduced drastically. However, line 00-847-25 can be considered "valorous" for any other uses if the random activity of big tubers is improved through other cultural methods.

REFERENCES

Blundy KS, Blundy MAC, Carter D, Wilson F, Park WD, Burrel MM (1991). The expression of class I patatin gene fusions in transgenic potato varies with both gene and cultivar. Plant Mol. Biol., 16: 153-160.

Bodea D (1994). Aspects of behavior in culture of potato varieties and lines in the environmental conditions in Suceava. Scientific papers, vol. 37 - Agronomic University, Iasi.

Ceapoiu N (1968). Statistical methods in agricultural and biological experiments. Agro House, Bucharest.

Ceapoiu N, Negulescu Floare (1983). Genetics and breeding plants resistance to diseases, Editura Academiei, Bucharest.

Grădinaru N, Macsim S, Siniavschi I (1986). Reaction of potato varieties stationed in Suceava County. Agronomic Research in Moldova, vol. 2.

Hawkes JG (1990). The potato. Evolution, biodiversity and genetic resources. London, Belhaven press, pp. 259-269.

Hijmans RJ, Spooner DM (2001). Geografic distribution of wild potato species. Am. J. Bot., 88: 2101-2112.

Tooly PW, Bunyard BA, Carras MM, Hatziloukas E (1997). Development of PCR Primers from Internal Transcribed Spacer Region 2 for Detection of Pytophthora species Infecting Potatoes. Appl. Environ. Microbiol., 63: 1467-1475.

16

Tissue engineered meat- Future meat

Z. F. Bhat[1]* and Hina Bhat[2]

[1]Division of Livestock Products Technology, Faculty of Veterinary Sciences and Animal Husbandry, Sher-e-Kashmir University of Agricultural Sciences and Technology of Jammu, R. S. Pura, India – 181 102.
[2]Department of Biotechnology, University of Kashmir, Hazratbal, Srinagar, Jammu and Kashmir, India-190006.

Current meat production methods have many health, environmental and other problems associated with them like high risk of infectious animal diseases, nutrition-related diseases, resource use and environmental pollution through green house gas emissions, decrease in the fresh water supply, erosion and subsequent habitat and biodiversity loss (Asner et al., 2004; Savadogo et al., 2007) besides the use of farm animals and non-sustainable meat supply. A new approach to produce meat and thereby reducing these risks is probably feasible with existing tissue engineering techniques and has been proposed as a humane, safe and environmentally beneficial alternative to slaughtered animal flesh. The growing demand for meat and the shrinking resources available to produce it by current methods also demand a new sustainable production system. *In vitro* meat production system ensures sustainable production of a new chemically safe and disease free meat besides reducing the animal suffering significantly. This review discusses the requirements that need to be met to increase the feasibility of *in vitro* meat production, which includes finding an appropriate stem cell source, their growth inside a bioreactor and providing essential cues for proliferation and differentiation.

Key words: *In vitro* meat, tissue engineering, future meat substitute.

INTRODUCTION

The present meat production methods have some serious consequences associated with them and the consumers are showing serious concern over them since last few years. Currently, 70% of all agricultural land, corresponding to 30% of the total global surface, is being used for livestock production; with 33% of arable land being used for growing livestock feed crops and 26% being used for grazing (FAO, 2006; Steinfeld et al., 2006). World meat production at present is contributing between 15 and 24% of total current greenhouse gas emissions, which is more than the total emission of the transportation sector, a great proportion of this percentage is due to deforestation to create grazing land (FAO, 2006, Steinfeld et al., 2006). Food production directly or indirectly involves about 70% of the fresh water

use and 20% of the energy consumption of mankind, of which a considerable proportion is used for the production of meat. The water use for livestock and accompanying feed crop production also has a dramatic effect on the environment such as a decrease in the fresh water supply, erosion and subsequent habitat and biodiversity loss (Asner et al., 2004; Savadogo et al., 2007). In addition, other consequences associated with current meat production methods are emergence of multi-drug-resistant strains of pathogenic bacteria (Sanders, 1999), animal disease epidemics or even pandemics, which can kill millions of people (Webster, 2002), nutritionally related diseases, such as cardiovascular diseases and diabetes, associated with the over-consumption of animal fats which are now responsible for a third of global mortality (WHO, 2001) and food-borne illnesses with a six fold increase in gastro-enteritis and food poisoning in industrialized countries in the last 20 years (Nicholson et al., 2000) with contaminated meats and animal products being the most common causes of

*Corresponding author. E-mail: zuhaibbhat@yahoo.co.in.

food borne diseases (Barnard et al., 1995; Mead et al., 1999; Nataro and Kaper, 1998; European Food Safety Authority, 2006; Fisher and Meakens, 2006).

Global population is anticipated to increase to 9 billion by the year 2050 and the demand for meat continues to grow worldwide (Steinfeld et al., 2006). Annual global meat production will rise to 465 million tonnes by the year 2050 accompanied by a rise in annual greenhouse gas emissions to 19.7 gigatonne of carbondioxide, carbon equivalent (Steinfeld et al., 2006). The animals themselves are mostly responsible for the emission of greenhouse gases (Williams et al., 2006) and therefore a reduction of the number of animals that could be achieved by in vitro meat production would result in an appreciable decline of greenhouse gas emission. There are many other reasons for promoting in vitro meat production including animal well fare, process monitoring, environmental considerations, efficiency of food production in terms of feedstock, decrease in intense land usage and greenhouse gas emissions (Stamp Dawkins and Bonney, 2008). Thus, continuing the production of meat by current methods is going to further aggravate the problems and in vitro meat production system seems to be an appealing alternative and is becoming increasingly justifiable in light of the sizable negative effects of current meat production system.

Advantages/need of *in vitro* meat

The first important advantage of producing cultured meat is better control over meat composition and quality by manipulating the flavor, fatty acid composition, fat content and ratio of saturated to poly-unsaturated fatty acids through composition of the culture medium or co-culturing with other cell types. Furthermore, health aspects of the meat can be enhanced by adding factors like certain types of vitamins to the culture medium which might have an advantageous effect on the health (van Eelen et al., 1999). Secondly due to strict quality control rules, such as Good Manufacturing Practice, that are impossible to be introduced in modern animal farms, slaughterhouses, or meat packing plants, the chance of meat contamination and incidence of food borne disease could be significantly reduced. In addition, the risks of exposure to pesticides, arsenic, dioxins, and hormones associated with conventional meat could also be significantly reduced. Third advantage may be the production of exotic cultured meats. In theory, cells from captive rare or endangered animals (or even cells from samples of extinct animals) could be used to produce exotic meats in cultures and thus a sustainable alternative to global trade of meats from rare and endangered animals will help in increasing wild populations of many species in many countries. Cultured meat also reduces animal use in the meat production system as theoretically a single farm animal may be used to produce the world's meat supply. Another advantage is the reduction in the amount of

nutrients and energy needed for their growth and maintenance as the biological structures in addition to muscle tissue are not required to produce meat in an *in vitro* system. Furthermore, *in vitro* system significantly lowers time to grow the meat and takes several weeks instead of months for chickens and pigs and years for beef cattle before the meat can be harvested and thus, the amount of feed and labor required per kilogram of *in vitro* cultured meat is much lower. Another advantage of cultured meat is that the bioreactors for *in vitro* meat production, unlike farm animals, do not need extra space and can be stacked up in a fabric hall. Thus, nutritional costs for *in vitro* cultured meat will be significantly lower and the decrease in costs of resources, labor, and land may be compensated by the extra costs of a stricter hygiene regime, stricter control, computer management, etc. Need for other protein sources also demands production of cultured meat and because it is, unlike the other products, animal-derived and with respect to composition most like meat, it may be the preferred alternative. A definite market available for meat substitutes and a small market comprising the vegetarians who do not eat meat for ethical reasons also demand the production of *in vitro* meat. The proteins produced using plants and fungi are animal friendly, sustainable and have been used to make a variety of good chief products but they lack a good texture and taste and such products are no solution for the craving for meat. Further cultured meat will be safer than conventional meat and due to the non-sustainability of traditional meat production there is a huge market for this. The comparatively minimal land requirement of an *in vitro* meat production system allows meat production and processing to take place domestically in countries which would normally rely on imported meats. By bringing the stages of the meat production process closer together spatially and temporally, meat supply can be better determined by demand.

Culturing of *in vitro* meat

The idea of cultured meat for human consumption in a lab *ex vivo* is not a new concept but was predicted long back by Winston Churchill in the 1920s. Alexis Carrel in 1912 managed to keep a piece of chick heart muscle alive and beating in a Petri dish demonstrating that it was possible to keep muscle tissue alive outside the body, provided that it was nourished with suitable nutrients. It was Willem van Eelen of Netherlands who independently had the idea of using tissue culture for the generation of meat products in the early 1950s and in 1999 van Eelen's theoretical idea was patented. SymbioticA harvested muscle biopsies from frogs and kept these tissues alive and growing in culture dishes (Catts and Zurr, 2002). A study involving the use of muscle tissue from the common goldfish (*Carassius auratus*) cultured in Petri dishes was published in 2002 (Benjaminson et al., 2002). Other research initiatives have also achieved keeping muscle tissue alive in a fungal medium, anticipating on the infection risk associated with serum-based media (Benjaminson et al., 2002).

Tissue engineering can be employed to produce cultured meat (Edelman et al., 2005) and a number of demands need to be met for using tissue engineering techniques for meat production. Firstly, a cell source is required that can proliferate indefinitely and also differentiate into functional skeletal muscle tissue. Secondly, these cells need to be embedded in a three dimensional matrix that allows for muscle growth, while keeping the delivery of nutrients and release of waste products undisturbed and lastly, muscle cells need to be conditioned adequately in a bioreactor to get mature, functional muscle fibers for processing to various meat products. The different design approaches for an *in vitro* meat production system can be roughly divided into scaffold/cell culture based and self organizing/tissue culture techniques.

There are two similar untested detailed proposals based on emerging field of tissue engineering (Boland et al., 2003, Zandonella, 2003) for using cell culture for producing *in vitro* meat on scaffold-based techniques. One of the two proposals has been written by Vladimir Mironov for the NASA (Wolfson, 2002) while the other proposal has been written by Willem van Eelen who also holds a worldwide patent for this system (van Eelen et al., 1999). Catts and Zurr (2002) however, appear to have been the first to have actually produced meat by this method. Both of these systems work by growing myoblasts in suspension in a culture medium. Embryonic myoblasts or adult skeletal muscle satellite cells are proliferated, attached to a scaffold or carrier such as a collagen meshwork or microcarrier beads, and then perfused with a culture medium in a stationary or rotating bioreactor. By introducing a variety of environmental cues, these cells fuse into myotubes, which can then differentiate into myofibers (Kosnik et al., 2003). The resulting myofibers may then be harvested, cooked, and consumed as meat (Figures 1 and 2). Mironov proposal (Wolfson, 2002) uses a bioreactor in which cells are grown together with collagen spheres to provide a substrate onto which the myoblasts can attach and differentiate whereas van Eelen's proposal (van Eelen et al., 1999) uses a collagen meshwork and the culture medium is refreshed from time to time or percolated through the meshwork. Once differentiated into myofibers, the mixture of collagen and muscle cells can be harvested and used as meat. While these kinds of techniques work for producing ground processed (boneless) meats with soft consistency, they do not lend themselves to highly structured meats like steaks.

In order to produce highly structured meats, one would need self organizing/tissue culture techniques, creating structured muscle tissue as self-organizing constructs

(Dennis and Kosnik, 2000) or proliferating existing muscle tissue *in vitro*. Benjaminson et al. (2002) cultured Gold fish (*Carassius auratus*) muscle explants. In this study muscle tissue cultured with crude cell extracts showed a limited increase in cell mass and the cultured muscle explants so obtained were washed, dipped in olive oil with spices, covered in breadcrumbs and fried. A test-panel judged these processed explants and agreed that the product was acceptable as food. Tissue culture techniques have the advantage that explants contain all the tissues which make up meat in the right proportions and closely mimics *in vivo* situation. However, lack of blood circulation in these explants makes substantial growth impossible, as cells become necrotic if separated for long periods by more than 0.5 mm from a nutrient supply (Dennis and Kosnik, 2000). According to Vladimir Mironov entirely artificial muscle can be created with tissue engineering techniques by a branching network of edible porous polymer through which nutrients are perfused and myoblasts and other cell types can attach (Wolfson, 2002).

Cell sources for tissue engineered meat

In vitro meat can be produced by culturing embryonic stem cells from farm animal species and are ideal for culturing since these cells have an almost infinite self-renewal capacity. But these cells must be specifically stimulated to differentiate into myoblasts and may inaccurately recapitulate myogenesis (Bach et al., 2003). However, different efforts invested into establishing ungulate stem-cell lines over the past two decades have been generally unsuccessful with difficulties arising in the recognition, isolation and differentiation of these cells (Keefer et al., 2007). Although embryonic stem cells have been cultured for many generations but so far it has not been possible to culture cell lines with unlimited self-renewal potential from pre-implantation embryos of farm animal species. Until now, true embryonic stem cell lines have only been generated from mouse, rhesus monkey, human and rat embryos (Talbot and Blomberg, 2008) but the social resistance to cultured meat obtained from mouse, rat or rhesus monkey will be considerable and will not result in a marketable product. Myosatellite cells isolated from different animal species have different benefits and limitations as a cell source and that isolated from different muscles have different capabilities to proliferate, differentiate, or be regulated by growth modifiers (Burton et al., 2000). Myosatellite cells have been isolated and characterized from the skeletal muscle tissue of cattle (Dodson et al., 1987), chicken (Yablonka-Reuveni, 1987), fish (Powell et al., 1989), lambs (Dodson et al., 1986), pigs (Blanton et al., 1999, Wilschut et al., 2008), and turkeys (McFarland et al., 1988). Porcine muscle progenitor cells have the potential for multilineage differentiation into adipogenic, osteogenic and

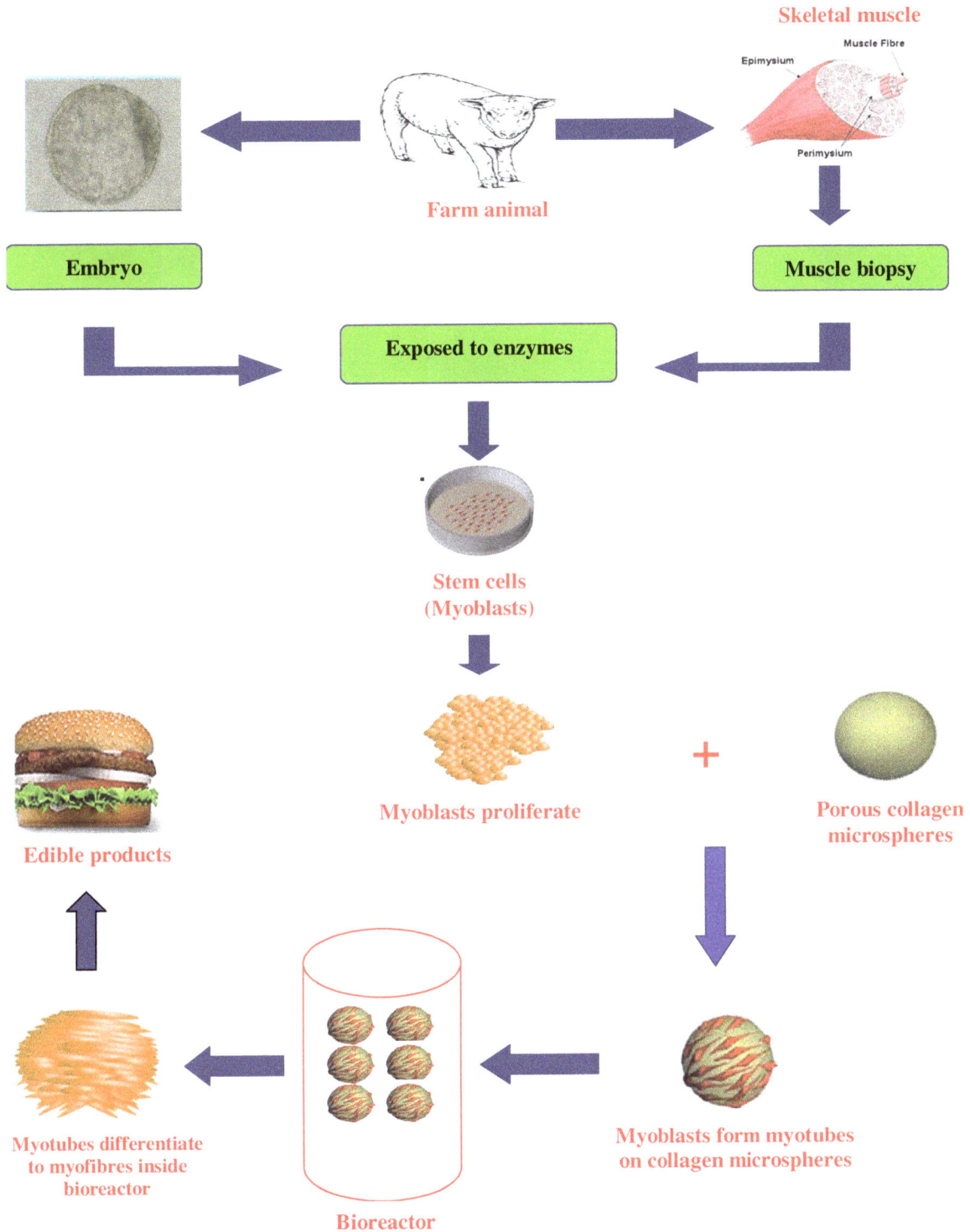

Figure1. Scaffold-based cultured meat production.

chondrogenic lineages, which may play a role in the development of co-cultures (Wilschut et al., 2008).

Adult stem cells from farm animal species can alternatively be used and myosatellite cells are one example of an adult stem-cell type with multilineage potential (Asakura et al., 2001). Adult stem cells have been isolated from several different adult tissues (Wagers and Weissman, 2004) but their *in vitro* proliferation capacity is not unlimited and can proliferate *in vitro* for several months at most. These cells also have the

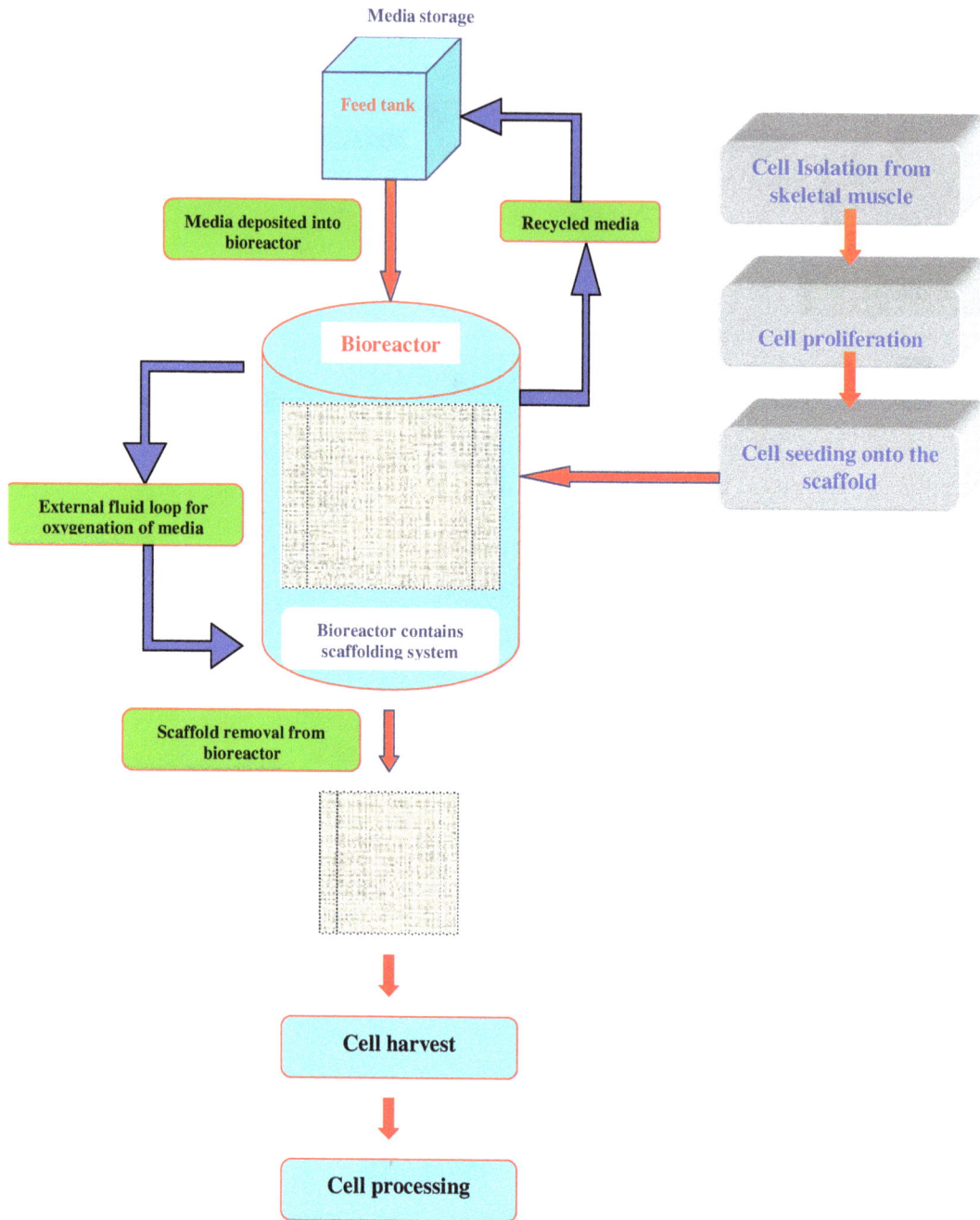

Figure 2. Possible *in vitro* meat production scheme.

capacity to differentiate into skeletal muscle cells, although not very efficiently but for now, these are the most promising cell type for use in the production of cultured meat. However, adult stem cells are prone to malignant transformation in long-term culture (Lazennec and Jorgensen, 2008) that is the greatest matter of debate.

A rare population of multipotent cells found in adipose tissue known as adipose tissue-derived adult stem cells (ADSCs) is another relevant cell type for *in vitro* meat production (Gimble et al., 2007) which can be obtained from subcutaneous fat and subsequently trans-differentiated to myogenic, osteogenic, chondrogenic or adipogenic cell lineages (Kim et al., 2006). It has been observed that adipose tissue-derived adult stem cells immortalize at high frequency and undergo spontaneous transformation in long-term (4 to 5 months) culturing (Rubio et al., 2005), while evidence of adult stem cells remaining untransformed have also been reported (Bernardo et al., 2007). To minimize the risk of spontaneous transformation, re-harvesting of adult stem cells may be necessary in an *in vitro* meat production system and as such obtaining ADSCs from subcutaneous fat is far less invasive than collection of myosatellite cells from muscle tissue. Matsumoto et al. (2007) reported that mature adipocytes can be dedifferentiated *in vitro* into a multipotent preadipocyte cell line known as dedifferentiated fat (DFAT) cells, reversion of a terminally differentiated cell into a multipotent cell type. These DFAT cells are capable of being transdifferentiated into skeletal myocytes (Kazama et al., 2008) and appear to be an attractive alternative to the use of stem cells. This process known as "ceiling culture method" certainly seems achievable on an industrial scale but Rizzino (2007) has put forth the argument that many of the claims of transdifferentiation, dedifferentiation and multipotency of once terminally differentiated cells may be due to abnormal processes resulting in cellular look-alikes.

Co-culturing

Myoblasts are specialized to produce contractile proteins but produce only little extracellular matrix and as such other cells likely need to be introduced to engineer muscle. Fibroblasts residing in the muscle are mainly responsible for the production of extracellular matrix which could be beneficial to add to the culture system (Brady et al., 2008). However, due to the difference in growth rate, co-culturing involves the risk of fibroblasts overgrowing the myoblasts. Meat also contains fat and a vasculature and possibly, co-culture with fat cells should also be considered (Edelman et al., 2005). The problem of vascularization is a general issue in tissue engineering and currently we can only produce thin tissues because of passive diffusion limitations. To overcome the tissue thickness limit of 100 to 200 μm, a vasculature needs to be created (Jain et al., 2005).

Scaffolds

A substratum or scaffold must be provided for proliferation and differentiation of myoblasts as they are anchorage-dependent cells (Stoker et al., 1968). Scaffolding mechanisms differ in shape, composition, characteristics and an ideal scaffold must have a large surface area for growth and attachment, be flexible to allow for contraction as myoblasts are capable of spontaneous contraction, maximize medium diffusion and be easily dissociated from the meat culture in order to optimize muscle cell and tissue morphology. A best scaffold is one that mimics the *in vivo* situation as myotubes differentiate optimally on scaffold with a tissue-like stiffness (Engler et al., 2004) and its by-products must be edible and natural and may be derived from non-animal sources, though inedible scaffold materials cannot

be disregarded. Scaffolds based on new biomaterials may be developed that may offer additional characteristics, such as fulfilling the requirement of contraction for proliferation and differentiation (De Deyne, 2000). Thus, challenge is to develop a scaffold that can mechanically stretch attached cells to stimulate differentiation and a flexible substratum to prevent detachment of developing myotubes that will normally undergo spontaneous contraction.

Edelman et al. (2005) proposed porous beads made of edible collagen as a substrate while as Van Eelen et al. (1999) proposed a collagen meshwork described as a "collagen sponge" of bovine origin. The tribeculate structure of the sponge allows for increased surface area and diffusion, but may impede harvesting of the tissue culture. Other possible scaffold forms include large elastic sheets or an array of long, thin filaments. Cytodex-3 micro-carrier beads have been used as scaffolds in rotary bioreactors but these beads have no stretching potential. One elegant approach to mechanically stretch myoblasts would be to use edible, stimuli-sensitive porous microspheres made from cellulose, alginate, chitosan, or collagen (Edelman et al., 2005) that undergo, at minimum, a 10% change in surface area following small changes in temperature or pH. Once myoblasts attach to the spheres, they could be stretched periodically provided such variation in the pH or temperature would not negatively affect cell proliferation, adhesion, and growth. Jun et al. (2009) have found that growing myoblasts on electrically conductive fibers induces their differentiation, forming more myotubes of greater length without the addition of electrical stimulation but use of such inedible scaffolding systems necessitates simple and nondestructive techniques for removal of the culture from the scaffold.

Bioreactors

Commercial production of *in vitro* meat based products requires large bioreactors for large-scale culturing for the generation of sufficient number of muscle cells. Development of new bioreactors that will maintain low shear and uniform perfusion at large volumes is required. The designing of a bioreactor is intended to promote the growth of tissue cultures which accurately resemble native tissue architecture and provides an environment which allows for increased culture volumes. A laminar flow of the medium is created in rotating wall vessel bioreactors by rotating the cylindrical wall at a speed that balances centrifugal force, drag force and gravitational force, leaving the three-dimensional culture submerged in the medium in a perpetual free fall state (Carrier et al., 1999) which improves diffusion with high mass transfer rates at minimal levels of shear stress, producing three dimensional tissues with structures very similar to those *in vivo* (Martin et al., 2004). Direct perfusion bioreactors appear more appropriate for scaffold based myocyte cultivation and flow medium through a porous scaffold with gas exchange taking place in an external fluid loop (Carrier et al., 2002). Besides offering high mass transfer they also offer significant shear stress, so determining an appropriate flow rate is essential (Martin et al., 2004). Direct perfusion bioreactors are also used for high-density, uniform myocyte cell seeding (Radisic et al., 2003). Another method of increasing medium perfusion is by vascularizing the tissue being grown. Levenberg et al., (2005) had induced endothelial vessel networks in skeletal muscle tissue constructs by using a co-culture of myoblasts, embryonic fibroblasts and endothelial cells co-seeded onto a highly porous biodegradable scaffold. Research size rotating bioreactors have been scaled up to three liters and, theoretically, scale up to industrial sizes should not affect the physics of the system.

As cell viability and density positively correlate with the oxygen gradient in statically grown tissue cultures, it is necessary to have adequate oxygen perfusion during cell seeding and cultivation on the scaffold (Radisic et al., 2008). Adequate oxygen perfusion is mediated by bioreactors which increase mass transport between culture medium and cells and by the use of oxygen carriers to mimic hemoglobin provided oxygen supply to maintain high oxygen concentrations in solution, similar to that of blood. Modified versions of hemoglobin or artificially produced perfluorochemicals (PFCs) that are chemically inert are used as oxygen carriers (Lowe, 2006) but their bovine or human source makes them an unfit candidate and alternatively, human hemoglobin has been produced by genetically modified plants (Dieryck et al., 1997) and microorganisms (Zuckerman et al., 1998).

Culture media and growth factors

To enjoy its potential advantages over conventional meat production, *in vitro* meat would need an affordable medium system containing the necessary nutritional components available in free form to myoblasts and accompanying cells. Myoblast culturing usually takes place in animal sera, a costly media that does not lend itself well to consumer acceptance or large-scale use. Animal sera are from adult, newborn or fetal source, with fetal bovine serum being the standard supplement for cell culture media (Coecke et al., 2005). Because of its *in vivo* source, it can have a large number of constituents in highly variable composition and potentially introduce pathogenic agents (Shah, 1999). The harvest of fetal bovine serum also raises ethical concern and for the generation of an animal-free protein product, the addition of fetal calf serum to the cells would not be an option and it is therefore essential to develop a serum-free culture medium. Commercially available serum replacements and serum-free culture media offer some more realistic options for culturing mammalian cells *in vitro*. Serum-free

media reduce operating costs and process variability while lessening the potential source of infectious agents (Froud, 1999).

Serum-free media have been developed to support *in vitro* myosatellite cell cultures from the turkey (McFarland et al., 1991), sheep (Dodson and Mathison, 1988) and pig (Doumit et al., 1993). Variations among different serum-free media outline the fact that satellite cells from different species have different requirements and respond differentially to certain additives (Dodson et al., 1996). Thus, an appropriate array of growth factors is also required to growing muscle cells in culture in addition to proper nutrition and these growth factors are synthesized and released by muscle cells themselves and, in tissues, are also provided by other cell types locally (paracrine effects) and non-locally (endocrine effects). The myosatellite cells of different species respond differentially to the same regulatory factors (Burton et al., 2000) and as such extrinsic regulatory factors must be specific to the chosen cell type and species. Furthermore, formulation may be required to change over the course of the culturing process from proliferation period to the differentiation and maturation period, requiring different set of factors. A multitude of regulatory factors have been identified as being capable of inducing myosatellite cell proliferation (Cheng et al., 2006), and the regulation of meat animal-derived myosatellite cells by hormones, polypeptide growth factors and extracellular matrix proteins has also been investigated (Dodson et al., 1996, Doumit et al., 1993). Purified growth factors or hormones may be supplemented into the media from an external source such as transgenic bacterial, plant or animal species which produce recombinant proteins (Houdebine, 2009). Alternatively, a sort of synthetic paracrine signalling system can be arranged so that co-cultured cell types can secrete growth factors, which can promote cell growth and proliferation in neighbouring cells. Appropriate co-culture systems like hepatocytes may be developed to provide growth factors necessary for cultured muscle production that provide insulin-like growth factors which stimulate myoblast proliferation and differentiation (Cen et al., 2008), as well as myosatellite cell proliferation in several meat-animal species *in vitro* (Dodson et al., 1996).

Atrophy and exercise

One of the potential problems associated with cultured meat is that of atrophy or muscle wasting due to a reduction of cell size (Fox, 1996) caused by lack of use, denervation, or one of a variety of diseases (Charge et al., 2002, Ohira et al., 2002). Regular contraction is a necessity for skeletal muscle and promotes differentiation and healthy myofiber morphology while preventing atrophy. Muscle *in vivo* is innervated, allowing for regular, controlled contraction whereas *in vitro* system would necessarily culture denervated muscle tissue, so contraction must be stimulated by alternate means. Proliferation and differentiation of myoblasts have been found to be affected by the mechanical, electromagnetic, gravitational, and fluid flow fields (Kosnik et al., 2003, De Deyne, 2000). Repetitive stretch and relaxation equal to 10% of length, six times per hour increase differentiation into myotubes (Powell et al., 2002). Myoblasts seeded with magnetic microparticles induced differentiation by placing them in a magnetic field without adding special growth factors or any conditioned medium (Yuge and Kataoka, 2000). Electrical stimulation also contributes to differentiation, as well as sarcomere formation within established myotubes (Kosnik et al., 2003).

Electrical stimulation

Neuronal activity can be mimicked by applying appropriate electrical stimuli *in vitro* cultures (Bach et al., 2004) and has proven to be pivotal in the development of mature muscle fibers (Wilson and Harris, 1993). It has been shown that induction of contractile activity promoted the differentiation of myotubes in culture by myosin heavy chain expression of different isoforms and sarcomere development (Fujita et al., 2007, Naumann and Pette, 1994). Electrical stimulation can provide a non-invasive and accurate tool to assess the functionality of engineered muscle constructs (Dennis et al., 2009). Functional muscle constructs will exert a force due to active contractions of the muscle cells by generating a homogeneous electrical field inside the bioreactor but so far, these forces generated by engineered muscle constructs only reach 2 to 8% of those generated by skeletal muscles of adult rodents (Dennis et al., 2001). Thus, functional properties of tissue engineered muscle constructs are still unsatisfactory at this moment.

Mechanical stimulation

Mechanotransduction is the process through which cells react to mechanical stimuli and is a complex mechanism (Burkholder, 2007; Hinz, 2006) that is another important biophysical stimulus in myogenesis (Vandenburgh and Karlisch, 1989). It is mainly by means of the family of integrin receptors that cells attach to the insoluble meshwork of extracellular matrix proteins (Juliano and Haskill, 1993) transmitting the applied force to the cytoskeleton. The resulting series of events shows parallels to growth factor receptor signaling pathways, which ultimately lead to changes in cell behavior, such as proliferation and differentiation (Burkholder, 2007). Muscle growth and maturation is affected by different mechanical stimulation regimes and the application of static mechanical stretch to myoblasts *in vitro* results in a facilitated alignment and fusion of myotubes, and also

results in hypertrophy of the myotubes (Vandenburgh and Karlisch, 1989). Furthermore, cyclic strain activates quiescent satellite cells (Tatsumi et al., 2001) and increases proliferation of myoblasts (Kook et al., 2008). Thus all these results indicate that mechanical stimulation protocols affect both proliferation and differentiation of muscle cells and different parameters that presumably influence the outcome of the given stimulus are percentage of applied stretch, frequency of the stimulus and timing in the differentiation process.

Food processing technology

Depending on the starting material utilized, new food processing technologies need to be developed to make *in vitro* meat based products attractive.

CONCLUSION

Being a sustainable, humane and safer system, cultured meat holds great promises as an alternative to traditionally livestock flesh provided consumer resistance can be overcome. *In vitro* meat production system can alleviate the ill consequences associated with current meat production methods by reducing the number of livestock animals, incidence of food borne disease, pollution level, emission of green house gases besides reducing the animal suffering significantly. Since crucial knowledge is still lacking on the biology and technology, it may be concluded that commercial production of cultured meat is as yet not possible and the focus must be on filling these gaps in knowledge.

REFERENCES

Asakura A, Komaki M, Rudnicki M (2001). Muscle satellite cells are multi-potential stem cells that exhibit myogenic, osteogenic, and adipogenic differentiation. Differentiation, 68(4-5): 245-253.

Asner GP, Elmore AJ, Olander LP, Martin RE and Harris AT (2004). Grazing systems, ecosystem responses and global change. Annu. Rev. Environ. Resour., 29: 261-299.

Bach AD, Stem-Straeter J, Beier JP, Bannasch H, Stark GB (2003). Engineering of muscle tissue. Clin. Plast. Surg., 30(4): 589-599.

Bach AD, Beier JP, Stern-Staeter J, Horch RE (2004). Skeletal muscle tissue engineering. J. Cell. Mol. Med., 8(4): 413-422.

Barnard ND, Nicholson A, Howard JL (1995). The medical costs attributable to meat consumption. Prev. Med., 24: 646-655.

Benjaminson MA, Gilchriest JA, Lorenz M (2002). *In vitro* edible muscle protein production system (MPPS): Stage 1, fish. Acta. Astronaut., 51(12): 879-889.

Bernardo ME, Zaffaroni N, Novara F, Cometa AM, Avanzini MA, Moretta A (2007). Human bone marrow-derived mesenchymal stem cells do not undergo transformation after long-term *in vitro* culture and do not exhibit telomere maintenance mechanisms. Cancer Res., 67(19): 9142-9149.

Blanton JR, Grand AL, McFarland DC, Robinson JP, Bidwell CA (1999). Isolation of two populations of myoblasts from porcine skeletal muscle. Muscle Nerve, 22(1): 43-50.

Boland T, Mironov V, Gutowska A, Roth E, Markwald R (2003). Cell and organ printing 2: Fusion of cell aggregates in three-dimensional gels. Anat. Rec., 272A(2): 497-502.

Brady MA, Lewis MP, Mudera V (2008). Synergy between myogenic and non myogenic cells in a 3D tissue-engineered craniofacial skeletal muscle construct. J. Tissue Eng. Regen. M., 2(7): 408-417.

Burkholder TJ (2007). Mechanotransduction in skeletal muscle. Front. Biosci., 12: 174-191.

Burton NM, Vierck JL, Krabbenhoft L, Byrne K, Dodson MV (2000). Methods for animal satellite cell culture under a variety of conditions. Methods Cell Sci., 22(1): 51–61.

Carrier RL, Papadaki M, Rupnick M, Schoen FJ, Bursac N, Langer R (1999). Cardiac tissue engineering: cell seeding, cultivation parameters and tissue construct characterization. Biotechnol. Bioeng., 64(5): 580-589.

Carrier RL, Rupnick M, Langer R, Schoen FJ, Freed LE, Vunjak-Novakovic G (2002). Perfusion improves tissue architecture of engineered cardiac muscle. Tissue Eng., 8(2):175-188.

Catts O, Zurr I (2002). Growing semi-living sculptures: The tissue culture project, 35(4): 365-370.

Cen S, Zhang J, Huang F, Yang Z, Xie H (2008). Effect of IGF-I on proliferation and differentiation of primary human embryonic myoblasts. Chinese J. Reparative Reconstr. Surg., 22(1): 84–87.

Charge S, Brack A, Hughes S (2002). Aging-related satellite cell differentiation defect occurs prematurely after Ski-induced muscle hypertrophy. Am. J. Physiol. Cell Physiol., 283(4): C1228-241.

Cheng L, Gu X, Sanderson JE, Wang X, Lee K, Yao X (2006). A new function of a previously isolated compound that stimulates activation and differentiation of myogenic precursor cells leading to efficient myofiber regeneration and muscle repair. Int. J. Biochem. Cell Biol., 38(7): 1123–1133.

Coecke S, Balls M, Bowe G, Davis J, Gstraunthaler G, Hartung T (2005). Guidance on good cell culture practice: A report of the second ECVAM Task Force on good cell culture practice. Altern. Lab. Anim., 33(3): 261-287.

De Deyne PG (2000). Formation of sarcomeres in developing myotubes: Role of mechanical stretch and contractile activation. Am. J. Physiol. Cell Physiol., 279(6): C1801–C1811.

Dennis R, Kosnik 2nd P (2000). Excitability and isometric contractile properties of mammalian skeletal muscle constructs engineered in vitro. In vitro Cell Dev. Biol. Anim., 36(5): 327-35.

Dennis R, Kosnik 2nd P, Gilbert M, Faulkner J (2001). Excitability and contractility of skeletal muscle engineered from primary cultures and cell lines. Am. J. Physiol. Cell Physiol., 280(2): C288–95.

Dennis RG, Smith B, Philp A, Donnelly K, Baar K (2009). Bioreactors for guiding muscle tissue growth and development. Adv. Biochem. Eng./Biotechnol., 112: 39-79.

Dieryck W, Pagnier J, Poyart C, Marden M, Gruber V, Bournat P (1997). Human haemoglobin from transgenic tobacco. Nature, 386(6620): 29-30.

Dodson MV, Martin EL, Brannon MA, Mathison BA, McFarland DC (1987). Optimization of bovine satellite cell derived myotube formation in vitro. Tissue Cell, 19(2):159-166.

Dodson MV, Mathison BA (1988). Comparison of ovine and rat muscle-derived satellite cells: Response to insulin. Tissue Cell, 20(6): 909-918.

Dodson MV, McFarland DC, Grant AL, Doumit ME, Velleman SG (1996). Extrinsic regulation of domestic animal-derived satellite cells. Domest. Anim. Endocrinol., 13(2): 107–126.

Dodson MV, McFarland DC, Martin EL, Brannon MA (1986). Isolation of satellite cells from ovine skeletal muscles. J. Tissue Cult. Meth. 10(4): 233–237.

Doumit ME, Cook DR, Merkel RA (1993). Fibroblast growth factor, epidermal growth factor, insulin-like growth factor and platelet-derived growth factor-BB stimulate proliferate of clonally derived porcine myogenic satellite cells. J. Cell Physiol., 157(2): 326-332.

Edelman PD, McFarland DC, Mironov VA, Matheny JG (2005). Commentary: In vitro-cultured meat production. Tissue Eng., 11(5): 659-662.

Engler AJ, Griffin MA, Sen S, Bönnemann CG, Sweeney HL, Discher DE (2004). Myotubes differentiate optimally on substrates with tissue-like stiffness, pathological implications for soft or stiff microenvironments. J. Cell Biol., 166(6): 877-887.

European Food Safety Authority (2006). The Community Summary Report on Trends and Sources of Zoonoses, Zoonotic Agents, Antimicrobial Resistance and Food borne Outbreaks in the European Union in 2005. EFSA J., 94: 2-288.

FAO (2006). Livestock's long shadow–environmental issues and options. Food and Agricultural Organization of the United Nations, Rome, pp. 1-176.

Fisher IS, Meakens S (2006). Surveillance of enteric pathogens in Europe and beyond: Enter-net annual report for 2004. Euro Surveillance: Bulletin Européen sur les Maladies Transmissibles 11, E060824.060823. Available at: http://www.hpa.org.uk/hpa/inter/enter-net/Enter-net%20annual%20report%202004.pdf, (26 March 2007 last date accessed).

Fox SI (1996). Human Physiology. Wim C, Brown Publishers, Boston.

Froud SJ (1999). The development, benefits and disadvantages of serum-free media. Dev. Biol. Stand., 99: 157-166.

Fujita H, Nedachi T, Kanzaki M (2007). Accelerated de novo sarcomere assembly by electric pulse stimulation in C2C12 myotubes. Exp. Cell Res., 313(9): 1853-1865.

Gimble JM, Katz AJ, Bunnell BA (2007). Adipose-derived stem cells for regenerative medicine. Circ. Res., 100(9): 1249–1260.

Hinz B (2006). Masters and servants of the force: the role of matrix adhesions in myofibroblast force perception and transmission. Eur. J. Cell Biol., 85(3-4): 175-181.

Houdebine LM (2009). Production of pharmaceutical proteins by transgenic animals. Comp. Immunol. Microbiol. Infect. Dis., 32(2): 107-121.

Jain RK, Au P, Tam J, Duda DG, Fukumura D (2005). Engineering vascularized tissue. Nat. Biotechnol., 23(7): 821-823.

Juliano RL, Haskill S (1993). Signal transduction from the extracellular matrix. J. Cell Biol., 120(3): 577-585.

Jun I, Jeong S, Shin H (2009). The stimulation of myoblast differentiation by electrically conductive sub-micron fibers. Biomaterials, 30(11): 2038-2047.

Kazama T, Fujie M, Endo T, Kano K (2008). Mature adipocyte-derived dedifferentiated fat cells can transdifferentiate into skeletalmyocytes in vitro. Biochem. Bioph. Res. Co., 377(3): 780-785.

Keefer CL, Pant D, Blomberg L, Talbot NC (2007). Challenges and prospects for the establishment of embryonic stem cell lines of domesticated ungulates. Anim. Reprod. Sci., 98(1-2): 147-168.

Kim MJ, Choi YS, Yang SH, Hong HN, Cho SW, Cha SM (2006). Muscle regeneration by adipose tissue-derived adult stem cells attached to injectable PLGA spheres. Biochem. Bioph. Res. Co., 348(2): 386-392.

Kook SH, Lee HJ, Chung WT, Hwang IH, Lee SA, Kim BS (2008). Cyclic mechanical stretch stimulates the proliferation of C2C12 myoblasts and inhibits their differentiation via prolonged activation of p38 MAPK. Mol. Cells, 25(4): 479-486.

Kosnik PE, Dennis RG, Vandenburgh H (2003). Tissue engineering skeletal muscle. In Guilak F, Butler DL, Goldstein SA and D. Mooney (Ed), Functional tissue engineering. New York: Springer-Verlag, pp. 377-392.

Lazennec G, Jorgensen C (2008). Concise review: Adult multipotent stromal cells and cancer: Risk or benefit? Stem Cells, 26(6): 1387-1394.

Levenberg S, Rouwkema J, Macdonald M, Garfein ES, Kohane DS, Darland DC (2005). Engineering vascularized skeletal muscle tissue. Nat. Biotechnol., 23(7): 879-884.

Lowe K (2006). Blood substitutes: From chemistry to clinic. J. Mater. Chem., 16(43): 4189-4196.

Martin I, Wendt D, Herberer M (2004). The role of bioreactors in tissue engineering. Trends Biotechnol., 22(2): 80-86.

Matsumoto T, Kano K, Kondo D, Fukuda N, Iribe Y, Tanaka N (2007). Mature adipocyte-derived dedifferentiated fat cells exhibit multilineage potential. J. Cell Physiol., 215(1): 210-222.

McFarland DC, Doumit ME, Minshall RD (1988). The turkey myogenic satellite cell: Optimization of in vitro proliferation and differentiation. Tissue Cell, 20(6): 899-908.

McFarland DC, Pesall JE, Norberg JM, Dvoracek MA (1991). Proliferation of the turkey myogenic satellite cell in a serum-free medium. Comp. Biochem. Physiol., 99(1-2): 163-167.

Mead P, Slutsker L, Dietz A, McCaig L, Bresee J, Shapiro C, Griffin P, Tauxe R (1999). Food-Related Illness and Death in the United

States. Emerg. Infect. Dis., 5(5): 607-625.

Nataro JP, Kaper JB (1998). Diarrheagenic *Escherichia coli.* Clin. Microbiol. Rev., 11: 142-201.

Naumann K, Pette D (1994). Effects of chronic stimulation with different impulse patterns on the expression of myosin isoforms in rat myotube cultures. Differ. Res. Biol. Divers., 55(3): 203-211.

Nicholson FA, Hutchison ML, Smith KA, Keevil CW, Chambers BJ, Moore A (2000). A Study on Farm Manure Applications to Agricultural Land and an Assessment of the Risks of Pathogen Transfer into the Food Chain. Project Number FS2526, Final report to the Ministry of Agriculture, Fisheries and Food, London.

Ohira Y, Yoshinaga T, Nomura T, Kawano F, Ishihara A, Nonaka I, Roy R, Edgerton V (2002). Gravitational unloading effects on muscle fiber size, phenotype and myonuclear number. Adv. Space Res., 30(4): 777-781.

Powell RE, Dodson MV, Cloud JG (1989). Cultivation and differentiation of satellite cells from skeletal muscle of the rainbow trout Salmo. gairdneri. J. Exp. Zool., 250(3): 333-338.

Powell CA, Smiley BL, Mills J, Vandenburgh HH (2002). Mechanical stimulation improves tissue-engineered human skeletal muscle. Am. J. Physiol. Cell Physiol., 283:C1557.

Radisic M, Euloth M, Yang L, Langer R, Freed LE, Vunjak-Novakovic G (2003). High-density seeding of myocyte cells for cardiac tissue engineering. Biotechnol. Bioeng., 82(4): 403–414.

Radisic M, Marsano A, Maidhof R, Wang Y, Vunjak-Novakovic G (2008). Cardiac tissue engineering using perfusion bioreactor systems. Nat. Protoc., 3(4): 719-738.

Rizzino A (2007). A challenge for regenerative medicine: Proper genetic programming, not cellular mimicry. Dev. Dyn., 236(12): 3199-3207.

Rubio D, Garcia-Castro J, Martin MC, de la Fuente R, Cigudosa JC, Lloyd AC (2005). Spontaneous human adult stem cell transformation. Cancer Res., 65(8): 3035-3039.

Sanders T (1999). The nutritional adequacy of plant-based diets. Proc. Nutr. Soc., 58(2): 265-269.

Savadogo P, Sawadogo L, Tiveau D (2007). Effects of grazing intensity and prescribed fire on soil physical and hydrological properties and pasture yield in the savanna woodlands of Burkina Faso. Agricult. Ecosys. Environ., 118: 80-92.

Shah G (1999). Why do we still use serum in production of biopharmaceuticals? Dev. Biol. Stand., 99: 17-22.

Stamp Dawkins M, Bonney R (2008). Future of Animal Farming; Renewing the Ancient Contract, Wiley-Blackwell.

Steinfeld H, Gerber P, Wassenaar T, Castel V, Rosales M, De Haan C (2006). Livestock's long shadow: environmental issues and options (Rome: Food and Agriculture Organization of the United Nations, p. xxi). Available at: www.virtualcentre.org/en/library/key_pub/longshad/A0701E00.pdf. Accessed March 7, 2008.

Stoker M, O'Neil C, Berryman S, Waxman V (1968). Anchorage and growth regulation in normal and virus-transformed cells. Int. J. Cancer, 3: 683-693.

Talbot NC, Blomberg LA (2008). The pursuit of ES cell lines of domesticated ungulates. Stem Cell Rev. Reports, 4(3): 235-254.

Tatsumi R, Sheehan SM, Iwasaki H, Hattori A, Allen RE (2001). Mechanical stretch induces activation of skeletal muscle satellite cells *in vitro.* Exp. Cell Res., 267(1): 107-114.

Van Eelen WF, van Kooten WJ, Westerhof W (1999). WO/1999/031223: Industrial production of meat from *in vitro* cell cultures. Patent Description http://www.wipo.int/pctdb/en/wo.jsp?wo=1999031223 Accessed Mar. 2009.

Vandenburgh HH, Karlisch P (1989). Longitudinal growth of skeletal myotubes *in vitro* in a new horizontal mechanical cell stimulator. *In vitro* Cell. Dev. Biol., 25(7): 607-616.

Wagers AJ, Weissman IL (2004). Plasticity of adult stem cells. Cell, 116(5): 639-648.

Webster R (2002). The importance of animal influenza for human disease. Vaccine, 20(2): S16-20.

WHO (2001). Global Burden of Disease estimates for 2001. Geneva, World Health Organization. http://www3.who.int/whosis/menu.cfm?path=evidence,burden,burden _estimates,burden_estimates_2001&language=english, accessed 9 April 2004.

Williams AG, Audsley E, Sandars DL (2006). Determining the environmental burdens and resource use in the production of agricultural and horticultural commodities. Main Report. Defra Research Project IS0205. Bedford: Cranfield University and Defra.

Wilschut KJ, Jaksani S, Van Den Dolder J, Haagsman HP, Roelen BAJ (2008). Isolation and characterization of porcine adult muscle-derived progenitor cells. J. Cell Biochem., 105(5): 1228-1239.

Wilson SJ, Harris AJ (1993). Formation of myotubes in aneural rat muscles. Dev. Biol., 156(2): 509-518.

Wolfson W (2002). Raising the steaks. New Scientist, 176: 60-63.

Yablonka-Reuveni Z, Quinn LS, Nameroff M (1987). Isolation and clonal analysis of satellite cells from chicken pectoralis muscle. Dev. Biol., 119(1): 252-259.

Yuge L, Kataoka K (2000). Differentiation of myoblasts accelerated in culture in a magnetic field. *In vitro* Cell Dev. Biol. Anim., 36: 383.

Zandonella C (2003). Tissue engineering: The beat goes on. Nature, 421(6926): 884-886.

Zuckerman SH, Doyle MP, Gorczynski R, Rosenthal GJ (1998). Preclinical biology of recombinant human hemoglobin, rHb1.1. Artif. Cells Blood Substit. Immobil. Biotechnol., 26(3): 231-257.

17

Rheological and sensory properties of four kinds of jams

Xin Gao, Tian Yu, Zhao-hui Zhang*, Jia-chao Xu and Xiao-ting Fu

College of Food Science and Engineering, Ocean University of China, Qingdao, Shandong 266003, China.

The rheological properties of four kinds of jams (Kewpie strawberry, Marmalade, Blueberry and ST. Dalfour) were determined by the rheometer of MCR101. It showed that all the jam samples belonged to non-Newtonian fluid model, with thixotropy and pseudoplastic. Three jams were fit for Herschel-Bulkley model, although yield stress existed in the samples at 25°C and was calculated as 25.07, 99.13 and 103.67 Pa for ST. Dalfour, Kewpie strawberry and Blueberry, respectively. More so, three jams were fit for Arrhenius model, and the flow activation energies were 13.79, 15.10 and 21.46 kJ/mol for Blueberry, ST. Dalfour and Kewpie strawberry, respectively. The oscillatory test data were revealed due to the increase in frequency, storage modules (G') and loss modules (G"), the decrease in viscosity (η) and the exhibition of shear thinning behavior by the jams. The G', G" and η of Blueberry were the highest, while those of ST. Dalfour were the lowest. According to the time dependent behavior, it was revealed that Kewpie strawberry had the most instability and ST. Dalfour had the most stability, which may be related to the flow activation energy. However, by the sensory properties, the lowest of the overall acceptability was Marmalade sample.

Key words: Jams, rheological properties, viscosity, storage modulus G', loss modulus G'.

INTRODUCTION

Jam is a mixture of sugars, pulp and a pure drop of one or more kinds of fruit and water brought to a suitable gelled consistency (Fugel et al., 2005). According to Bureau of Indian Standards (BIS) and Prevention of Food Adulteration (PFA) specifications, jam should contain more than 68.5% total soluble solids (TSS) and at least 45% fruit (The Prevention of Food Adulteration Rules, 1955; Santanu 2010), whereas, the Codex Alimentarius Commission specifies that the finished jam should contain more than 65% TSS. Sugar constitutes more than 40% of total weight and 80% of total solids in jam (Lal et al., 1998, Santanu 2010). Commercial jam usually has an extremely variable composition. Nearly all manufacturers have a formula of their own which differs in some respects from those of other manufacturers. The ingredients affect the jam quality in terms of both subjective (sensory) and objective (textural and rheological) attributes.

China is a large agricultural country, rich in fruits and vegetables. It shows that fruits provide a sufficient source of jam. Both the labor cost of the U.S.A, Europe and

Japan's growth and the people who engaged in agricultural production becoming old have resulted to the increase in the price of fruits and vegetables. So the purchasing agent turns his attention towards the China market; as such, the Japanese demand for jam increase rapidly. According to the investigation carried out on Tokyo Customs, as compared to that carried out 7 years ago, only jam imports increased by 3 times, but the amount of imports increased by just 1.5 times. In Europe and America, jam is also the main food taken by consumers. On the one hand, owing to the food culture for centuries, bread with jam has been their most common and popular way of life; on the other hand, fast-paced modern city has given them even less time to properly prepare and enjoy their breakfast and lunch. So the convenient food has become their first choice and it has also met nutritional needs.

Rheological research in food science is therefore closely linked to the development of food products and could address the industrial production of food (stirring, pumping, dosing, dispersing and spraying), home based cooking as well as consumption of food (oral perception, digestion and well-being). For the processed food, the composition and the addition of ingredients to obtain a

*Corresponding author. E-mail: zhangzhh@ouc.edu.cn.

certain food quality and product performance require profound rheological understanding of individual ingredients related to food processing and final perception (Peter and Erich, 2010). Variation in ingredients or their concentration levels usually lead to changes in gel structure in jam that are often perceived by consumers through texture or mouthfeel. Texture influences the mouthfeel of a product, whereas taste is the sensory experience derived from the sensation in the mouth or on the tongue after ingestion of a food material.

The consumer judges the quality (fresh, stale, tender, ripe) when the food produces a physical sensation (hard, soft, crisp, moist and dry) in the mouth (Szczesniak, 1963; Kokini and Plutchok, 1987; Santanu 2010). Therefore, four kinds of jam products were chosen to determine the rheological properties including the static flow (Herschel-Bulkley parameters, thixotropy, and the determination of the apparent activation energy) and dynamic flow patterns (frequency, temperature and time sweep). More so, the sensory evaluation was also determined.

MATERIALS AND METHODS

The following four brands of commercial jams were purchased from JUSCO supermarket in Qingdao, China, and were preserved in refrigerator at 4°C:

(1) Kewpie strawberry jam was manufactured by Beijing Kewpie Corporation;
(2) Marmalade was manufactured by Shanghai Liang Food Corporation;
(3) Hero Blueberry was produced in Germany by Hero-Group for Hero AG, CH-5600 Lenzburg, Switzerland;
(4) ST. Dalfour was manufactured by Chambord et Cie Chateau St. Dalfour 82 Route de Bracieux Cour Cheverrny, France.

The composition of jams' samples

Moisture content, the soluble solids and ash content of the jams were determined according to the methods of Weixuan et al. (1989).

Rheological measurements

Rheological properties of the four jams were measured using a rheometer (MCR101, Austria).

Hysteresis area

The hysteresis area presents the area between two curves, which is commonly the up and down curve of a shear rate sweep (Sharoba et al., 2005). However, the steady-state rheological behavior of the four jam samples was studied at 25°C, where the shear rate was increased linearly from 0.1 to 150 s^{-1}. It was observed that the steady-state relationship between shear stress and shear rate of food materials is expressed in terms of Herschel–Bulkey model:

$$\tau = \tau_0 + K\gamma^n \tag{1}$$

where, τ is the shear stress (Pa), τ_0 is the yield stress (Pa), γ is the

shear rate (s-1), K is the consistency index (Pa.sn) and n is the flow behavior index (dimensionless) signifying the extent of deviation from Newtonian behavior (Santanu and Shinhare, 2010).

Dependence of the flow behavior of fluid foods on temperature

The dependence of the flow behavior of fluid foods on temperature is taken to satisfy the Arrhenius relationship:

$$K = A_k \exp(E_k/RT) \tag{2}$$

Where, A_k is the frequency factor (Pa·sn), E_k represents the activation energy (kJ/mol), R is the gas law constant (R = 8.314 J/mol·K), and T is the absolute temperature (K). However, the flow activation energy was calculated using the Arrhenius-type equation.

Oscillatory measurement analysis

The oscillatory test, also called the dynamic rheological experiment, can be used to determine viscoelastic properties of food. The storage modulus G' expresses the magnitude of the energy that is stored in the material or that is recoverable per cycle of deformation. G" is a measure of the energy that is lost as viscous dissipation per cycle of deformation. Therefore, for a perfectly elastic solid, all the energy are stored, that is, G" is zero and the stress and strain will be in phase. In contrast, for a liquid with no elastic properties, all the energy are dissipated as heat; that is, G' is zero and the stress and strain will be out of phase by -90°C (Sharoba et al., 2005). Each measurement was performed with one point every 2 s for 60 points. However, measurements were performed at least in triplicate and the phase angle (δ) was calculated from the measurements of G' and G" as tanδ=G"/G' (Karin et al., 2009).

Sensory evaluation

The samples stored at 4°C were taken out 3 h before serving. Color, taste, odor, texture,and overall acceptability of the four jam samples were evaluated by eight panelists following nine point hedonic scale (9 = like extremely, 8 = like very much, 7 = like moderately, 6 = like slightly, 5 = neither like nor dislike,4 = dislike slightly, 3 = dislike moderately, 2 = dislike very much, 1 = dislike extremely) Santanu 2010; Lawless and Heymann 1998. All samples were presented before the panelists at room temperature under normal lighting conditions in 50 mL cups coded with random,three-digit numbers.

Statistical analysis

All determinations were made in triplicate, unless otherwise specified. Data were analysed by analysis of variance (ANOVA). The mean comparison was carried out by using SPSS software version 16.0 at P < 0.05 level.

RESULTS

The composition of jam samples

The main composition of the four jam samples is shown in Table 1. There was an insignificant difference in the ash content between the four jam samples. The lowest ash content was 0.13% of Blueberry, while the highest content was 0.25% of ST. Dalfour samples. Also, the moisture

Table 1. The main composition of jam samples.

Samples	Moisture content (%)	Ash content (%)	Soluble solid content (%)
ST. Dalfour	44.04 ± 0.27^a	0.25 ± 0.01^a	53.97 ± 0.32^d
Blueberry	30.55 ± 0.53^c	0.13 ± 0.01^a	64.73 ± 0.21^b
Marmalade	28.47 ± 0.43^d	0.24 ± 0.01^a	66.40 ± 0.10^a
Kewpie strawberry	34.40 ± 0.34^b	0.19 ± 0.03^a	58.83 ± 0.76^c

a,b,c,d Mean with different letters in the same column are significantly different (P < 0.05).

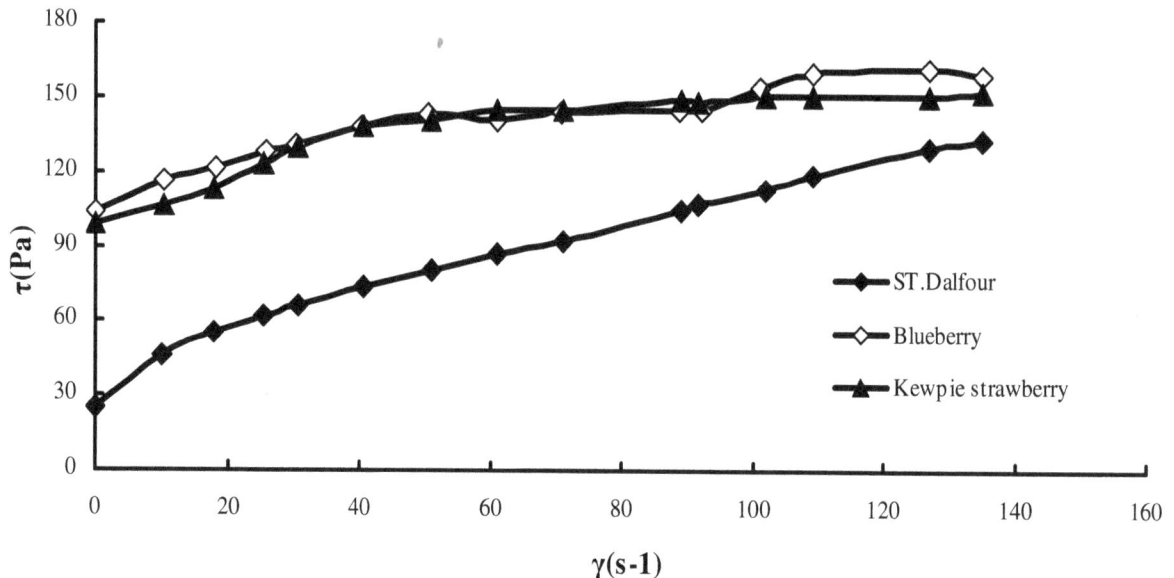

Figure 1. Changes of the shear stress with the shear rate increase.

Table 2. The Herschel-Bulkley parameters of jam samples.

Samples	σ_0/Pa	K/(Pa.s)	n	R^2
ST. Dalfour	25.07	4.77	0.63	0.9860
Blueberry	103.67	3.61	0.57	0.9509
Kewpie strawberry	99.14	2.02	0.71	0.9492

Table 3. The Arrhenius relationship parameters of jam samples.

Parameter	ST. Dalfour	Blueberry	Kewpie strawberry
Ea(KJ/mol)	15.10	13.79	21.46
R^2	0.9917	0.9678	0.9808

content and the soluble solid content of the four jam samples were presented in Table 1. The opposite variation tendency of the four jam samples appeared from the moisture content and the soluble solid content. The ST. Dalfour sample had the highest moisture content (44.04%) and the lowest soluble solid content (53.97%), while Marmalade had the highest soluble solid content (66.40%) and the lowest moisture content (28.47%).

Steady-state rheology analysis

As shown in Figure 1, the steady flow curves of ST. Dalfour, Blueberry and Kewpie strawberry were well described by the Herschel-Bulkley model, while the Marmalade sample was not fit for the Herschel-Bulkey model. The experimental values of the shear stress and the shear rate were fitted by Equation (1). The rheological parameters ($\sigma 0$, K and n) of Dalfour, Blueberry and Kewpie strawberry were calculated and shown in Table 2. At 25°C, the highest value of the yield point ($\sigma 0$) was found in Blueberry sample, while the lowest was found in ST. Dalfour sample. At 25°C, the K value was higher for ST. Dalfour sample and lower for Kewpie strawberry sample. All the flow index values n for the jam samples are given in Table 2, with n<1 indicating that the rheological behaviour is pseudoplastic.

As the temperature of the jam was increased, the viscosity of the jam samples decreased. The flow activation energies of the jam samples as listed in Table 3 were calculated at a constant shear rate of 100 1/s. The viscosity decreased with temperature. This effect of temperature on the flow behaviour of Dalfour, Blueberry and Kewpie strawberry can be described by the Arrhenius

Table 4. The sensory evaluation of jam samples.

Samples	Consistency	Colour	Taste	Odour	Overall acceptability
ST. Dalfour	5.88±0.64	6.25±1.49	5.88±1.13	5.63±1.30	6.38±1.19
Kewpie strawberry	6.00±1.20	5.50±1.07	6.13±0.99	5.63±1.19	6.00±1.07
Blueberry	6.00±0.76	5.38±1.51	5.75±1.04	5.50±1.07	6.38±0.74
Marmalade	5.88±0.99	6.25±1.49	5.75±1.58	5.50±1.20	5.63±0.92

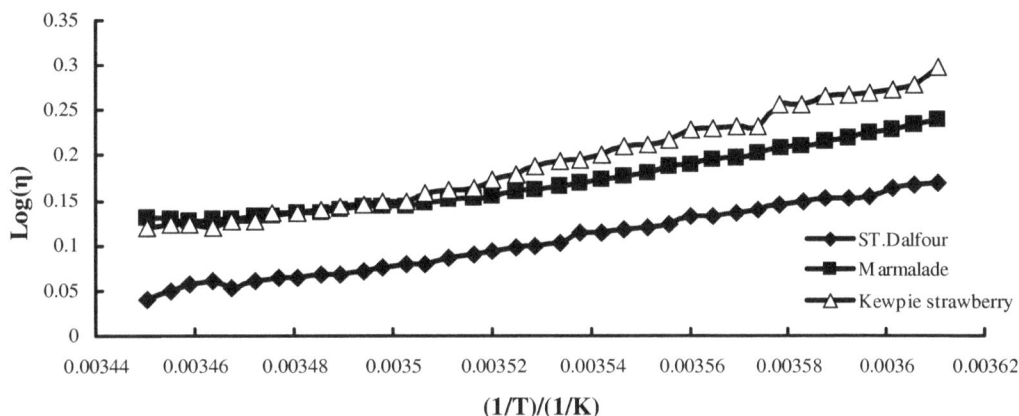

Figure 2. Effects of temperature on viscosity of the jams' samples.

relationship.

The largest of hysteresis loop area was Kewpie strawberry, while the smallest was Marmalade. The value assigned to a certain hysteresis area depends on the loop contour and the shear resistance of the sample. It can also vary according to the experimental conditions of the test (total shearing time and range of shear rates applied).

Dynamic measurement analysis

As Figure 4 shows the strain sweep, when the strain is 0.5%, the complex module (G*) is very steady. So, 0.5% strain at 3 Hz frequency is taken as the linear visco-elastic (LVE) region. The frequency sweep measurements agree with the results obtained previously. All the jam samples show G'>G" in the LVE region, which means that they all have semisolid characteristics at very low deformation. As the frequency was increased, tanδ of the jam samples were increased as well. The tanδ values decreased with an increase in temperature as shown in Figure 6. Of the jam samples, Marmalade declined most, while Blueberry was the steadiest jam. However, the value of tanδ for the jam samples ranged between 0.30 and 0.32, 0.28 and 0.30, 0.27 and 0.32, and 0.29 and 0.33 for ST. Dalfour, Blueberry, Marmalade and Kewpie strawberry, respectively.

Figure 7 shows different characters of the time dependent behavior of the four jam samples. With the exception of ST. Dalfour sample, the complex module (G*)

increased rapidly with time of shearing, while the most significant increase was observed for Kewpie strawberry. The results comply with the measurements of the hysteresis loop.

Analysis of the sensory properties of jams

The results of the sensory evaluation scored are shown in Table 4. There was no significant difference for the four jam samples with regards to consistency, colour, taste, odour and overall acceptability. It was observed that the colour of Marmalade sample had the highest score, while the other three items had the lowest. Moreover, the highest score of taste and odour was observed for Kewpie strawberry.

DISCUSSION

In this study, we determined the rheological behavior of four jam samples. Some studies on the steady-shear and dynamic-shear rheological characterization of one kind of jam had been done (Sharoba et al., 2005; Santanu and Shinhare, 2010; Sesmero et al., 2009). However, the dynamic-shear rheological characterization is not complete and the differences of several jams have not been reported. Generally, the greater the intermolecular distances required, the higher the flow activation energy. So, the effect of viscosity on temperature is greater. As

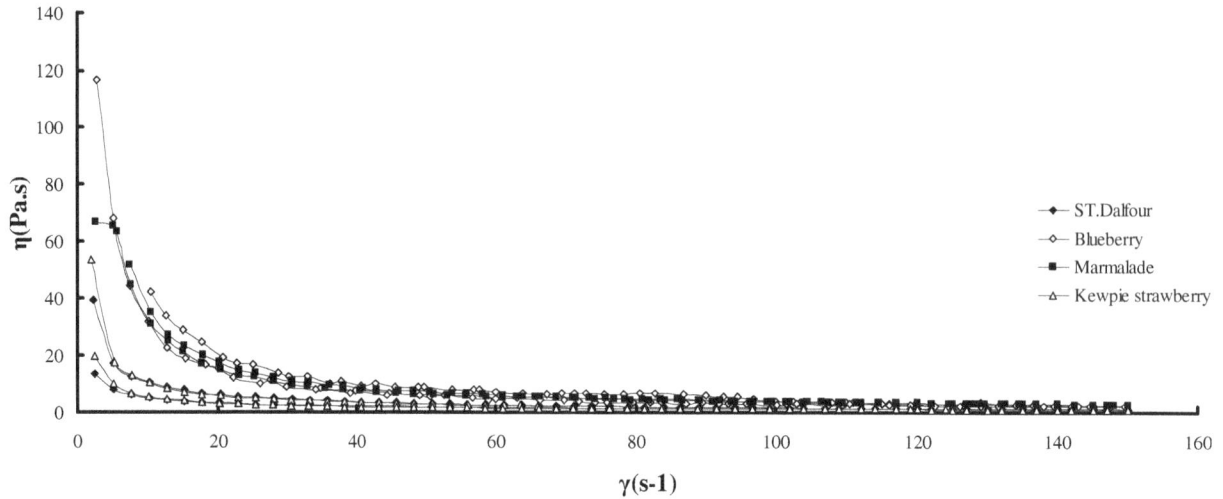

Figure 3. Hysteresis loop area of the four jams' samples.

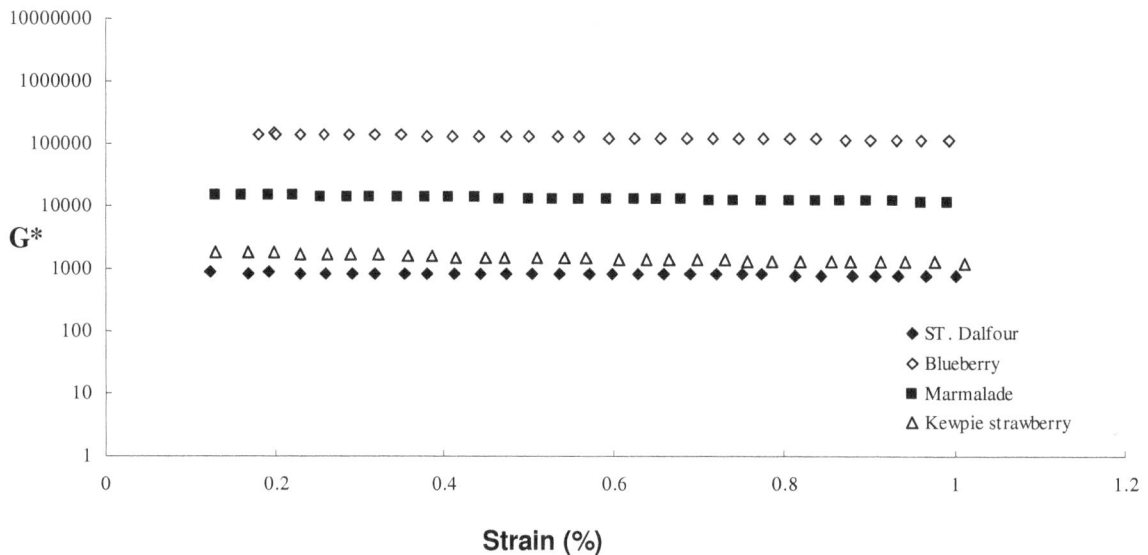

Figure 4. The linear viscoelasticity region of the four jams' samples.

the temperature increased, the intermolecular distances increased and therefore the viscosity decreased. The temperature dependence of K on Dalfour, Blueberry and Kewpie strawberry jams was well described by Arrhenius relationship (Figure 2). So far, Santanu Basu (Santanu and Shinhare, 2010) reported the flow activation energy of mogoo jam. The flow activation energies show the energies that the samples are trying to overcome. It is due to the fact that jam is a physical gel that it gave rise to more resistance to deformation.

As reported by Hernandez, a high-viscosity thixotropic fluid may show a larger hysteresis area than that with lower viscosity even if the latter undergoes a stronger structural destruction. Comparison of straight loop areas between differently viscous systems may not validate the conclusions on the extension of time-dependent structural

breakdown (Tarrega et al., 2004). In this study as shown by Figure 3, Kewpie strawberry shows the largest loop area, indicating the highest resistance to flow. Assuming that a hysteresis loop area is an index of the energy needed to destroy the structure responsible for flow time dependence, the experimental data showed that Kewpie strawberry was the one needing the highest energy to breakdown, while Marmalade was the one needing the lowest energy to breakdown. For this type of soft solids, rheological properties change not only with the shear rate, but also with the time of shearing. Therefore, it is important to determine time dependent behavior. Typical time dependent rheograms revealed the thixotropic behavior and stability indicating continuous breakdown or rearrangement of structure with time of shearing. The results of time dependent behavior in Figure 7 were

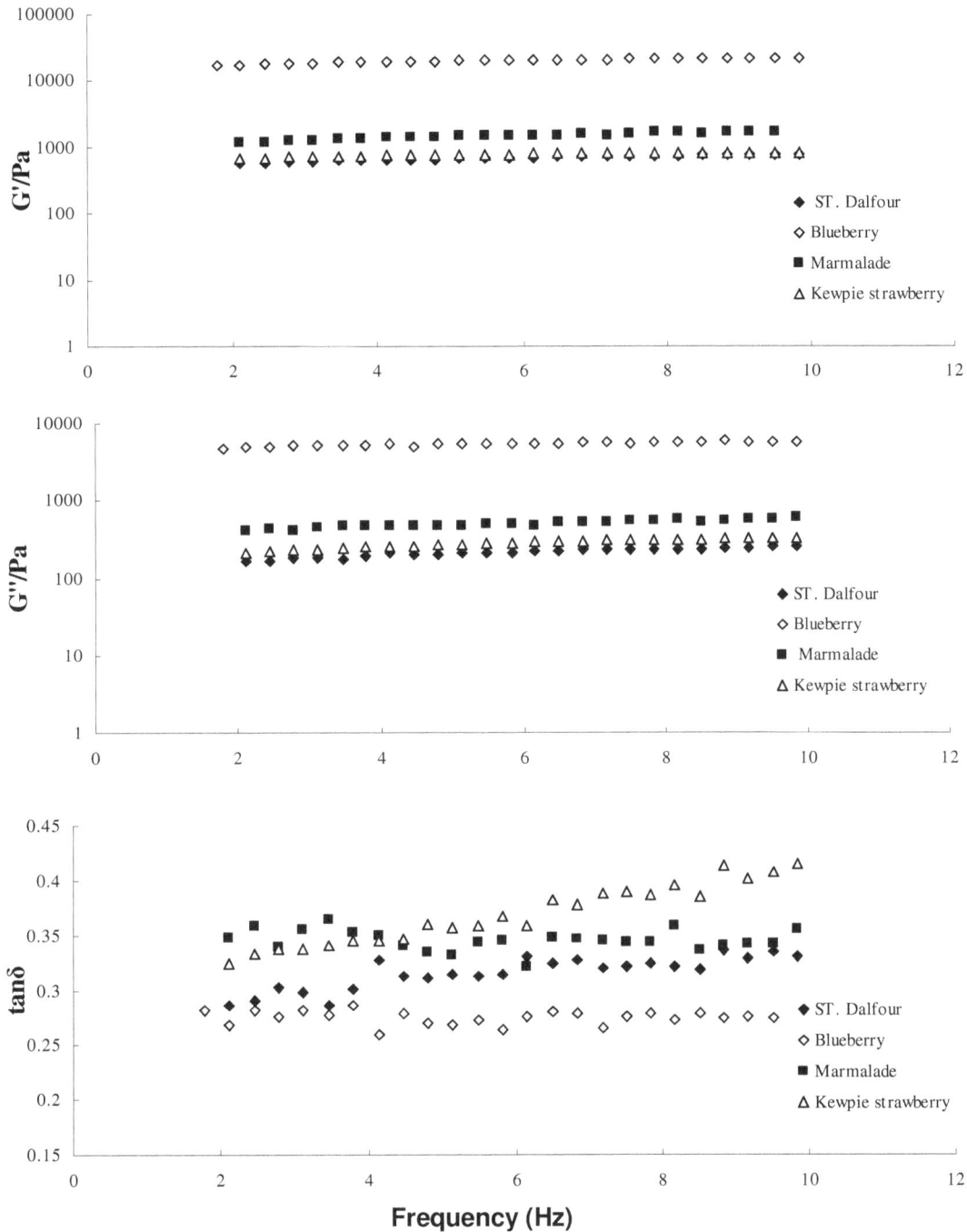

Figure 5. Frequency sweep of the four jams' samples at 25°C.

obviously consistent with that shown in Figure 3.

Dynamic rheology measure (Figure 5) revealed that the value of G' was higher than that of G", which showed that the jam samples were more elastic than viscous. Typical weak gel characteristics were observed, that is, G' was greater than G" throughout the frequency range, and the moduli showed a slight dependence on frequency (Talip and Sevim, 2003). The results of the frequency sweep analyzed showed that the jam samples are within the category of weak gel (Clark and Ross-Murphy, 1987;

Sundaram and Mehmet 2000). For unlinked polymers, the values of both G' and G" fall constantly towards lower frequencies. At low frequencies (lower than 2 Hz), the bonds in the structural flow units were stretched and relaxed with very little breakage taking place. In these samples, very large particles act as an energy storing agent. Therefore, long molecules govern the behaviour of the system at low frequencies in places where the entanglement which seemed to be the mechanical interactions is probably neither chemical bonds nor

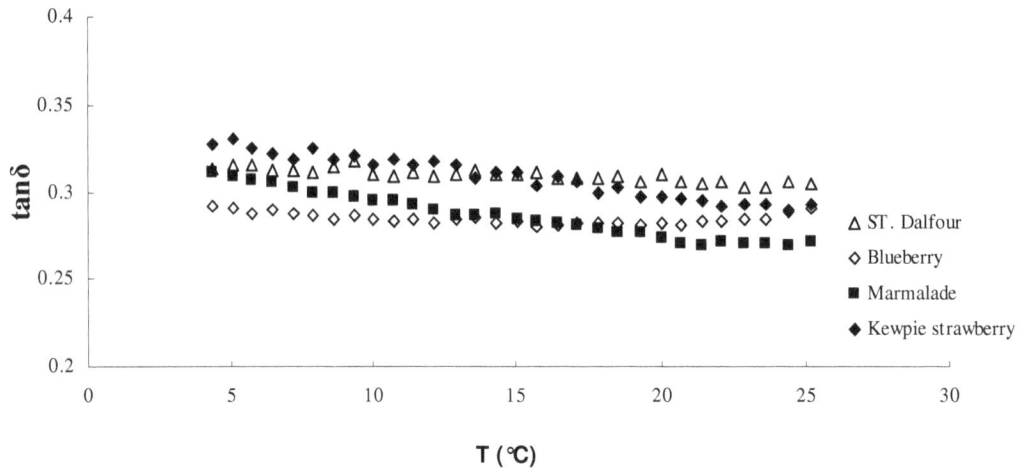

Figure 6. Temperature sweep of the four jams' samples at 3 Hz.

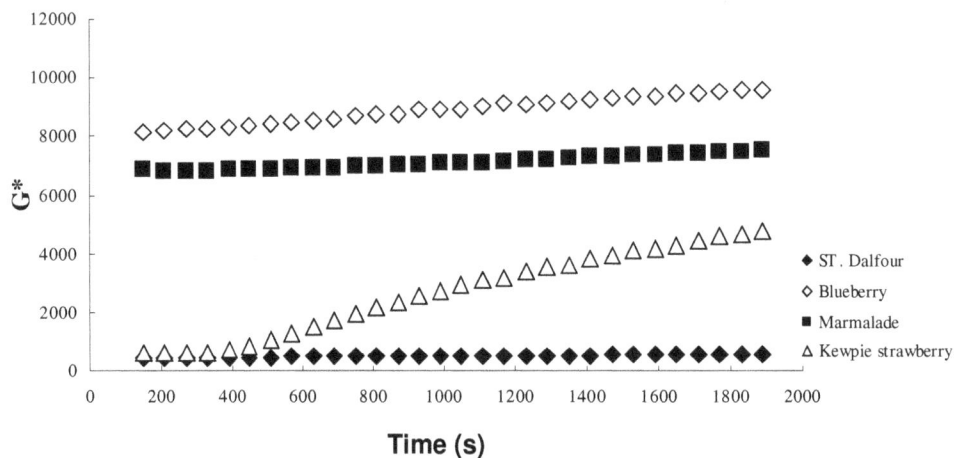

Figure 7. Time dependent behavior of the four jams' samples.

physical-chemical bonds. Hence, it has not been broken in places where solid-like behaviour predominates (G' > G") (Sima et al., 2011). The tangent of this loss angle [tan δ= G"/G'] denotes the relative effects of viscous and elastic components on viscoelastic behavior (Sundaram and Mehmet, 2000).

As the frequency was increased, the viscous character increased. By the tendency of tanδ, the solid-like characteristics of the Marmalade and Blueberry were greater than the other two jams. This may be caused by the lower moisture content of the former two jam samples, which could result in less freedom of movement for molecules (Clark and Ross-Murphy, 1987). Loss factor, tanδ= G"/G', is a dimensionless value that compares the amount of energy lost during a test cycle to the amount of energy stored during this time. Observations of polymer systems give the following numerical ranges for tanδ: very high for dilute solutions, 0.2 to 0.3 for amorphous polymers, and low (near 0.01) for glassy crystalline polymers and gels (Sharoba et al., 2005; Steffe, 1989;

James and Steffe 1996). The value of tanδ for the jam samples ranged between 0.30 and 0.32, 0.28 and 0.30, 0.27 and 0.32, and 0.29 and 0.33 for ST. Dalfour, Blueberry, Marmalade and Kewpie strawberry, respectively. These differences might be due to the different particles or three-dimensional networks. The difference in loss factor could be considered as an index to distinguish the fine structure. The quantitative relation among tanδ, G' and G" and the network structure needs further investigation. It was observed that when the temperature was increased, both G' and G" decreased. However, the shape of the cures is more or less similar, showing that all the jam samples exhibit similar viscoelastic properties.

As in all foods, the organoleptic tests are generally the final guide of the quality from the consumer's point of view (Sharoba et al., 2005; Jimenez et al., 1989; David and Stuart, 2002). Thus, it is beneficial to make a comparison between the four jam samples. There was no significant difference for the four jam samples with regards to consistency, colour, taste, odour and overall acceptability.

It was observed that the colour of Marmalade sample had the highest score, while the other three items had the lowest. Moreover, the highest score of taste and odour was observed for Kewpie strawberry.

Conclusion

This study determined the detailed rheological characteristics of jams. Thus, the obtained rheological parameters could be used for processing calculations and product development. All the four jam samples belonged to non-Newtonian fluid model, with thixotropy and pseudoplastic. It was observed that three jams (ST. Dalfour, Kewpie strawberry and Blueberry) were fit for Herschel–Bulkley model, in that yield stress existed in the model. The flow activation energies were calculated for Blueberry, ST. Dalfour and Kewpie strawberry as 13.79, 15.10 and 21.46 kJ/mol, respectively. The oscillatory test data revealed the time the frequency, storage modules (G′) and loss modules (G″) increased and the time the jams exhibited shear thinning behavior.

According to the time dependent behavior, Blueberry and Marmalade increased more gently than the Kewpie strawberry, so Kewpie strawberry had the most instability and ST. Dalfour had the most stability, which may be related to the flow activation energy. By the sensory properties, there was no significant difference for the four jam samples as regards consistency, colour, taste, odour and overall acceptability. However, while the lowest of the overall acceptability was Marmalade, the hysteresis loop of Marmalade was the smallest.

ACKNOWLEDGMENTS

This study was partly supported by grants from the Natural Science Foundation of China (No. 31071631) and 948 of Chinese Ministry of Agriculture (2011-Z26).

REFERENCES

Fugel R, Carle R, Schieber A (2005). Quality and authenticity control of fruit purees, fruit preparations and jams-a review, Trends Food Sci. Technol., 16: 433-441.

The Prevention of Food Adulteration Rules (1955). A.16.07.287.

Lal G, Siddappaa GS, Tandon GL (1998). Preservation of Fruit and Vegetables. ICAR Publication, New Delhi, India, pp. 51-58.

Peter F, Erich JW (2010). Rheology of food materials. Curr. Opin. Colloid Interface Sci., 7, 5-9.

Szczesniak AS (1963). Objective measurement of food texture. Food Sci., 28: 410-420.

Kokini JL, Cussler EL (1987). The psycho physics of fluid texture. In: Moskowitz,H.R. (Ed.), Food Texture. Marcel Dekker, New York, USA.

Kokini JL, Plutchok GJ (1987). Viscoelastic properties of semisolid foods and their biopolymers components. Food Technol., 41(3): 89-95.

Weixuan L, Shujuan J, Yansong M (1989). Food analysis, pp. 46-52.

Sharoba AM, Senge B, El-Mansy HA, EIM. Bahlol H, Blochwitz R (2005). Chemical, Sensory and rheological properties of some commercial German and Egyptian tomato ketchups. Eur. Food Res. Technol., 220: 142-151.

Santanu B, Shinhare US (2010). Rheological, textural, micro-structural and sensory properties of mango jam. J. Food Eng., 100: 357-365.

Karin H, Karin W, Anne-Marie H (2009). Sweetness and texture perception in mixed pectin gels with 30% sugar and a designed rheology. LWT - Food Sci. Tech., 42: 788-795.

Lawless HT, Heymann H (1998). Sensory Evaluation of Food: Principles and Practices. Chapman and Hall, New York, USA.

Sesmero R, Mitchell JR, Mercado JA, Quesad MA (2009). Rheological characterisation of juices obtained from transgenic pectate lyase-silenced strawberry fruit, 116: 426-432.

Tarrega A, Duran L, Costell E (2004). Flow behaviour of semi-solid dairy desserts. Effect of temperature. Int. Dairy J., 14: 345-353.

Talip K, Sevim K (2003). Effects of heat treatment and fat reduction on the rheological and functional properties of Gaziantep cheese, 13, 867-875.

Clark AH, Ross-Murphy SB (1987). Structural and mechanical properties of biopolymer gels, 83, 57-192.

Sima BA, Mohammad AMA, Azizollaah ZA, Hassan AGB, Mehrdad MA (2011). Compositional analysis and rheological characterization of gum tragacanth exudates from six species of Iranian Astragalus. Food Hydrocolloids, 25: 1775-1784.

Sundaram G, Mehmet MAK (2000). Dynamic oscillatory shear testingof foods—Selected applications. Trends Food Sci. Technol., 11: 115-127.

Steffe JF (1996). Rheological methods in food process engineering, 2nd edn. Freeman, East Lansing, MI.

Jimenez L, Ferrer L, Paniego ML (1989). J Food Eng., 9: 119-128.

David K, Stuart C (2002). Sensory perception of creaminess and its relationship with food structure. Food Qual. Prefer., 13: 609-623.

James F, Steffe PE (1996). Rheological methods in food process engineering,2nd edn., pp. 312-323.

18

Application of oven drying method on moisture content of ungrounded and grounded (long and short) rice for storage

M. A. Talpur[1], J Changying[1]*, F. A. Chandio[1,2], S. A. Junejo[3,4] and I. A. Mari[1]

[1]Department of Agricultural Mechanization, College of Engineering, Nanjing Agricultural University, Post Code 210031, Nanjing, Peoples Republic of China.
[2]Department of Farm Power and Machinery, Faculty of Agricultural Engineering, Sindh Agriculture University, Tandojam, Pakistan.
[3]Department of Hydrology, School of Earth Sciences and Engineering, Nanjing University, P. R. China.
[4]Department of Geography, Sindh University Jamshoro, Sindh Pakistan.

This study was conducted on ungrounded, grounded, long and short rice grains to determine the moisture content for storage. The rice samples were dried in an oven at 105°C; in this regard, every sample was divided in 6 parts with equal volume. The moisture contents were measured in six different ways, such as 1st part with 1 h interval, 2nd with 2 h, 3rd with 4 h, 4th with 6 h, 5th with 12 h and 6th part after 24 h. It is observed that an ungrounded grain sample with weight of 28.9 g showed in 1, 2, 4, 6, 12, and 24 h moisture release 5.81, 7.82, 9.10, 9.62, 10.48, and 11.11%, respectively. However, long grains weighing 44.86 g released moisture in 1, 2, 4, 6, 12, 24 h as 4.41, 6.87, 8.76, 9.59, 10.63 and 11.39%, respectively. While the short grains with weight of 45.68 g showed moisture release in the interval of 1, 2, 4, 6, 12 and 24 h as 3.96, 6.17, 7.99, 8.76, 9.68 and 10.29% respectively. It is evident from the study that ungrounded grains may be stored for long time as the moisture is easily released from them and they may retain the quality as compared to grounded short grains.

Key words: Ungrounded, grounded, moisture content.

INTRODUCTION

The moisture content plays an important role in the storage of rice for long period in terms of maintaining of its quality. There are several practical air-oven procedures that have been standardized to determine moisture content of grains (Hart et al., 1959; United states Department of Agriculture (USDA), 1971; Association of Official Analytical Chemists (AOAC), 1980; American Society of Agricultural Engineers, 1982; Jindal and Siebenmorgen, 1987; De Datta Surajit (1981) These methods are based on drying whole or ground grains in an oven over a fixed period of time. The cleaning, drying, and storage of grains are postharvest operations required to maintain their product quality (Bakker-Arkema et al., 1999). The drying temperature and time are usually

specified for a particular type of grain on the basis of moisture content comparison with the reference method.

Moisture content determinations made with different oven methods and different grains may not be the same due to the empirical nature of the methods. Oven exposure time depends upon the type of grain and the method used (Hart and Neustadt, 1957; Warner and Browne, 1963; Young et al., 1982; Bowden, 1984).

Mechanical systems, especially those using hot air for rapid drying of high moisture grain are becoming increasingly popular throughout the region (Soponronnarit et al., 1996; Wiset et al., 2001; Huang-Nguyen et al., 1999; Nguyen et al., 1999). Fluidised and spouted bed dryers are examples of high temperature dryers. Due to the high air temperatures used, residence time of grain in the dryer must be short to prevent heat damage.

*Corresponding author. E-mail: chyji@njau.edu.cn.

Table 1. Shows the trend/regression trends.

S/N	Type of grain	Formulae (Logarithmic)	R^2 value
1	Ungrounded Grains	y = 1.6185Ln(x) + 6.4189	0.9496
2	Grounded Long Grains	y = 2.1714Ln(x) + 5.1579	0.9496
3	Grounded Short Grains	y = 1.9809Ln(x) + 4.6602	0.9429

Experiment was expressed by the aforestated equations.

Though air-oven procedures have been standardized for moisture determination of several common whole grains, there exists no such standard for rough rice. Noomhorm and Verma (1982) have compared rough rice moisture content determinations using the I30DC-16 h whole-grain method based on the work of Matthews (1962). They used the AOAC (1980) method as a standard, incorporating two-stage drying over the moisture content range of approximately 10 to 19% w.b. They concluded that the whole-grain oven method gave significantly higher moisture contents compared to the AOAC method. Thus, there is a need to develop a standard oven procedure for whole-grain rough rice moisture content determination that would be accurate, rapid and easy to use.

Grains are among the major commodities for feeding mankind. The cleaning, drying and storage of grains are post harvest operations required to maintain their product quality. Grain drying is a process of simultaneous heat and moisture transfer. When air is moved through grain two things happen, first of all, the grain will cool/warm until at equilibrium with the air being blown through. Secondly, the grain will dry/moisten until at equilibrium with the air being blown through.

The medium of drying is air. The major physical properties of air that affect the drying rate of grains are the relative humidity or humidity ratio, the dry bulb temperature, the specific volume, and the enthalpy. The current study is based on the assessment of moisture content in ungrounded and grounded (long and short) rice, to determine the effects of oven drying temperature with different times on its moisture content.

MATERIALS AND METHODS

This study was carried as a laboratory exercise to explore the overall effects of oven drying temperature in different timings, to determining the moisture content of ungrounded, long and short rough rice grain. A preliminary study on moisture content of three samples was taken in account by using oven drying box model DHG-101. The temperature of drying box was kept constant at 105°C. The fresh harvested rice (T-259) was purchased in 2009 for experiments. The packet of the ungrounded rice sample and were grounded with machine and long grains were obtained. After that, an ungrounded long sample was cut into two pieces to get the short grain. The three samples with equal volume of each sample were kept in the sampler with the weight of, Ungrounded Grain 28.9 g, Long Grain 44.8 g, and Short Grain 45.6 g, six different ways and time intervals (Table 1).

Measurement of moisture drying

The weight of empty samplers (pots) was measured by electronic weight balance model G&G. The same volume of grains of ungrounded, long and short rice samples were kept separately in each sampler (pots) and were measured again; it was placed in the drying box on fixed temperature on the timings that is, 1, 2, 4, 6, and 24 h; after the given time interval its weight was again measured. The results of the study revealed a trend that indicated a possible relationship between the moisture contents attained with the various drying temperature/time combinations.

RESULTS AND DISCUSSION

The experiment was conducted to analyze the moisture content variations in three different samples in order to see the ability for long time storage. The perceived data showed that Ungrounded Grain sample (weight 28.9 g) showed 5.81% moisture release in 1 h, 7.82% moisture release in 2 h, 9.10% moisture release in 4 h, 9.62% moisture release in 6 h, 10.48% moisture release in 12 h, and 11.11% moisture release in 24 h (Figure 1) However long grains (weight 44.86 g) showed 4.41% moisture release in 1 h, 6.87% moisture release in 2 h, 8.76% moisture release in 4 h, 9.59% moisture release in 6 h, 10.63% moisture release in 12 h, and 11.39% moisture release in 24 h (Figure 1). While the short grains (weight 45.68 g) showed 3.96%, moisture release in 1 h, 6.17% moisture release in 2 h, 7.99% moisture release in 4 h, 8.76% moisture release in 6 h, 9.68% moisture release in 12 h and 10.29% in 24 h (Figure 1). It is observed from the data that ungrounded grains released more moisture as compared to grounded grains because of voids. By measuring the equal volume of ungrounded and grounded grains, it was found that 40.1% more space is required to store ungrounded grains as compared to grounded long grains and 41.16% more than short grains of same volume. However, it is found that the trend of moisture releases in all samples is nearly the same (Table 1).

Conclusion

It is concluded from the perceived data that the moisture release of all grain samples decreased with deferent time intervals. However, moisture release in long grains is higher than smaller grains. Trend of moisture releases in

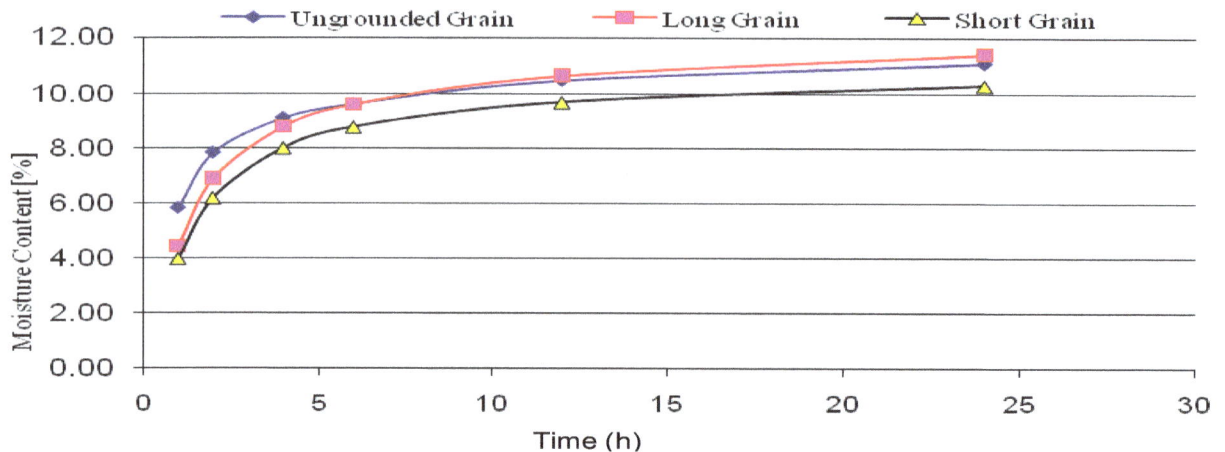

Figure 1. Shows moisture content trend in different rice grain samples.

all three types is almost the same. It is recommended that the same kind of study be conducted on full spikes.

REFERENCES

Association of Official Analytical Chemists (1980). Official methods of analysis. 13th Edition, AOAC, Washington, D.C.

American Society of Agricultural Engineers (1982). Standard: ASAE S352.1. Moisture measurement-Grains and seeds.

Jindal VK and Siebenmorgen TJ (1987). Effects of Oven Drying Temperature and Drying Time on Rough Rice Moisture Content Determination. Am. Soc. Agric. Eng., 30(4): 1185-1192.

American Society of Agricultural Engineer (1999). CIGR Handbook of Agricultural Engineering-Agro Processing Engineering volume-IV.

Bakker-Arkema FW, DeBaerdemaeker J, Amirante P, Ruiz-Altisent M Studman CJ (1999). CIGR Handbook of Agricultural Engineering, Volume 4, Agro-Processing Engineering. ASAE, St. Joseph, MI.

Bowden PJ (1984). Comparison of three routine oven methods for grain moisture content determination. J. Stored Prod. Res., 20(2):97-106.

De Datta S (1981). Principles and Practices of Rice Production, A Wiley-Inter science publication.

Hart JR, Feinstein L, Golumbic C (1959). Oven methods for precise measurement of moisture content of seeds. Marketing Research Report No. 304 (USDA-AMS), US Government Printing Otlice, Washington, D.C.

Hart JR, Neustadt MH (1957). Application of the Karl Fischer method of grain moisture determination. Cereal Chem., 34: 26-37.

HungWiset L, Srzednicki G, Driscoll R, Nimmuntavin C, Siwapornrak P (2001). "Effects of High Temperature Drying on Rice Quality". Agricultural Engineering International: the CIGR J. Scientific Res. Develop. Manuscript FP 01 003. May, 3: 2.

Matthews J (1962). The accuracy of measurement of known changes in moisture content of cereals by typical oven methods. J. Agric. Engr. Res., 7(3): 185-191.

Noomhorm A, Verma LR (1982). A comparison of microwave, air oven and moisture meters with the standard method for rough rice moisture determination. Trans. ASAE, 25(5): 1464-1470.

Soponronnarit S, Prachayawarakorn S, Wangji M (1996). Commercial Fluidised Bed Paddy Dryer. In: Strumillo, C. and Pakowski, Z. (Eds.), Proc. The 10th International Drying Symposium, Krakow, Poland, 30 July-2 August, A: 638-64.

Warner MGR, Browne DA (1963). Investigations into oven methods of moisture content measurement of grain. J. Agric. Engr. Res., 8(4): 289-305.

Wiset L, Srzednicki G, Driscoll R, Nimmuntavin C, Siwapornrak P (2001). "Effects of High Temperature Drying on Rice Quality". Agricultural Engineering International: the CIGR J. Scientific Res. Develop. Manuscript FP 01 003. May Vol. III.

Young JH, Whitaker TB, Blankenship PD, Brusewitz GH, Troeger JM, Steele JL, Person NZ Jr. (1982). Effect of oven drying time on peanut moisture determination. Trans. ASAE, 25(2): 491-496.

19

The use of two indigenous medicinal plant leaf powders (*Cymbopogon citratus* and *Ocimum suave*) applied as mixed and individual powders to evaluate the reproductive fitness of F_1 generation (eggs laid and F_2 adult) of *Callosobruchus maculatus* (cowpea bruchid)

Ojianwuna, C. C.[1]* and Umoru, P. A.[2]

[1]Department of Animal and Environmental Biology, Delta State University, Abraka, Delta State, Nigeria.
[2]Department of Zoology, Ambrose Alli University, Ekpoma, Edo State, Nigeria.

It has been estimated that over 90% cowpea is lost to cowpea bruchids especially *Callosobruchus maculatus*. A mixture of plant containing *Cymbopogon citratus* (Lemon grass) (L) and *Ocimum suave* (Wild basil) (W) powders in the ratios (Lemon grass: Wild basil): 100:0, 80:20, 60:40, 50:50, 40:60, 20:80, 0:100 and 0:0 were used under ambient laboratory conditions with the aim of evaluating the effects of these plant powders on the reproductive fitness of F_1 generation of *C. maculatus* (F.) (Cowpea bruchids). The mixed powders were each applied at 1, 2, and 3 g concentration to 20 g of cowpea seeds. Number of eggs laid by F_1 and subsequent F_2 adult that emerged from the untreated combination were compared with the untreated control (0L: 0W). The plant mixture 60L: 40W had the least mean number of egg counts and significantly ($P < 0.05$) suppressed F_2 adult emergence. This plant material mixture had the most knock on effect from the original parent and this shows that it can serve as grain protectants against *C. maculatus*.

Key words: *Cymbopogon citratus*, *Ocimum suave*, *Callosobruchus maculatus*, F_1, F_2 generation adults.

INTRODUCTION

It is estimated that between 60 and 80% of all grains produced in the tropics is stored at farm level. The main purpose of storing grains is to ensure house hold food supplies to cover future cash needs though sales by taking advantage of seasonal price increase. The largest quantity of food in the tropics is stored in traditional farmer's granaries and in most cases under one roof (Stathers et al., 2002). Developing countries are hit by great losses during storage of cereals and durable commodities such as pulses and oil seeds by storage insect pests. *Callosobruchus* species are major pest of stored grains and grain products in the tropics (Ofuya, 2003). Over 90% of the insect damage to cowpea seeds is caused by *Callosobruchus maculatus* (F) (Caswell and Akibu 1981). The insect lays its eggs on the seeds of cowpea, which hatch and produce larvae that bore hole into the seed cotyledons on which they feed (Onuh and Onyenekwe, 2008). Infestation may reach 100% within 3 to 5 months of storage. Control of *C. maculatus* relies heavily on the use of synthetic insecticides. Owing to the problems of synthetic organic chemicals, there is renewed interest on plants as alternative materials for use as stored grain protectants because they have been found to have broad spectrum insecticidal properties with reduced persistence compared to the organochlorines and organophosphates, carbamates and pyrethroids.

*Corresponding author. E-mail: iloh4u@yahoo.com.

They are easily available and can be produced within the farmers' vicinity, thus providing a more sustainable approach to pest control (FAO, 1985). They are also cheap to purchase and have no negative impact on the environment because they are easily biodegradable. Therefore, many scientists have been conducting researches over the last three decades aimed at identifying botanicals that would replace synthetic organic chemicals but the efficacies of the plant mixtures have been less investigated (Emeasor et al., 2007). This work is to evaluate the effects of *Cymbopogon citratus* (Lemon grass) and *Ocimum suave* (Wild basil) applied as mixed and individual powders to access the reproductive fitness of F_1 adults of *C. maculatus* (eggs laid and F_2 generation).

MATERIALS AND METHODS

Insect stock culture

Adult bruchids were obtained from already infested cowpea and identified as *C. maculatus* by the assistance of the Nigerian Stored Product Research Institute (NSPRI) Sapele, Delta State Nigeria. The *C. maculatus* adults obtained were introduced into undamaged cowpea (*Vigna unguiculata*) L seeds of the Kano white variety 1696 and maintained in large specimen bottles with fine mesh gauze covering the opened end, for them to mate and oviposit under laboratory conditions. Adult emergence was checked daily and the newly emerged adults were then used for the experiment.

Experimental cowpea

Undamaged and clean cowpea seeds that were used were purchased from Abraka market, Delta state. Each seed was examined under microscope to make sure there were no damages, eggs laid and exit holes on them. They were then kept by deep freezing for one week and left for 24 h under ambient conditions (Ofuya et al., 2007) with slight modifications (deep freezed for 2 weeks).

Preparation of insecticidal plant powder

Two researched plants identified as *C. citratus* and *O. suave* by Botany Department of Delta state University were used for the experiments. The plants were obtained from Issele-Azagba (Aniocha North) Local Government Area of Delta State. Fresh leaves from each plant were slowly dried for 3 weeks in an open wooden cabinet (1.0 × 0.5 × 1.0 m) under room temperature before pulverization in a motorized high speed grinder. The powder was passed through a sieve of 0.1 mm mesh size. The particles were then put in an air tight container to prevent active components from evaporating. This method was adopted by Denloye et al. (2007), with slight modification and constructed the cabinet with a 100 watt bulb.

Formulation of insecticidal plant powders into treatment combinations and efficacy test

The powders obtained from 3.3 were mixed in the following ratios (Lemon grass: Wild basil): 100:0, 80:20, 60:40, 50:50, 40:60, 20:80 and 0:100. Each combination was replicated two times. The untreated combination (0:0) served as the control for the experiment. Each combination was admixed with 20 g of experimental cowpea seed (3.2) at 1, 2 and 3 g concentrations or rates of application. Each formulated treatment combination were admixed with 20 g of cowpea seed of various sizes at different concentration of 1, 2 and 3 g and put into Petri dishes with lid. 3 pairs of adult *C. maculatus* (3 males and 3 females) were introduced into each Petri dishes and kept under laboratory conditions for 12 days. The adults were allowed to lay eggs for 5 days after which they were removed and eggs were counted between 5 and 8 days after infestation (DAI) with the aid of a magnifying lens. Adult emerged 21 DAI and counted for three consecutive weeks. Each adult that emerged was removed and put into a Petri dish. Three (3) pairs of F_1 adult were put into 20 g of cowpea seed. The number of eggs laid was counted with a binocular microscope and F_2 generation adults were counted after 21 days for 3 consecutive weeks.

Statistical analysis

All data collected were subjected to Analysis of Variance (ANOVA) and Multiple Comparism using Duncan's Multiple Range Test and LSD.

RESULTS

Oviposition (eggs laid) and F_1 adult emergence of *C. maculatus* in treated cowpea grain

Oviposition (number of eggs laid)

The mean number of eggs laid (oviposition) by the adult *C. maculatus* in cowpea seeds treated with different mixed proportions of *C. citratus* (Lemon grass, L) and *O. suave* (Wild basil, W) and the untreated seeds are shown in Table 1. There was significant difference (P <0.05) between the concentrations of the plant materials in the number of eggs laid. However, the Duncans' multiple range test could not differentiate where the difference lies. There was significant difference (P <0.05) between the treatment concentrations of the plant materials and the untreated in the number of eggs laid (Ojianwuna and Umoru, 2010).

Number of adult emergence (F_1 generation)

The mean number of adult *C. maculatus* that emerged from cowpea seeds treated with different concentrations of mixed (%) combinations of *C. citratus* (Lemon grass, L) and *O. suave* (Wild basil, W) and the untreated seeds are shown in Table 2. There was significant different (P<0.05) between the concentrations of the plant materials in the number of F_1 adults that emerged. However, the Duncans' multiple range test could not differentiate where the significant lies. There was significant difference (P <0.05) between treatment combinations of the plant materials in the number of F1 adult that emerged (Ojianwuna and Umoru, 2010).

Table 1. Mean number of eggs laid by the adult parents *Callosobruchus maculatus* on cowpea seeds treated with different mixed proportions of *Cymbopogon citratus* (lemon grass, L) and *Ocimum suave* (Wild basil, W) plant powders and the untreated (0:0) seeds.

Mixed proportions (%) of plant materials (L:W)	Concentration of plant material (grams)			$\overline{X}$
	1	2	3	
100:0	62.0	55.0	44.0	53.7[b]
80:20	58.5	41.0	35.5	45.0[bc]
60:40	41.0	37.0	21.5	33.2[c]
50:50	45.5	42.0	27.5	38.3[bc]
40:60	56.5	44.0	38.5	46.3[bc]
20:80	44.5	39.5	33.5	39.2[bc]
0:100	58.0	51.5	39.0	49.5[b]
0:0 (Control)	95.5	97.5	92.0	95.0[a]
$\overline{X}$	57.7[x]	50.9[x]	41.4[x]	

Means with the same superscript letters do not differ significantly (P <0.05) using Duncans Multiple Range Test (Ojianwuna and Umoru, 2010).

Table 2. The mean number of adults (F_1 generation) *Callosobruchus maculatus* adult that emerged from cowpea seeds treated with different mixed proportions of *Cymbopogon citratus* (Lemon grass, L) and *Ocimum suave* (Wild basil, W) plant powder on the untreated (0:0).

Mixed proportions (%) of plant materials (L:W)	Concentration of plant material (grams)			$\overline{X}$
	1	2	3	
100:0	21.5	16.0	11.0	16.2[b]
80:20	24.0	19.0	13.0	18.7[b]
60:40	16.5	11.0	9.0	12.2[b]
50:50	22.5	11.0	12.0	15.2[b]
40:60	29.0	14.5	11.0	18.2[b]
20:80	25.0	18.5	11.0	18.2[b]
0:100	24.5	16.5	12.5	17.8[b]
0:0 (Control)	56.0	57.0	55.0	56.0[a]
$\overline{X}$	27.4[x]	20.4[x]	16.8[x]	

Means with the same superscript letters do not differ significantly (P <0.05) using Duncans multiple range test (Ojianwuna and Umoru, 2010).

Reproductive fitness

Oviposition (eggs laid) and F_2 adult emergence of C. maculatus in untreated cowpea grains

Oviposition (number of eggs laid): The mean number of eggs laid by the adult *C. maculatus* of the first filial generation (F_1) on the treated and untreated cowpea seeds significantly (P <0.05) reduced as the concentration of plant material increased from 1 to 3 g per 20 g cowpea seeds (Table 3). However, the Duncans' multiple range test could not differentiate where the significance lies. There was also significant difference (P <0.05) between the treatment combination of the plant materials and the untreated in the number of eggs laid.

Number of adult emergence (F_2 generation): The

mean number of adult *C. maculatus* (F_2) that emerged from the treated and untreated cowpea seeds significantly reduced (P <0.05) as the concentration of plant materials *C. citratus* (Lemon grass) and *O. suave* (Wild basil) increased from 1 to 3 g per 20 g cowpea seeds. However, the Duncans' multiple range test could not differentiate where the significance lies. There was also significant difference (P <0.05) between the treatment combinations of plant material and the untreated in the adult F_2 (Second Filial Generation) *C. maculatus* emergence (Table 4).

DISCUSSION

The plant materials were toxic to the insects as the concentration increased from 1 to 3 g/20 g cowpea

Table 3. Mean number of eggs laid by F_1 adult *Callosobruchus maculatus* that emerged from cowpea seeds treated with different concentrations and mixed combinations of *Cymbopogon citratus* and *Ocimum suave* compared with the untreated cowpea seeds.

Mixed proportions (%) of plant materials (L:W)	Concentration of plant material (g)			$\overline{\text{X}}$
	1	2	3	
100:0	31.5	30.5	24.0	28.7[b]
80:20	27.5	22.0	20.0	23.2[bc]
60:40	21.5	19.5	14.5	18.5[c]
50:50	22.5	20.5	15.5	19.5[bc]
40:60	31.0	27.5	21.5	26.7[bc]
20:80	30.0	26.0	19.0	25.0[bc]
0:100	29.5	27.5	19.0	103.3[a]
0:0 (Control)	106.5	111.0	92.5	95.0[a]
$\overline{\text{X}}$	37.5[x]	35.6[x]	28.3[x]	

Means with the same superscript letters do not differ significantly (P <0.05) using Duncans multiple range test.

Table 4. The mean number of adults (F_2 generation) *Callosobruchus maculatus* that emerged from the treated and untreated cowpea grains.

Mixed proportions (%) of plant materials (L:W)	Concentration of plant material (g)			$\overline{\text{X}}$
	1	2	3	
100:0	13.5	9.5	6.0	9.7[b]
80:20	11.5	10.0	5.5	9.0[b]
60:40	9.0	7.0	3.5	6.5[b]
50:50	10.5	8.0	4.0	7.5[b]
40:60	13.0	9.5	5.5	9.3[b]
20:80	13.5	10.5	5.5	9.8[b]
0:100	13.5	8.5	5.0	9.0[b]
0:0 (Control)	57.0	56.0	50.0	54.3[a]
$\overline{\text{X}}$	17.7[x]	14.9[x]	10.6[x]	

Means with the same superscript letters do not differ significantly (P <0.05) using Duncans multiple range test.

seeds. The toxicity of the mixtures of plant materials in their different proportions could be as a result of the volatile compounds in the plant materials which acted as fumigants with insecticidal effects on the bruchids. In evaluating the reproductive fitness of *C. maculatus*, it was observed in this study that the number of eggs laid by the F_1 adults that emerged from treated seeds was significantly (P <0.05) reduced when compared to the eggs laid by the adults F_1 parents from the untreated control. It may be that the toxic material in the plant mixtures had a knock-on effect from the original parents of the F_1 generation adults. This could have caused reduction in the number of eggs laid in a number of ways; it may be that the males were sterile, or physiological changes that occurred in the female made it to lay a few non-viable eggs. Dike and Mbah (1992) suggested that *C. citratus* may be ovicidal or larvicidal in their action. Notably in the 60L: 40W and 50L: 50W, the toxicity of the

plant material was most retained in the F_1 progeny as seen from the number of eggs laid.

If the insecticidal materials in the plant mixture are systemic, they could move from the outside of the seeds into the inside to affect the feeding and developing F_1 larvae that would subsequently become F_1 adults. In this way, the mixture of plant materials in various proportions, especially in the 60L: 40W and 50L: 50W affected the number of F_2 adults that emerged. The number of F_2 adults that emerged could have further been reduced if the F_1 parents laid non-viable eggs compared to the F_1 adults from the untreated cowpea seeds. In an earlier study by Ojianwuna and Umoru (2010), it was reported that the number of eggs laid by adult *C. maculatus* in the proportions of plant mixtures significantly reduced as the concentration increased from 1 to 3 g / 20 g cowpea seeds with 60L: 40W having the least number of eggs laid.

In conclusion, this study has shown that mixed proportions of *C. citratus* (lemon grass) and *O. suave* (Wild basil) especially at 60L: 40W could be most effective in the control of bruchids. However, there is need for more investigation to identify the use of other local plant material mixtures to access their effects on the bruchids- *C. maculatus.*

REFERENCES

Caswell GH, Akibu S (1981). The use of pirimipho-methyl to control bruchids attacking selected varieties of stored cowpea. Trop. Grain Legume Bull., 17/18:9-11.

Denloye AA, Makanjuola WA, Don-Pedro KN, Negbenebor HE (2007). Insecticidal effects of *Tephrosia vogelii* Hook (Leguminosae) leaf powder and extracts on sitophilus zeamais motsch, *Callosobruchus maculatus F.* and *Tribolium castaneun* Herbst. Niger. J. Entomol., 24: 91-97.

Dike MC, Mbah OT (1992). Evaluation of the lemon grass *{Cymbopogon citratus)* products on control of *Callosobruchus maculatus F* (Coleoptera: Bruchidae) on stored cowpea. Niger. J. Plant Prot., 14: 81-91.

Emeasor KC, Emosairue SO, Ogbuji RO (2007). Comparative susceptibility of some cowpea (*Vigna unguiculata* (L) Walp) Accessions, Breeding lines and Winged Bean to the Cowpea Bruchid, *Caollosobruchus maculates* (F) (Coleoptera: Bruchidae). Niger. J. Entomol., 24: 131-136.

FAO (1985). Handling and storage of food grains. Food and Agriculture Organization of the United Nations, Rome. http://www.blackherbals.com/using natural pesticides. htm. Assessed 07/03/2008Ofuya.

Ofuya TI (2003). Beans, insects and man. Inaugural lecture series 35, The Federal University of Technology Akure Nigeria, 45 pp.

Ofuya TI., Olotuah OF, Ala desanwa RD (2007). Potential of Dusts of *Eugenia aromatica* Baill. Dry Flower Buds and *Piper guineense* Schum and Thonn Dry fruits formulated with three organic flours for controlling *Callosobruchus maculatus* Fabrieus (Coleoptera: Bruchidae). Nig. J. Bnt., 24: 98-106.

Ojianwuna CC, Umoru PA (2010). Effects of *Cymbopogon citratus* (Lemon grass) and *Ocimum suave* (Wild basil) applied as mixed and individual powders on the eggs laid and emergence of adult *Callosobruchus maculatus* (Cowpea bruchids). Afr. J. Agric. Res., 5(20): 2837-2840.

Onuh MO, Onyenekwe RI (2008). Accessment of Inhibitory substances in the seed coat of some cowpea cultivars for resistance against *Callosobruchus maculatus.* Sci. World J., 3: 2.

Stathers TE, Chigariro J, Mudiwa M, Mvumi BM, Golob P (2002). Small-scale farmer perceptions of diatomaceous earth products as potential stored grain protectants in Zimbabwe. Crop Prot., 21: 1049-1060.

20

Physical properties of honey products in Algeria

Bendeddouche Badis[1]* and Dahmani Kheira[2]

[1]Ecole Nationale Supérieure Vétérinaire BP 161 El Harrach, Algiers, Algeria.
[2]Inspection vétérinaire de la wilaya d'Alger. BHCA El Harrach, rue Benyoucef Khettab. El Harrach, Algiers, Algeria.

Honey is considered as the sweetened natural substance produced by mellifluous bees from the flowers nectar, plants parts or the insects excretions generally left on the same parts. Although Algeria possesses floral and climatic potentialities; it remains a small honey producer. Among analyzed samples,74% correspond to codex standards .This latter may increase if certain measures such as a national regulation including standards for Algerian honey, a referential beekeeping and hive products laboratory are implemented. A labeling system followed by a guide for a good practice of beekeeping and production should be implemented as well.

Key words: Honey, physicochemical analysis, quality, Algeria.

INTRODUCTION

In the Algerian collective consciousness, it is evident that honey produced in our country follows the common rule: The honey quality justifies the price. Indeed, this noble and food scarce, remains expensive and reserved for patients and special occasions. Yet, in absence of specific national rules in this matter "quality" remains theoretical and poorly studied.

Algerians beekeepers who have always sought to preserve and protect the natural qualities of honey hope to find other markets than the local ones to sell their surplus production. However, to achieve this aim, they must first comply with the conventional guidelines and regulations. Great quantities of honey of doubtful quality proposed on the market means that the consumer is often misled. The absence of national regulation, insufficient quality control laboratory and the lack of a professional organization make the honey industry unstructured.

The purpose of the present study is to evaluate the quality of Algerian honey and verify its compliance with the standards of codex. Eight parameters were selected: water content, rate of hydroxy-methyl furfural (HMF), free acidity, pH, electrical conductivity, color, falsification search and detecting residues of antibiotics.

MATERIALS AND METHODS

Sampling

50 honey samples were collected randomly from beekeepers of trade fairs in the central region of Algeria between 2009 and 2010. Each sample weighed 250 g and clearly identified by date of harvest, and floral (Figure 1) and geographical origin (Figure 2).

Technical analysis

In our study the physical and chemical parameters were studied using the methods harmonized by European Commission Directive /110/CE (2001): water content (index of refraction), electrical conductivity (conductometer), pH and free acidity, the HMF (White method), color (Pfund - Lovibond), falsification of research (Fiehe reaction) and the search for antibiotics: chloramphenicol, streptomycin and oxytetracycline by HPLC (Shimadzu LC/10AVP type).

RESULTS

Out of the 50 samples analyzed, 48 were compliant to the codex standard (96%) with an average of 17.38. It is worth mentioning that 36% of honey collected has moisture content between 17 and 17.5% (Figure 3) which make these honeys safe from fermentation (Prost, 2005). Table 1 presents the results of physicochemical analysis obtained by comparison with the limits of the codex (Codex standard 12-19811).

*Corresponding author. E-mail: bendeddouchebadis@hotmail.com.

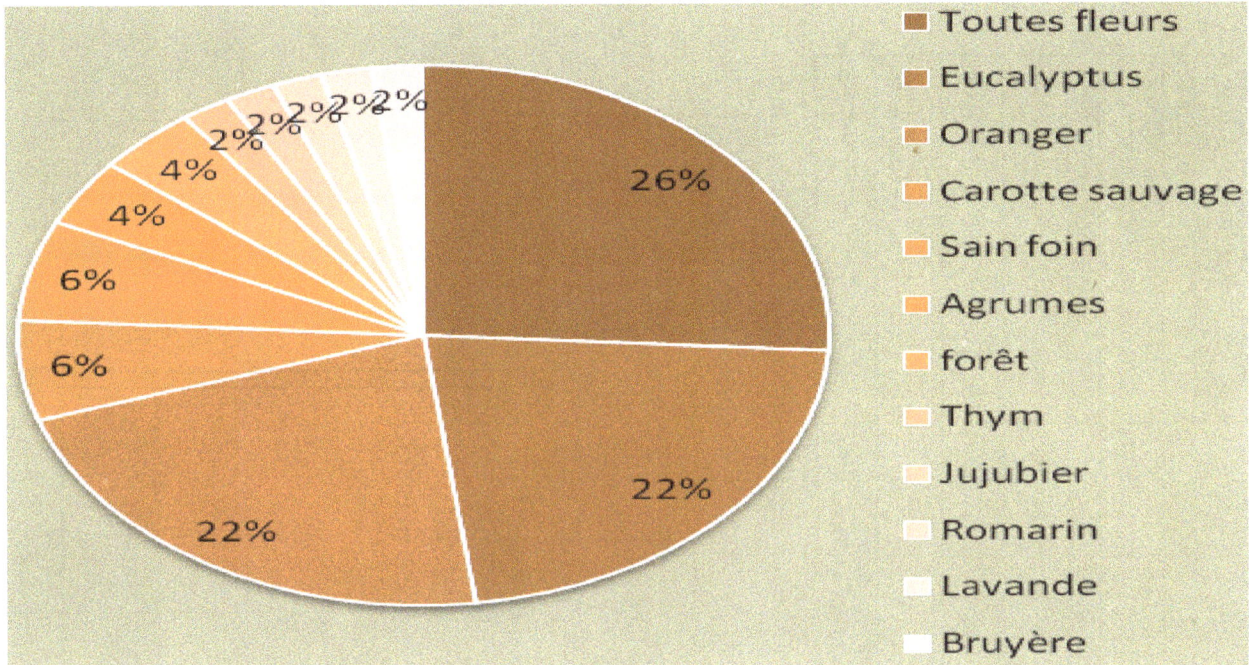

Figure 1. Distribution of samples of honey by floral source.

Figure 2. Regional origin of 50 samples of honey.

One sample n° 42 Heather floral origin of the Wilaya of Tizi Ouzou (or 02%) has an electrical conductivity of 0.82 mS/cm. The origin of this honey is presumed from heather, therefore normally characterized by high electrical conductivity and therefore considered by the codex standards. Since other samples have electrical conductivity less than 0.8 mS/cm, we concluded that all our samples are nectar honeys (Figure4) (Makhloufi, 2007).

43 samples (86%) correspond to the codex standard (≤ 50 meq/kg), 14% samples have a high acidity probably a sign of fermentation. These honeys might have lost some

Figure 3. Distribution of samples according to values of water content.

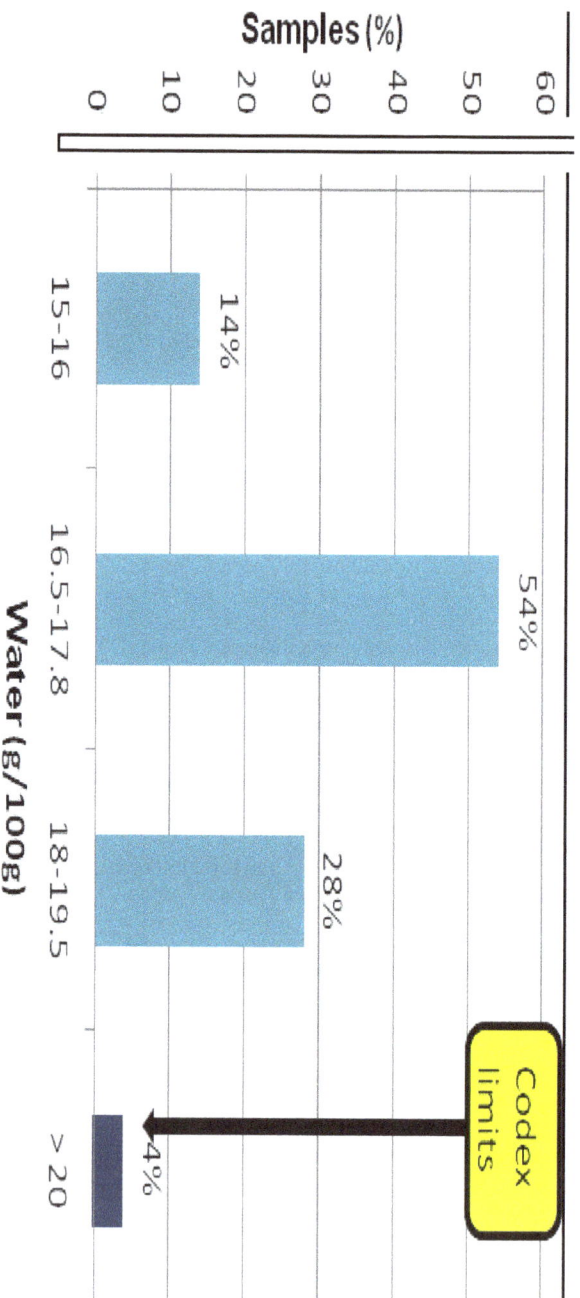

Table 1. Results of physicochemical analysis of 50 samples.

Parameter	Average value ± standard deviation (écart-type)	Minimum and maximum value	Limits of international standards (codex)	Number of samples exceeding international standards
Water content (%)	17.38 ± 1.26	15 – 21	< 20 %	2 samples
Electrical conductivity (mS/cm)	3.63 ±1.83	1.1 – 8.21	Honey nectar ≤0.8 mS/cm Honey dew ≥0.8 mS/cm	0
pH	3.74 ±0.38	2.97 – 5.35	No limits	
Free acidity (meq/kg)	37.41±15.00	13 – 66	50 meq/kg	7 samples
HMF (mg/kg)	37.80 ±111.69	0 – 598.8	<40 mg/kg	6 samples
Coloring (mm Pfund)	62.7 ±27	11– 119	No limits	
Search falsification	Positive reaction : 36.66 %			Negative reaction : 63.34%
Search antibiotiques residues	No residues of oxytetracycline No residues of chloramphenicol No residues of streptomycin			

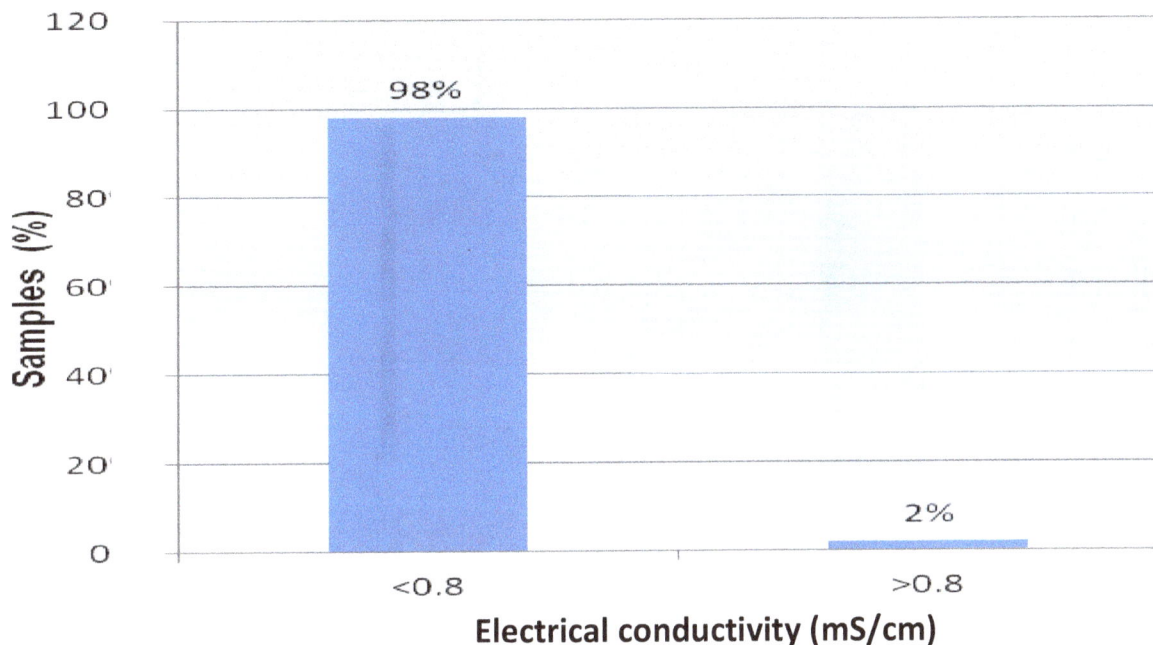

Figure 4. Distribution of honey samples according to values electrical conductivity.

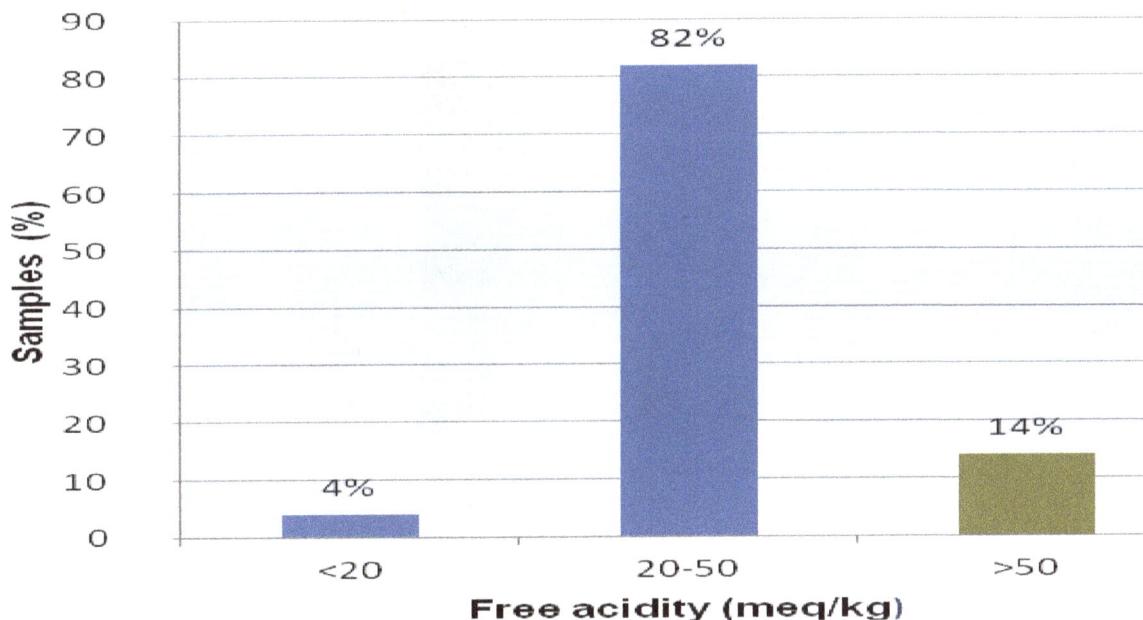

Figure 5. Distribution of samples according to the free acidity.

of their quality. All honeys are acidic, but those with high acidity will deteriorate even more rapidly with particularly adverse consequences on taste and shelf life (Schweitzer, 1998). Two samples (04%) had a rate of acidity below 20 meq/kg. These honeys are coded under numbers 15 (citrus floral origin of the wilaya of Blida) and No. 44 (original orange flower of the wilaya of

Blida) both collected between April and May 2010. They do not appear to have undergone fermentation (Figure 5).

The HMF is a quality criterion (Marceau, 1994). The spectrophotometric analysis of our samples revealed that 44 of them (88%) had levels below the limits HMF codex, 8% have a rate of HMF between 40 and 80 mg/kg and 02

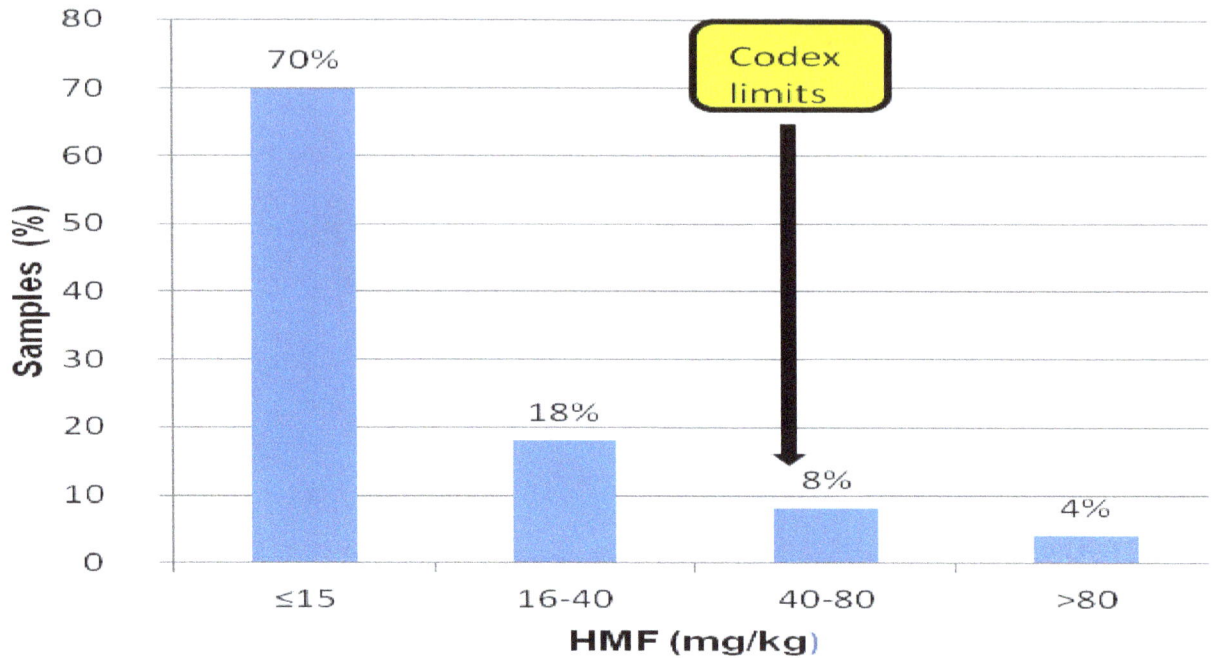

Figure 6. Distribution of samples according to the values of HMF.

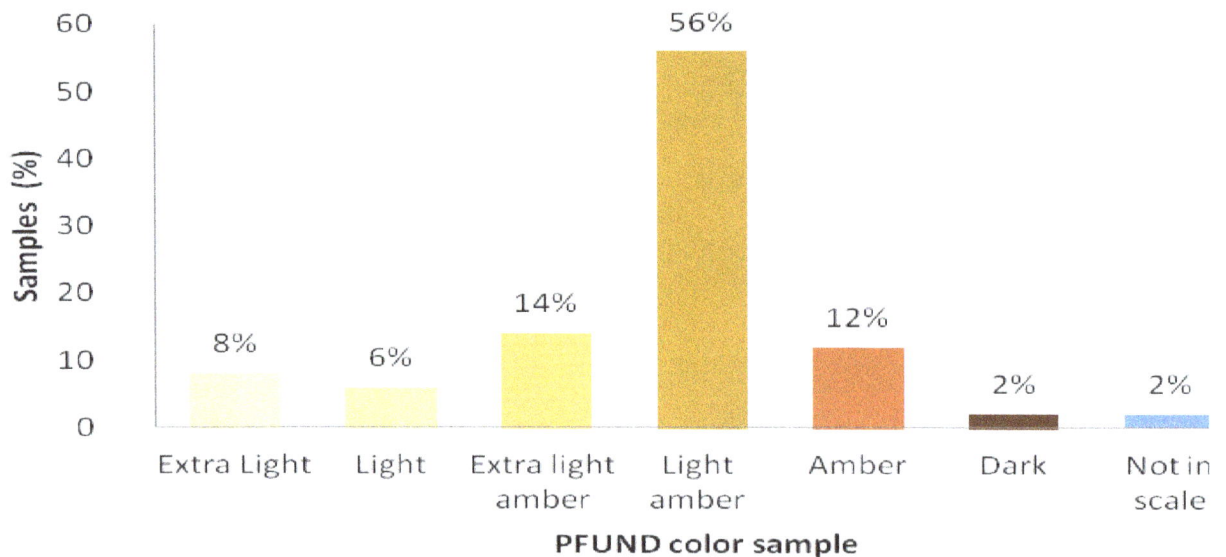

Figure 7. Distribution of samples by staining.

samples are considered of unacceptable quality with very high values of HMF (Figure 6). These values relate to samples n° 3 eucalyptus floral origin of the Wilaya of Tizi Ouzou and n°21 flower honey origin of the Wilaya of Bouira and can be explained by inadequate treatment of these honeys (Hadorn et al., 2003) probably overheating (White, 1964).The average color corresponding to our honey samples is 63 mm Pfund (amber light) with a standard deviation of 27 mm Pfund, or values between

36 and 90 mm Pfund (extra light amber to amber), which on the scale Pfund covers the full range of honey colors (Gonnet, 1982). For the sample No. 3 wide color could not be determined while the sample 42 is 119 (dark) this is probably the result of unsatisfactory conditions of processing or storage (Schweitzer, 2001) (Figure 7). We have had eleven positive reactions from 30 samples submitted to the reaction FIEHE (36.66%) indicating a probable forgery by adding synthetic sugar. This number

FIEHE reaction

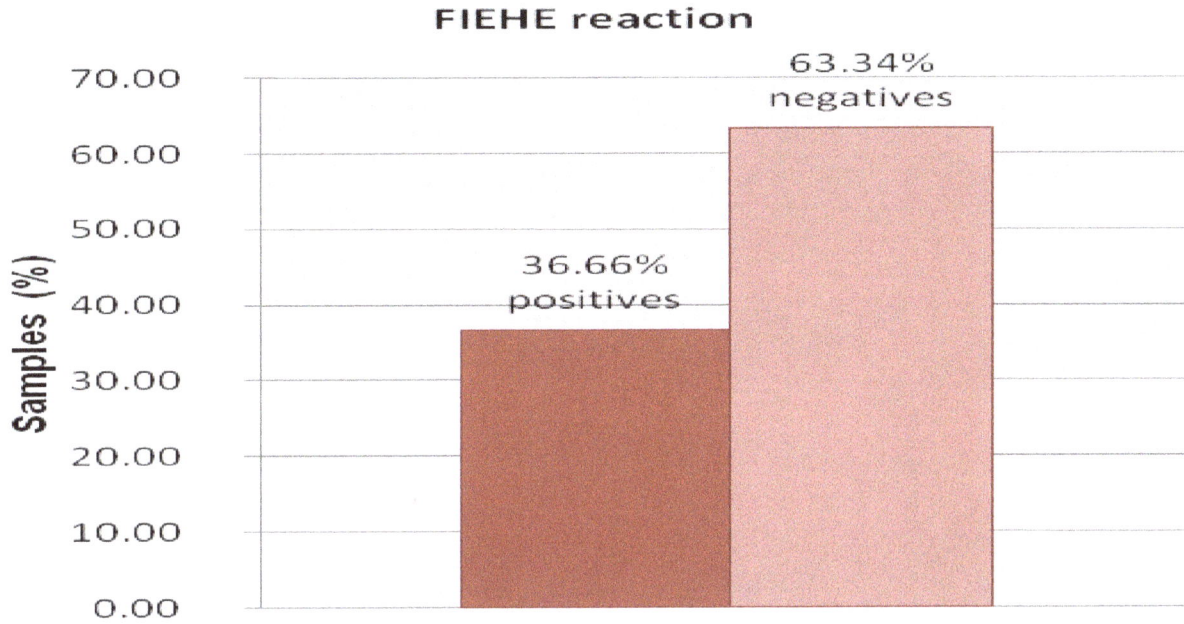

Figure 8. Distribution of samples according to research forgery.

Detector A -
1 (250nm)

Pk #	Name	Retention Time	Area	Area Percent	Height	Height Percent
1	? temoin streptomycine	2.567	30505	21.85	1694	23.17
2	? ?	3.085	82483	59.09	4274	58.45
Totals			112988	80.94	5968	81.62

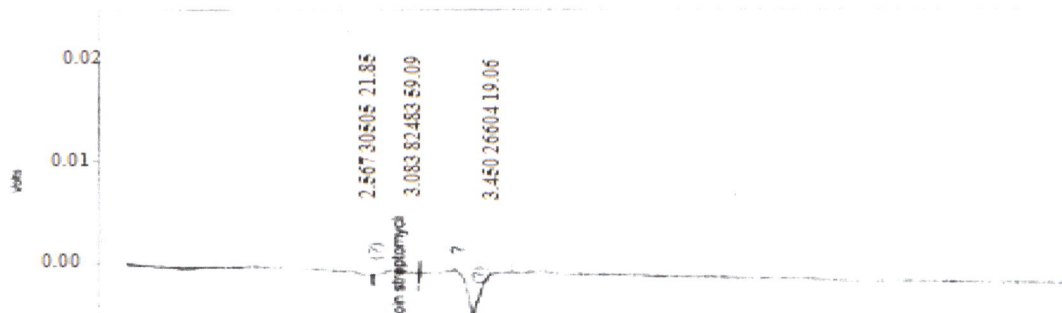

Figure 9. Chromatogram streptomycin.

seems too high as noted by LE COQ (1965). Weak reactions are also obtained with some honey heated as well as unheated. It is important to compare the color obtained with that provided from control sample with negative reaction in which a 2% of sugar is incorporated (Sweitzer, 1998) (Figure 8).

Given the absence of maximum residue limits (MRLs) set for honey; the detection threshold was considered the threshold of positivity: 15 mg/kg for tetracycline and 10 mg/kg for streptomycin (Hirsh, 2002). Zero tolerance is applied to chloramphenicol since it is banned from use in food producing animals in EU since 1990 (Regulation 2377/90/UE). Of the samples analyzed none was positive for antibiotics sought (Figures 9, 10 and 11).

CONCLUSION AND RECOMMENDATIONS

74% of samples analyzed correspond to codex standards (Codex standard 12-19811). In the light of these results

Detector A - 1
(325nm)

Name	Retention Time	Area	Area Percent	Height	Height Percent
temoin	1.783	13288	38.25	21.04	56.90
temoin	2.000	15258	43.92	1016	27.47
Totals		28546	82.16	3120	84.37

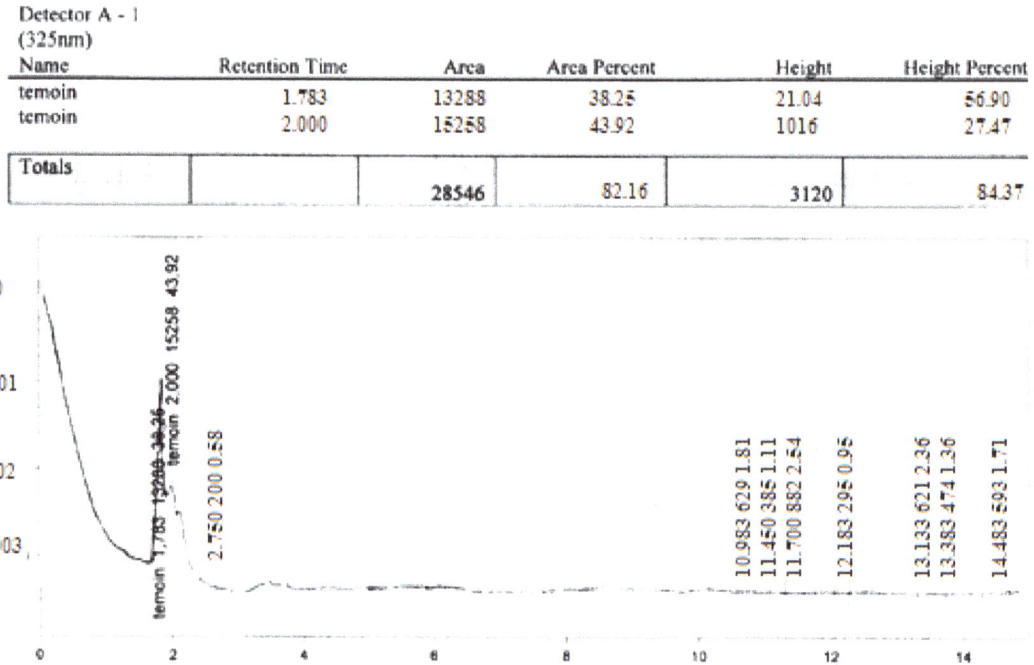

Figure 10. Chromatogram oxytétracycline.

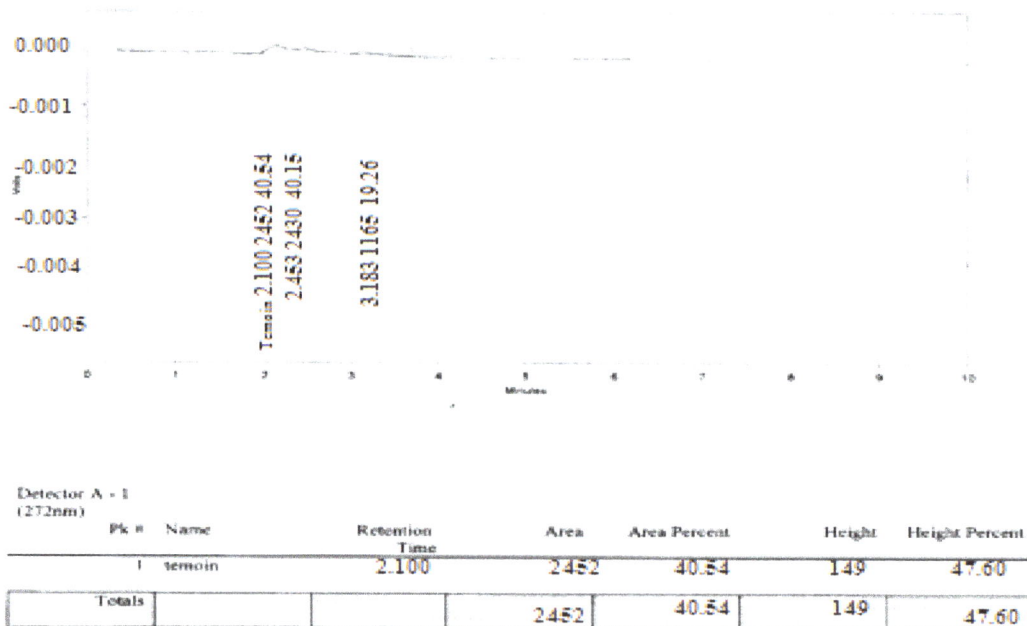

Detector A - 1
(272nm)

Pk #	Name	Retention Time	Area	Area Percent	Height	Height Percent
1	temoin	2.100	2452	40.54	149	47.60
	Totals		2452	40.54	149	47.60

Figure 11. Chromatogram chloramphénicol.

these honeys can be exported to foreign countries. 26% of them have one or more defects. The rate free acid (14% of non-compliance) and Hydroxy-Methyl-Furfural (12% of non-compliance), should be subject to more attention from beekeepers looking to improve the quality of their product. We found interesting to compare our results with the standards required in some competitions of honey. In fact this practice is increasingly common especially in Western countries. It has the double benefit of enhancing the product and a mean of advertizing and informing the media about beekeeping and honey. We have taken for example Apimondia 2009 contest held in France, where the requirements of water content should be lower than 2% of the threshold set by European legislation, that is, <18% and a rate of HMF <15 mg / kg. 60% of our samples meet these criterias and therefore can theoretically enter/participate in/take part of a competition. The economic benefits driven from such advertising can be of great benefit for the honey industry in the region. Antibiotic research conducted by HPLC showed the absence of antibiotics such as streptomycin, oxytetracycline and chloramphenicol.

For detection and quantification of residues of antibiotics, several methods are applied such as HPLC, or the method of Sharm screening test. Some international laboratories are using modern technologies such as the measurement system Biacore new immunochemical technique for optical sensors suitable for organic honey (Gatermann, 2004).

It is obvious that the usefulness of a highly developed beekeeping industry in Algeria is not debatable. Algeria is a country with a traditionally large community of consumers of honey and its production possibilities are very important due to its climate and resources. It is important to mention that the physico-chemical analysis showed that 74% of samples meet the codex standards, while 26% had one or more defects. The rate of free acid (14% of non-compliance) and HMF (12% of non-compliance) are the most important parameters that must be considered by the beekeepers to ensure compliance of honey produced. During the long journey of honey production, some steps are crucial for the quality of the final product. Therefore, beekeepers must comply with the rules of modern beekeeping practices to deliver to the consumer a honey that meets criteria quality. Qualitatively Algerian honey can compete with major producing countries, but this requires an upgrade of the industry incorporating all the food chain, focusing primarily on:

1. Rules in conformity with international standards;
2. Implementation of control laboratories to standardize honey analysis;
3. Training of all actors involved in honey production;
4. Establishment of a professional organization;
5. Improving quality through the labeling of certain products and promoting the export of local products.

ACKNOWLEDGEMENTS

The authors wish to thank the laboratory of technical institute of small breeding, as well as the Algerian Ministry of Agriculture and Rural Development.

REFERENCES

Codex standard 12-19811 Norme adoptée en 1981. Révisions en 1987 et 2001.

Directive 2001/110/CE du conseil européen du 20 décembre 2001 relative au miel.

Hadorn H, Zurcher K, Doevelaar F (2003). Produits apicoles 23 A miel, pp. 155-168.

Hirsh M (2002). Avis de l'Agence Française de Sécurité Sanitaire des Aliments relatif à l'évaluation du risque éventuel lié à la présence de résidus de tétracyclines et de streptomycine dans le miel Saisine n° 2002-SA-0126.

Gatermann R (2004). Revue eurofins n°15 (05)0. [En ligne] Adresse URL : http://www.eurofins.fr .Page consultée le 04/01/2008.

Gonnet M (1982). Le miel. Composition, propriétés, conservation. OPIDA. France, p. 31.

Le coq R (1965). Manuel d'analyses alimentaires et d'expertises usuelles. Tome 2 DOIN. Paris, p. 2185.

Makhloufi C (2007). Quelques propriétés de miels algériens, Apiacta, 42: 73-80.

Marceau J (1994). Les HMF et la qualité du miel .L'abeille. Volume 15 numéros 2 .Service de zootechnie, MAPAQ.[En ligne] Adresse URL : www.agrireseau.qc.ca Page consultée le 24/10/2007.

Prost PJ (2005). Apiculture, connaitre l'abeille –conduire le rucher. Lavoisier, Paris, p. 382.

Règlement UE n°2377/90 du 26 juin 1990 relative aux limites maximales de résidus de médicaments vétérinaires dans les aliments d'origine animale

Schweitzer P (1998). Sur les sentiers des miels de France. L'Abeille de France. Novembre.IN : PERDRIX L. Critères de qualité du miel .février.[En ligne] Adresse URL : http://www.zoo-logique.org. Page consultée le 18/09/2007.

Schweitzer P (2001). La couleur des miels. [En ligne] Adresse URL : http://www.beekeeping.com/abeille de France. Page consultée le 04/01/2008.

White J (1964). Effect of storage and processing temperatures on honey quality. Food technology 18,153-156 (164). IN BOGDANOV S.et al .Produits apicoles 23 A miel[En ligne] Adresse URL : www.alp.admin.ch.Page consultée le 28/02/2009.

21

Quality evaluation of segments-in-syrup as affected by steeping preservation of aonla fruits

Priyanka Nayak[1], Dharmendra Kumar Shukla[2], Devendra Kumar Bhatt[3] and Dileep Kumar Tandon[2]

[1]Manyavar Shri Kashi Ram Ji University of Agriculture and Technology, Banda, India.
[2]Division of Post Harvest Management, CISH, Lucknow, India.
[3]Department of Food Technology, Bundelkhand University, Jhansi, India.

Steeping preservation is one of the technologies comparatively economical and easy to follow for processing fruits for a prolonged period. In the present study, aonla fruits were preserved in solutions, viz: water (control), water + 500 ppm SO_2, water + 2% salt solution + 500 ppm SO_2, water + 5% sugar solution + 500 ppm SO_2 and water + 2% salt solution + 5% sugar solution + 500 ppm SO_2, up to 90 days and fruits were withdrawn at 0, 15, 30, 60 and 90 days of storage for the preparation of product segments-in-syrup. For the preparation of product, aonla fruits were blanched in boiling water for 10 min and segments were separated. The segments were dipped in successive increasing concentration (55, 65 and 72°B) of sugar syrup at room temperature till equilibrium was reached at 72°B. The segments were packed in syrup at 72°B in PET jars and quality parameters were evaluated in segments as well as in syrup. The analyses of fruits showed that the contents of TSS, titratable acidity, ascorbic acid, polyphenols and sugars decreased continuously as the period of steeping preservation of fruits increased. In general, the quality of the product decreased as the storage of fruits in different preservation solutions increased. The fruits preserved in water (control) spoiled up on 90 days of storage. The prepared product from fruits steeped preserved in water up to 60 days was acceptable. Highly acceptable product was obtained from fruits preserved in water + 2% salt solution + 5% sugar solution + 500 ppm SO_2, up to 90 days.

Key words: Aonla, segments-in-syrup, steeping, preservation, blanching.

INTRODUCTION

Aonla or Indian gooseberry (*Emblica officinalis* Gaertn.), a versatile fruit tree, belongs to the family Ephorbiaceae. Aonla has been cultivated in India since time immemorial. Besides India, naturally growing aonla trees are also found in different parts of the world, viz: Sri Lanka, Cuba, Puerto Rico, Chiana, Thiland and Japan (Hooker, 1973; Baileri, 1917). The fruit has high medicinal and nutritional value and is one of the richest known sources of ascorbic acid (300 to 1000 mg per 100 g edible portion) depending upon the cultivar and location (Kalra, 1988; Manny and Swamy, 1997). The fruit also contains polyphenols which have antioxidant property and, thus, have good free radical scavenging activity (Shanker, 1969). Polyphenols found in fruits contain gallic acid and ellagic acid, and glucose in the moiety, prevent and/or retard oxidation of ascorbic acid. Therefore, even after processing, aonla fruits retain a major part of ascorbic acid and polyphenols. Its regular use increases body resistance against diseases, prevents aging, improves vitality, stimulates digestive system, cures piles, urinary diseases and diabetes and, therefore, used for manufacturing drugs, cosmetics and herbal products (Prabhu et al., 2003). The *brahma rasayan*, *chyavanprash* and *triphala* are famous ayurvedic preparations in which aonla fruit is a major constituent. However, aonla fruit is not consumed freely in fresh form because of its astringent taste. It is, therefore, not popular as a dessert fruit and due to this fact it is processed into value added products (Nayak et al., 2011).

*Corresponding author. E-mail: dkbhatt_2003@ rediffmail.com.

Presently, a number of products like juice, preserve, candy, powder, pickle, etc., are prepared from aonla fruits. Aonla segments-in-syrup, a new product developed at the Institute, has good nutritional quality as compared to preserve (Nayak and Tandon, 2006; Bhattacharjee et al., 2012). Aonla is a highly perishable fruit and its shelf life is poor (Singh et al., 2005). Hence, its preservation is necessary for the preparation of products for a longer period. Steeping preservation of fruits and vegetables involving permissible chemical preservatives is one of the methods to enhance their storability without much deterioration in the quality (Sethi and Maini, 2000). Therefore, the present investigation was undertaken to assess the loss in quality of fruits as well as segments-in-syrup prepared from aonla fruits steeped preserved in water and different solutions.

MATERIALS AND METHODS

Mature aonla fruits of aonla cv. Chakaiya, obtained from the experimental farm of CISH, Lucknow, were washed thoroughly under tap water. The fruits were steeped preserved in different solutions, viz: water (control, T1), water + 500 ppm SO_2 (T2), water + 2% salt solution + 500 ppm SO_2 (T3), water + 5% sugar solution + 500 ppm SO_2 (T4) and water + 2% salt solution + 5% sugar solution + 500 ppm SO_2 (T5) in glass jars of 1 kg capacity at room temperature. There were four replications having 500 g fruits each. The water (control) was changed twice a week. The fruits were withdrawn at 0, 15, 30, 60 and 90 days of storage for the preparation of product segments-in-syrup. The fruits were blanched in 2% alum solution for 10 min and then segments were separated manually. The separated segments were dipped in sugar syrup of 55 °B having 500 ppm SO_2 as potassium metabisulphite and kept for one day at room temperature. Next day, the segments were taken out from the syrup and the syrup was concentrated to 65 °B by boiling and adding extra sugar. The segments were dipped again in sugar syrup and kept for two days at room temperature. Thereafter, TSS of syrup was increased to 72 °B by boiling and adding extra sugar after removing the segments. The segments were dipped again in sugar syrup (72 °B) and kept at room temperature for another two days. The segments were then packed with sugar syrup (72 °B) by adjusting the TSS in airtight 500 g capacity PET jars for quality evaluation.

The quality parameters, viz: TSS, titratable acidity, ascorbic acid, polyphenols, total sugars and reducing sugars at each withdrawl stage in steeped preserved fruits and products (both segments and syrups) were analyzed as per methods described by Ranganna (1997). The sensory quality of the product on the basis of colour, texture and taste was assessed by 5 semi skilled judges on Hedonic scale. The data was analyzed statistically and reported at 5% significance level (Panse and Sukhatme, 1961).

RESULTS

Changes in biochemical characters of steeped preserved fruits

The data on biochemical characters of fresh and steeped preserved aonla fruits are presented in Table 1. Fresh aonla fruit contained 9.7 °B TSS, 1.7% titratable acidity, 309 mg 100 g^{-1} ascorbic acid, 1.73% polyphenols, 7.4% total sugars and 1.6% reducing sugars. The fruits steeped in water (T1) were spoiled up on 90 days of storage. A continuous decrease in TSS of preserved fruits was noticed except in fruits stored in sugar solution where it increased. After 60 days of storage treatment, T1 showed minimum content of TSS (4.5 °B). Maximum TSS content (8.7 °B) was noted in fruits of treatment T4 (water + 2% sugar solution + 500 ppm SO_2) at 15 days of storage which increased gradually to 9.3 °B after 90 days of storage. Titratable acidity decreased in all the treatments as the storage period prolonged. Minimum (0.83%) titratable acidity was observed in treatment T4 after 15 days of storage, which further declined to 0.46% after 90 days of storage. On the other hand, maximum acid content (0.95%) was recorded in treatment T3 (water + 2% salt solution + 500 ppm SO_2) in fruits after 90 days of storage. The ascorbic acid declined from 309 mg 100 g^{-1} in fresh fruits to 172 mg 100 g^{-1} in fruits steeped preserved in treatment T3 after 90 days of storage. Similarly, polyphenols content decreased in steeped preserved fruits. Maximum content (1.50%) of polyphenols was noticed in treatment T3, whereas minimum (1.36%) in treatment T1 after 15 days of fruit storage. After 90 days of fruit preservation, highest (0.82%) polyphenols content was recorded in treatment T3 and lowest (0.66%) in treatments water + 2% sugar solution + 500 ppm SO_2 (T4) and water + 2% salt solution + 5% sugar + 500 ppm (T5).

The contents of total and reducing sugars decreased as the preservation of fruits in different solution increased. Maximum (3.6%) total sugars content was noted in T4, while minimum (2.3%) in T2 after 90 days of storage. Reducing sugars content was maximum (0.90%) in T4 and minimum (0.76%) in T2 after 90 days of fruit storage.

Changes in biochemical characters of the prepared product

The data on biochemical characters of segments and syrup are presented in Tables 2 and 3, respectively. The segments of the product contained 72.0 °B TSS, 1.2% titratable acidity, 113 mg 100 g^{-1} ascorbic acid, 0.90% polyphenols, 48.7% total sugars and 18.9% reducing sugars, while syrup contained 72.0 °B TSS, 1.08% titratable acidity, 98 mg 100 g^{-1} ascorbic acid, 0.51% polyphenols, 46.4% total sugars and 17.6% reducing sugars when prepared from fresh fruits. The maximum (92 mg 100 g^{-1}) ascorbic acid content in segments was recorded in treatment T4, while minimum (65 mg 100 g^{-1}) in treatment T5 when the product was prepared from 15 days steeped preserved fruits. Highest (56 mg 100 g^{-1}) ascorbic acid content in segments was found in treatment T4, while lowest (37 mg 100 g^{-1}) in treatment T5 when the product was prepared from fruits stored for 90 days. Similarly, maximum (40 mg 100 g^{-1}) ascorbic acid content in syrup was recorded in treatment T4, while minimum (25 mg 100 g^{-1}) in treatment T5 when the product was prepared from 90 days stored fruits.

Table 1. Quality evaluation of steeped preserved aonla fruits.

Treatments	Storage period (days)	T1	T2	T3	T4	T5	CD at 5%		
							Treatments	Storage period	Treatments × period
TSS (°B)	0	9.7	9.7	9.7	9.7	9.7			
	15	5.7	6.1	6.5	8.7	8.0			
	30	5.0	5.5	5.7	8.9	8.4	0.30	0.30	0.67
	60	4.5	5.1	5.3	9.0	8.5			
	90	Spoiled	4.6	5.0	9.3	8.7			
Titratable acidity (%)	0	1.7	1.7	1.7	1.7	1.7			
	15	1.33	1.36	1.30	0.83	0.95			
	30	1.18	1.22	1.15	0.70	0.83	0.97	0.97	0.21
	60	1.10	1.10	1.00	0.58	0.66			
	90	Spoiled	0.90	0.95	0.46	0.60			
Ascorbic acid (mg/100 g)	0	309	309	309	309	309			
	15	279	284	272	270	261			
	30	233	248	240	243	240	NS	32.4	73.5
	60	214	216	203	229	221			
	90	Spoiled	179	172	208	206			
Polyphenols (%)	0	1.73	1.73	1.73	1.73	1.73			
	15	1.36	1.46	1.50	1.40	1.40			
	30	1.05	1.08	1.12	1.06	1.03	0.026	0.026	0.059
	60	0.85	0.87	0.93	0.86	0.83			
	90	Spoiled	0.75	0.82	0.66	0.66			
Total sugar (%)	0	7.4	7.4	7.4	7.4	7.4			
	15	3.4	3.7	3.9	5.0	4.5			
	30	3.2	3.3	3.4	4.6	4.2	0.22	0.22	0.50
	60	2.7	2.9	3.0	3.8	3.8			
	90	Spoiled	2.3	2.5	3.6	3.3			
Reducing sugar (%)	0	1.6	1.6	1.6	1.6	1.6			
	15	1.06	1.03	1.10	1.36	1.20			
	30	0.86	0.96	0.98	1.26	1.00	0.46	0.46	0.10
	60	0.73	0.80	0.86	1.06	0.96			
	90	Spoiled	0.76	0.78	0.90	0.86			

T1- water, T2- water + 500 ppm SO_2, T3- water + 2% salt solution + 500 ppm SO_2, T4- water + 2% sugar solution + 500 ppm SO_2 and T5- water + 2% salt solution + 5% sugar + 500 ppm.

The polyphenols content in segments as well as in syrup decreased as the steeping preservation of fruits prolonged. The polyphenols content in segments was minimum (0.41%) in T2, while maximum (0.47%) in treatments T4 and T5 and in syrup it was minimum (0.25%) in T4 and maximum (0.32%) in water + 500 ppm SO_2 (T2) when the product was prepared after 90 days of preserved fruits.

Maximum (51.2%) total sugars was noted in segments of the product in treatment T3, prepared from 90 days stored fruits, while lowest (45.0%) in T1, prepared from 15 days stored fruits. Maximum (50.6%) total sugars was noted in syrup of the product T3, prepared from 90 days stored fruits, while lowest (46.5%) in T1, prepared from 15 days stored fruits. Similarly, reducing sugars content was maximum (20.6%) in segments of treatment T2, prepared from 90 days stored fruits, while lowest (17.2%) in T2, prepared from 15 days stored fruits. Maximum (18.7%) reducing sugars was noted in syrup of the product in treatment T5, prepared from 90 days stored

Table 2. Quality evaluation of the segments of the product prepared from steeped preserved aonla fruits.

Treatments	Storage period (days)	T1	T2	T3	T4	T5	CD at 5%		
							Treatments	Storage period	Treatments × period
TSS (°B)	0	72.0	72.0	72.0	72.0	72.0			
	15	72.0	72.0	72.0	72.0	72.0			
	30	72.0	72.0	72.0	72.0	72.0	0.80	0.71	1.60
	60	72.0	72.0	72.0	72.0	72.0			
	90	Spoiled	72.0	72.0	72.0	72.0			
Titratable acidity (%)	0	1.20	1.20	1.20	1.20	1.20			
	15	1.00	1.03	1.06	1.00	1.00			
	30	0.96	0.93	0.96	0.93	0.90	0.53	0.53	0.11
	60	0.83	0.90	0.90	0.86	0.86			
	90	Spoiled	0.80	0.80	0.76	0.76			
Ascorbic acid (mg/100 g)	0	113	113	113	113	113			
	15	76	84	83	92	65			
	30	58	74	67	87	51	0.91	0.91	2.04
	60	42	57	46	66	42			
	90	Spoiled	39	38	56	37			
Polyphenols (%)	0	0.90	0.90	0.90	0.90	0.90			
	15	0.72	0.74	0.70	0.73	0.77			
	30	0.56	0.62	0.62	0.65	0.70	0.41	0.41	0.09
	60	0.48	0.50	0.53	0.57	0.59			
	90	Spoiled	0.41	0.43	0.47	0.47			
Total sugars (%)	0	48.7	48.7	48.7	48.7	48.7			
	15	45.0	48.3	48.2	49.9	47.3			
	30	46.4	49.6	50.2	50.5	48.3	0.16	0.16	0.35
	60	46.6	48.9	50.6	50.4	48.9			
	90	Spoiled	47.4	51.2	49.6	50.4			
Reducing sugars (%)	0	18.9	18.9	18.9	18.9	18.9			
	15	18.7	17.2	18.3	15.7	17.9			
	30	19.3	19.7	19.7	18.9	19.0	0.16	0.16	0.35
	60	19.2	20.2	19.3	19.0	19.3			
	90	Spoiled	20.6	19.6	19.2	19.6			

fruits, while lowest (15.7%) in T1, prepared from 15 days stored fruits. The sensory evaluation of the product was assessed on the basis of colour, appearance, texture and taste and overall average score was worked out on a 9 point Hedonic scale. The data are presented in Table 4. The product prepared from 15 days preserved fruits from treatment T4 scored highest (8.1) followed by treatment T2 (7.3). The treatment T4 scored maximum (7.6) followed by treatment T2 (7.5) when the product was prepared from 90 days preserved fruits. However, the texture and taste of the product prepared from treatment T4 (7.8 and 7.2, respectively) was the best along with

treatment T2 (7.6 and 7.6, respectively), while that prepared from treatments T3 (4.8 and 4.6, respectively) and T5 (4.8 and 4.6, respectively) were less acceptable when prepared from 90 days stored fruits.

In general, the organoleptic quality of the product decreased with the prolongation of the preservation period of fruits.

DISCUSSION

Aonla cv. Chakaiya was steeped preserved in different

Table 3. Quality evaluation of syrup of the product prepared from steeped preserved aonla fruits.

Treatments	Storage period (days)	T1	T2	T3	T4	T5	CD at 5%		
							Treatments	Storage period	Treatments × period
TSS (°B)	0	72.0	72.0	72.0	72.0	72.0	0.80	0.71	1.6
	15	72.0	72.0	72.0	72.0	72.0			
	30	72.0	72.0	72.0	72.0	72.0			
	60	72.0	72.0	72.0	72.0	72.0			
	90	Spoiled	72.0	72.0	72.0	72.0			
Titratable acidity (%)	0	1.08	1.08	1.08	1.08	1.08	0.42	0.41	0.08
	15	0.90	0.86	0.90	0.90	0.80			
	30	0.88	0.83	0.93	0.90	0.76			
	60	0.78	0.80	0.80	0.80	0.73			
	90	Spoiled	0.76	0.86	0.83	0.66			
Ascorbic acid (mg/100 g)	0	98	98	98	98	98	0.88	0.80	1.92
	15	63	68	63	82	56			
	30	48	59	46	75	44			
	60	37	44	35	48	35			
	90	Spoiled	30	28	40	25			
Polyphenols (%)	0	0.51	0.51	0.51	0.51	0.51	0.22	0.20	0.04
	15	0.49	0.41	0.44	0.41	0.45			
	30	0.43	0.39	0.37	0.38	0.40			
	60	0.37	0.37	0.33	0.34	0.34			
	90	Spoiled	0.32	0.27	0.25	0.28			
Total sugars (%)	0	46.4	46.4	46.4	46.4	46.4	0.14	0.14	0.29
	15	46.5	49.1	48.5	48.1	47.4			
	30	48.8	50.3	50.0	47.8	47.2			
	60	46.6	49.4	50.4	46.6	48.4			
	90	Spoiled	49.0	50.6	47.2	46.6			
Reducing sugars (%)	0	17.6	17.6	17.6	17.6	17.6	0.12	0.14	0.31
	15	15.7	16.2	17.3	17.2	17.3			
	30	16.9	17.7	17.6	17.9	17.7			
	60	16.4	17.3	17.2	17.8	17.3			
	90	Spoiled	17.0	17.4	17.5	18.7			

solutions, viz: water (control, T1), water + 500 ppm SO_2 (T2), water + 2% salt solution + 500 ppm SO_2 (T3), water + 5% sugar solution + 500 ppm SO_2 (T4) and water + 2% salt solution + 5% sugar solution + 500 ppm SO_2 (T5) and the product segments-in-syrup was prepared at 0, 15, 30, 60 and 90 days of storage. The fruits stored in water (T1) were spoiled after 60 days of steeped preservation. A continuous decrease in TSS of the fruits was recorded in all the treatments except T4 and T5 where sugar was in the preservation solution. The titrarable acidity also decreased continuously in the fruits steeped in different solutions. The maximum ascorbic acid content in fruits was recorded in treatment T4 (208 mg 100 g⁻¹) followed by T5 (206 mg 100 g⁻¹) after 90 days of preservation. The maximum poyphenols content in fruits was recorded in treatment T3 followed by T2 after 90 days of preservation. The data indicated that the fruits could be preserved with minimum loss in quality parameters and without spoilage containing SO_2 as an added preservative up to 90 days. Sethi and Maini (2000) have also reported that fruits could be stored safely for fairly long periods in chemical solution consisting of chemical preservatives (potassium metabisulphite) along with salt, sugar and spices in water. Kumar and Shukla, (2009)

Table 4. Sensory evaluation of the product segments-in-syrup prepared from steeped preserved aonla fruits.

Treatments	Storage period (days)	Colour	Appearance	Texture	Taste	Overall average (out of 9)
Fresh fruit	0	8.2	8.4	8.1	8.5	8.3
T1	15	7.6	7.6	6.4	6.4	7.0
	30	7.8	7.6	6.4	6.6	7.1
	60	7.6	7.4	6.4	6.2	6.9
	90	Spoiled	Spoiled	Spoiled	Spoiled	
T2	15	7.6	7.0	7.2	7.4	7.3
	30	8.2	8.0	7.4	6.4	7.5
	60	8.0	7.8	7.8	7.6	7.8
	90	7.4	7.4	7.6	7.6	7.5
T3	15	8.0	8.2	5.5	5.1	6.7
	30	5.0	5.0	5.2	5.2	5.0
	60	5.1	5.0	5.0	4.5	4.9
	90	4.6	4.4	4.8	4.6	4.6
T4	15	7.8	8.2	8.4	8.0	8.1
	30	8.0	8.0	8.0	8.0	8.0
	60	8.0	7.8	7.8	7.6	7.8
	90	8.2	7.4	7.8	7.2	7.6
T5	15	7.8	8.0	5.5	5.5	6.7
	30	5.2	5.1	5.4	5.1	5.2
	60	5.0	5.2	5.0	4.8	5.0
	90	4.8	4.6	4.8	4.6	4.7

CD at 5% treatments 0.30, storage period 0.30 and treatments x period 0.67.

have stated that aonla fruits could be stored in 15% steeping salt solution with minimum quality loss for 3 months. Premi et al. (1998) successfully preserved aonla fruits in a steeping solution (10% salt, 0.5% acetic acid and 200 ppm sulphur dioxide) up to 3 months. The product segments-in-syrup, prepared from steeped preserved fruits was evaluated for their quality parameters.

The analysis of segments of the product showed that the contents of titratable acidity, ascorbic acid and polyphenols decreased continuously in all the treatments, while total and reducing sugars varied slightly. The maximum ascorbic acid content in segments was recorded in treatment T4 followed by treatment T1 after 90 days of preserved fruits. The maximum ascorbic acid content in syrup of the product was recorded in treatment T4 followed by treatment T2 after 90 days fruit preservation. The maximum contents of polyphenols in segments and syrup were recorded in treatments T4 and T2, respectively, from 90 days of fruit preservation. The sensory quality of the product segments-in-syrup was highly acceptable even after it was prepared from 90 days of preserved fruits of treatments T2 and T4. The

product prepared from treatments T3 and T5 were not acceptable. It was inferred from the study that segments-in-syrup of good quality could be prepared from the steeped preserved fruits up to 90 days of storage in a solution containing water + 500 ppm SO_2 (T2) and water + 2% sugar solution + 500 ppm SO_2 (T4).

ACKNOWLEDGEMENT

Authors acknowledge with thanks the help provided by Mr. Abhay Dixit, Central Institute for Subtropical Horticulture, Lucknow during the course of investigation.

REFERENCES

Baileri LH (1917). "The standard cyclopedia of Horticulture", The Mac Million Co., London.

Bhattacharjee AK, Dikshit A, Kumar S, Tandon DK (2012). "Steeping preservation of aonla and quality of products", Indian J. Nat. Prod. Resourc. (In Press).

Hooker JD (1973). "The flora of British India", vol. V, 1st Indian Reprint, Periodical Experts, Delhi.

Kalra CL (1988). "The chemistry and technology of aonla (*Phyllanthus emblica* L.) a resume", Indian Food Packer 38(4):67.

Kumar S, Shukla SK (2009). "Aonla, the book of under utilized subtropical fruits", International book dist. Co. pp. 1-10.

Manny NS, Swamy MS (1997). "Food facts and principles", New age International (P) ltd., New Delhi, p. 190.

Nayak P, Bhatt DK, Shukla DK, Tandon DK (2011). "Evaluation of aonla (*Emblica Officinalis* G.) segments-in-syrup prepared from stored fruits", Res. J Agric. Sci. 43(2):252-257.

Nayak P, Tandon DK (2006). "Standardization of pretreatment for preparation of aonla (*Emblica officinalis Gaertn.*) segments in syrup", National seminar on Production and Processing of Aonla (*Emblica Officinalis G.*), Amdavad – Gujrat 21–23 November 35.

Panse VG, Sukhatme PV (1961). "Statistical methods for agricultural workers" ICAR, New Delhi.

Prabhu T, Kumar S, Shanthakumar P, Kalyansundaram P (2003). "Uses of aonla" 8-10 August, Aonla Growers Association of India, Salem, Tamilnadu.

Premi BR, Sethi V, Saxena DB (1998). "Studies on the identification of white specks in cured aonla fruits", Food Chem. 61(1-2):11.

Ranganna S (1997). "Handbook of analysis and quality control for fruit and vegetable products", IInd Ed. Tata McGraw Hill Pub. Co. Ltd., New Delhi, pp. 81-82.

Sethi V, Maini SB (2000). "Steeped preserved products", Post harvest technology of fruits & vegetables, Indian Pub Co, New Delhi, 941-961.

Shanker G (1969). "Aonla for your daily requirement of vitamin C", Indian Horticulture pp. 11-19, 35.

Singh BP, Pandey G, Pandey MK, Pathak RK (2005). "Shelf life evaluation of aonla cultivars", Indian J. Hort. 62(2):137-140.

22

Storage and consumer acceptability of fruit: Ginger based drinks for combating micronutrient deficiency

Omodamiro R. M.*, Aniedu C., Chijoke U. and Oti E.

National Root Crops Research Institute Umudike, P. M. B. 7006, Umuahia, Abia State, Nigeria.

Fruits are good sources of micronutrients especially mineral and vitamin C. Ginger satisfies the function of color and flavour enhancers. Matured healthy pineapple, orange and paw-paw fruits were used to prepare pineapple-ginger, orange-ginger and paw-paw-ginger based drinks. The drinks were distributed into the same type of plastic container with lid (50-ml capacity filled up to about 45 ml). They were kept under ambient condition (shelves in the laboratory) at 30 to 32°C while some were kept in a refrigerator (24 to 26°C). Visual examinations of the drinks were carried out and they were evaluated for taste/flavour. By 96th-h of storage, all the drinks under ambient condition got spoilt. All the fruits-ginger drinks kept well under refrigeration as they remained acceptable to the panelists for 21 days with pawpaw-gingered drink more acceptable ($P<0.05$) to the panelists than orange-ginger and pineapple-ginger drinks. The pH of the drinks stored under ambient conditions decreased as storage progressed thus encourages proliferation of fermentation microorganisms while those in refrigeration did not fluctuate.

Key words: Spices, herbs, fruits, ginger drinks, consumer acceptability, preservative.

INTRODUCTION

The high rate of micronutrient deficiencies in Nigeria has been attributed partly to poor dietary habits. Increased production and consumption of micronutrient rich foods would improve the micronutrient status of the Nigerian population. Fruits are good sources of micronutrients especially mineral and vitamin C (Egbekun and Akubor, 2006). High post harvest losses are usually recorded in most of our fruits. The use of chemical preservative such as sodium benzoate used to preserve fruit juices or drinks have been found to be toxic, hence the use of spices and herbs for this purpose. Many preservatives are readily available for many diverse uses. Some foods, however, because of their delicate balance of flavours require the utmost care in selecting preservatives (Giese, 1994).

Sweet orange (*Citrus sinensis*) is the most widely grown of all the citrus crops. The juice contains 86 to 91% water, 5 to 9% carbohydrate, 0.70% protein and 1 to 2% fiber; 1,200 J of energy is provided by 1 kg of the edible materials. Pineapple and paw-paw are available in almost all parts of Nigeria. However, high post harvest losses are usually recorded coupled with problem of off-season. Proper handling of these fruits reduces the losses, thus making them also increase availability in off-season. Spice and herbs act as colour and flavour enhancers, ginger satisfies these functions. Products with pH less than 4.2 are considered to be safe from food poisoning (Ihekoronye, 1998).

Food security exists when all people, at all times, have physical and economic access to sufficient, safe and nutritious food that meets their dietary needs and food preferences for an active and healthy life (WHO, 1995). Food security is assumed to occur when all people at all times have access to enough food that is affordable, safe and healthy, culturally acceptable, meets specific dietary needs, obtained in a dignified manner, produced in ways that are environmentally sound and socially just (WHO, 1995). The prevention of hypovitaminosis A can be achieved through dietary diversification, food and beverage fortification, as well as periodic and specific supplementation.

The objectives of the study were to determine consumer acceptance and sensory changes occurring during storage of pineapple-ginger, orange-ginger and paw-paw

*Corresponding author. E-mail: majekdamiro@yahoo.com.

Table 1. Sensory evaluation of fruit: ginger drinks stored at ambient temperature / refrigerated conditions.

Samples / Storage period	Acceptability scores*					
	Ambient temperature (h)			Refrigerated conditions (days)		
	0	48	96	7 days	14 days	21 days
Ginger-pineapple	5.27[b]	5.04[b]	Spoilt	5.52[b]	5.25[b]	5.25[b]
Ginger-orange	5.72[b]	5.12[b]	Spoilt	5.70[b]	5.69[b]	5.71[b]
Ginger-pawpaw	6.54[a]	5.89[b]	5.56	6.35[a]	5.90[a]	6.02[a]
Bitter lemon (control)	5.27[b]	5.25[b]	5.26	5.23[b]	5.25[b]	5.24[b]
LSD	**0.82**	**0.64**	**-**	**0.65**	**0.21**	**0.31**

*7-point hedonic scale: 1 = dislike extremely, 4 = neither like nor dislike and 7 = dislike extremely.

Table 2. The pH of the fruit: ginger drinks during storage.

Sample	Code	Storage period (days)						
		0	2	4	6	7	14	21
Ginger-pineapple	A_A	3.40	2.94	2.93	ND*	ND	ND	ND
Ginger-pineapple	A_R	3.40	3.35	3.33	3.32	3. 05	2.98	2.96
Ginger-orange	B_A	3.98	2.80	2.77	ND	ND	ND	ND
Ginger-orange	B_R	3.98	3.76	3.70	3.17	3.69	3.62	3.59
Ginger-pawpaw	C_A	5.10	3.18	3.16	ND	ND	ND	ND
Ginger-pawpaw	C_R	5.10	5.24	5.23	4.27	4.17	4.10	3.98
Ginger drink	D_A	3.90	3.90	3.92	3.95	3.95	3.96	3.95
Ginger drink	D_R	3.90	3.90	3.90	3.91	3.19	3.90	3.90

*ND-not determine.

-ginger drinks towards increasing micronutrient intake of the populace especially the low income earners.

MATERIALS AND METHODS

Matured and healthy pineapple, orange and paw-paw fruits were purchased from the Umudike central market while the ginger was obtained from the ginger programme of the National Root Crops Research Institute (NRCRI) Umudike. Pineapple-ginger; orange-ginger and paw-paw-ginger drinks were prepared using standard method described by Omodamiro et al. (2008).

The drinks were distributed into the same type of plastic kegs with cover (50-ml container filled up to about 45 ml). Four kegs kept under ambient condition (on shelves in the laboratory) at a temperature of 30 to 32°C and the other 4 were kept in the refrigerator (24 to 26°C). Kegs under ambient conditions were coded as A_A, B_A, C_A, C_A and D_A and those under refrigerated coded as A_R, B_R, C_R D_R and D_R and D_A represent the control. Sensory evaluation was done for the drinks using 7-point Hedonic scale where 7 = like extremely; 4 = neither like nor dislike; 1 = dislike extremely (Iwe, 2002).

The pH of the stored drinks was monitored for 3 weeks using methods of AOAC (1990). Visual examination of the drinks was carried out and they were evaluated for their taste/flavour.

RESULTS AND DISCUSSION

Table 1 shows that all the fruit: ginger drinks were acceptable to the panelists up to 48 h after preparation and storage under ambient conditions there was significant difference (P<0.05) between the pawpaw; ginger drink and the other fruit; ginger drinks in acceptance before storage and after 48 h storage; the paw-paw; ginger drink was found most acceptable by the panelists (Table 1).

By the 96th-h of storage all the drinks got spoilt. Table 1 show that all the fruit-ginger kept well under refrigeration as they remained acceptable to the panelists for 21 days. The pawpaw: ginger based drink was more acceptable (P<0.05) to the panelists than either the orange: ginger of pineapple: ginger drink.

The pH (Table 2) of the drinks stored under ambient conditions decreased as storage progressed thus encouraged proliferation of fermentative micro-organisms. The pH of the refrigerated drinks did not fluctuate and they remained acceptable to the panelists for as long as 21 days after preparation.

The results obtained from this work suggest that consumers accepts the gingered based drinks-pineapple-ginger, orange-ginger and paw-paw-ginger drinks. Changes observed during the storage shows that it could be produced and kept up to 48 h without the use of refrigerator, which the poor populace may not be able to afford. The adoption of this processing method in the production of micronutrient drinks will not only bring about increasing micronutrient intake of the populace especially the low income earners but will also serve as income generation for the rural women and the unemployed youth with little capital.

REFERENCES

AOAC (1990). Official Methods of Analysis. Washington DC: Association of official Analytical Chemists.

Egbekun NK, Akubar PI (2006). Chemical composition and sensory properties of melon seed milk-orange juice beverage. Food Sci. Technol., 24(1): 42-49.

Ihekoronye AI (1998). Manual on small-scale food processing. The academic publishers. p. 5.

Giese J (1994). Antimicrobials: Assuring Food Safety. Food Technol., 48(6): 102-109.

Iwe MO (2002). Handbook of Sensory Methods and Analysis. Rojoint Communication   Services Ltd, 65 Adelabu St. Uwani-Enugu. AJFS

Omodamiro RM, Aniedu C, Chijoke U, Oti E (2008). Preliminary studies on the consumer acceptability of fruit: ginger drinks Stored under ambient and refrigerated conditions. Nutrition Society of Nigeria. 39th Annual Conference.

World Health Organisation (1995). The global prevalence of vitamin A deficiency. Micronutrient series document WHO/NUT/95.3.Geneva: WHO, 1995.

23

Identification of compounds characterizing the aroma of oblate-peach fruit during storage by GC-MS

Lin-lin Cheng[1], Li-mei Xiao[1], Wei-Xin Chen[2], Ji-de Wang[1], Feng-bin Che[3] and Bin Wu [1,3*]

[1]College of Chemistry and Chemical Engineering, XinJiang University,Urumqi, 830046, China.
[2]College of Horticulture, South China Agricultural University, Guangzhou 510642, China.
[3]Farm Product Storage and Freshening Institute, Xinjiang Academy of Agricultural Sciences, Urumqi, 830091, China.

Three analytical methods, including headspace solid-phase microextraction (HS-SPME), liquid–liquid extraction (LLE) and steam distillation extraction (SDE), were utilized to investigate the aroma profile characteristics of Xinjiang oblate-peach fruit during storage, and the characterizing compounds were detected by gas chromatography mass spectrometry (GC-MS). Silica fiber coated with (DVB-CAR-PDMS) was found to be more efficient for collecting the SPME headspace volatile compounds, and the extraction time of 40 min was preferred in this study. The SPME headspace volatile constituents present in oblate-peach before and after 4 weeks of cold storage (4±1°C) have been analysed, while some physical characteristics such as fruit weight, firmness, soluble solids content (SSC), and titratable acid were monitored during storage. The results show that the volatile compounds displayed different composition during storage, a total of 58 volatiles were identified, 52 prior to storage and 45 after 4 week's storage, the content of lactones and esters, characteristic compounds of peach aroma, were much lower after 4 weeks of storage. Thirteen of the pre-storage volatiles were not found after storage.

Key words: Oblate-peaches, aroma, gas chromatography mass spectrometry (GC–MS), headspace solid-phase microextraction (HS-SPME), liquid–liquid extraction (LLE), steam distillation extraction (SDE).

INTRODUCTION

Flavour, besides other parameters such as texture and appearance, plays a very important role in the quality assessment of fruits and vegetables. As such, this parameter affects the appreciation and the acceptance of horticultural products and, as a consequence, the purchasing behaviour of consumers (Harker et al., 2003; Kühn and Thybo, 2001; Peneau et al., 2006). Flavour is defined as the interaction of individual taste and aroma components with the human sensory system (Vermeir et al., 2009). Oblate-peach *(Prunus persica* L.*)*, an economically important variety plant grown in Xinjiang of China, which is famous for its crisp, juicy texture and sweet flavour, along with important nutrient contributions from its phytochemical constituents is highly favoured by consumers worldwide. However, oblate-peach has a very

short shelf-life under normal ambient conditions. Low temperature and modified atmosphere storage is commonly used for peaches to delay fruit ripening, senescence, and textural changes, reduce activity of certain enzymes and other positive benefits, however, they often have little aroma which upon removal from storage, greatly diminish consumer acceptability (Hardenburg et al., 1986; Biale, 1960; Aubert et al., 2003).

Conventional sampling methods for fruit aromas in previous studies are mainly liquid-liquid extraction (LLE) and steam distillation extraction (SDE) (Aubert et al., 2007; Baldry et al., 1972). These conventional methods always require long extraction times, large amounts of solvents and multiple steps. Moreover, many unstable aroma volatiles may be thermally decomposed and degraded during thermal extraction or distillation. However, being simple and straightforward procedures, they were still extensively applied for fragrance-and-aroma characterization, either alone or combined with

---

*Corresponding author. E-mail: xjuwubin0320@sina.com.

other sample-preparation procedures (Moser et al., 1980). Solid-phase microextracton (SPME) is a simple, solvent-free method for concentration of volatiles present in the headspace. This technique had been used to analyse the volatile compounds of different fruits (Zhang et al., 2007).

In this study, HS-SPME combined with LLE and SDE were used to sample aroma volatiles of oblate-peach fruit emanating quantitatively and qualitatively with 4 weeks storage (4±1°C), followed by GC-MS analysis. The type of SPME fibre coating of various polarities and extraction time were carried out to select the optimum condition for the analysis of the volatile compounds. The objective of our research was to evaluate the effectiveness of three different sampling methods combined with GC–MS, to identify the volatile compounds of oblate-peaches. In addition, fruit quality was evaluated from a consumer perspective. It is hoped that the study would provide a method for investigating the aroma profile characteristics of oblate-peach pulp during storage and some helpful clues for quality discrimination.

MATERIALS AND METHODS

Plant materials

Oblate-peach (*P. persica* L.) CV. 'Ruipan' at commercial harvest maturity (SSC 12.5±1.0% and firmness 11.8±2.5 kg/m^2) was obtained from the orchard at Shihezi, Xinjiang Province, China, and quickly transported to the laboratory on the day of harvest. The fruits were selected for uniform size, maturity and appearance and free from defects and mechanical damage, transported to the laboratory and stored at 8 to 10°C overnight before processing.

Treatments

Oblate-peach were dipped for 2 min in 1.0 g/L sodium hypochlorite solution to control disease and dried at room temperature with air. The fruits were kept in hermetically sealed plastic drums (20 L) of low-density polyethylene (LDPE, 0.02 mm thickness). Three replicates per treatment were used and evaluations were made every 7 days for up to 28 days during cold storage (4±1°C)

HS-SPME procedure and sample preparation

Duplicate measurements of each sample were performed for the optimization of the HS-SPME conditions. The silica fibers and the manual SPME holder were purchased from Supelco (Bellefonte, Inc., PA, USA) Three fibers were tested and compared: polydimethylsiloxane (PDMS, 100 μm), carboxen-polydimethylsiloxane (CAR-PDMS, 65 μm) and divinylbenzene-carboxen-polydimethyl-siloxane (DVB-CAR-PDMS, 50/30 μm). The fibers were conditioned prior use according to the supplier's prescriptions. In this experiment, the different extraction times (20, 30, 40, 50 and 60 min) were evaluated to obtain the optimized sampling efficiency.

For sampling, seeds were removed, and 500 g of peaches were homogenized using a commercial blender, to which was added 10 g NaCl. For each measure, 8 g of puree was transferred into a 15 ml capped solid-phase microextraction vial. Vials were equilibrated during the equilibrium time (depending on the experimental design)

in a thermostatic bath at desired temperature (depending on the experimental design). Experimental conditions were set before fiber screening study, as follows: extraction temperature 50°C, equilibrium time 15 min, extraction time 30 min. After sampling, desorption of the analytes from the fiber coating was made in the injection port of GC at 250°C during 3 min in splitless mode. Before sampling, each fiber was reconditioned for 1 min in the GC injector port at 250°C.

LLE and SDE procedure

In the LLE analysis, the extraction procedure was a modification of the method described by Aubert et al. (2005), 200 g fresh peach pulp free of seeds was cut, homogenized, and centrifuged at 800 rpm for 30 min. The filtered supernatant was subjected to continuous liquid-liquid extraction with 3×25 ml of dichloromethane.

In the SDE procedure, as described by Zhang et al. (2007), 200 g fresh peach pulp free of seeds was cut, homogenized, and subjected to hydrodistillation for 120 min. The obtained fraction was extracted using 3×25 ml of dichloromethane. The organic phase was dried over anhydrous sodium sulfate, concentrated to 1 ml using a rotor evaporator (Yarong Instrument, Shanghai, China); 0.5 μL concentrated organic phase from the SDE and SD samples were introduced to the GC-MS for subsequent analysis.

GC–MS analysis

Desorption and analysis of volatile components was carried out on a QP2010 GC-MS (Shimadzu, Tokyo, Japan). Chromatographic separation was performed with a DB-5MS capillary column (30.0 m ×0.25 mm i.d., 0.5 μm film thickness) under the following instrumental conditions. The carrier gas was helium with a flow rate of 1.0 ml/min. The injector temperature was set at 250°C. The GC oven temperature was maintained at 40°C for 2 min after injection, then programmed at 8°C /min to 200°C for 5 min, and then at 10°C /min to 250°C which was maintained for 5 min. The temperature of mass spectrometer was 230°C. The ionizing energy was 70 eV. All data were obtained by collecting the full-scan mass spectra within the scan range 20 to 600 amu.

Quality evaluation

Fruit firmness was measured on two paired sides of 10 fruits from each replicate (skin removed) with a hand-held firm meter (GY-B, Jilin, P.R. China) with a 10 mm diameter probe at a speed of 1 mm s^{-1}. Data were expressed as kg/cm^2. Soluble solids content (SSC) was determined with a digital refractometer (Atago, Japan) and expressed as a percentage. Percentage of titratable acidity (TA) was determined by titration with 0.01 M NaOH and calculated as citric acid equivalents from 10 g of pulp obtained from 6 fruits. In each treatment, 50 fruits were selected for investigating the rot index of fruits. All fruits were classified in four ranks by the extent of rot: 0, fruits were not rotten; 1, the rotten surface was less than 1/3; 2, the rotten surface was between 1/3 and 2/3; 3, the rotten surface was more than 2/3. The rot index was expressed as the following equation:

Rot index= $\sum$ (Rank × Quantity) / (4×50) ×100%.

Fruit appearance (visible structural integrity, off-aroma, color and flavor) was evaluated by six experienced panelists. The visual quality score was based on the following scale: 5, excellent; 4, very good; 3, good, limit of marketability; 2, fair, limit of usability; 1, poor, inedible.

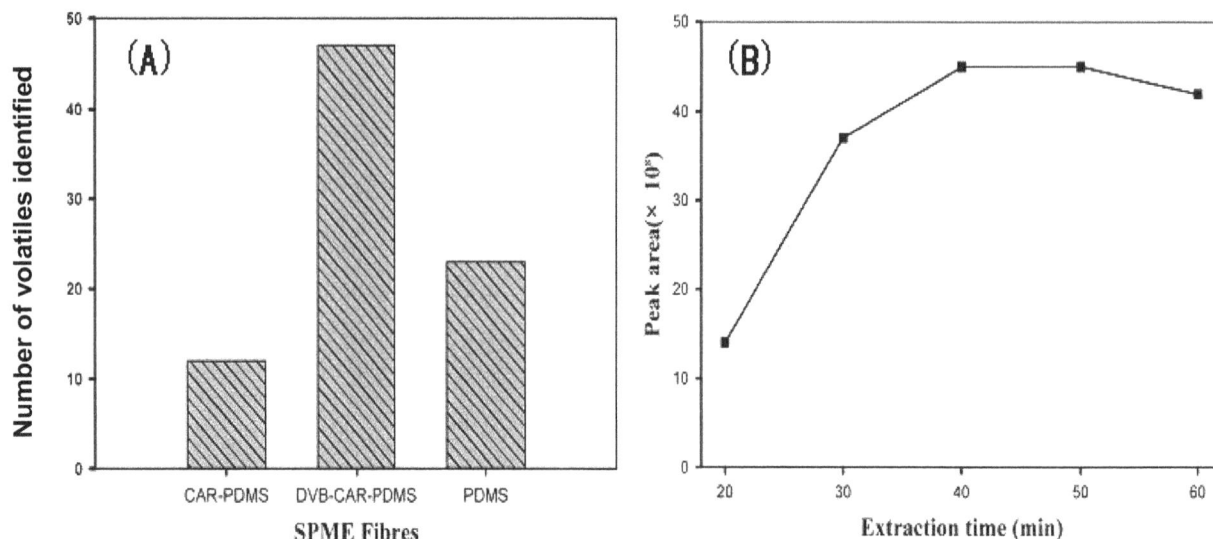

Figure 1. Effect of SPME fibre coatings (A) and extraction time (B) on sampling efficiency.

Statistical analysis

Each experiment was repeated three times and statistical analyses were performed using Microsoft Excel software and SPSS 13.0 software. P-values were determined by t-test. Data are presented as the mean ± standard error of the mean (S.E.M.). Data were treated for multiple comparisons by analysis of variance with least significant difference (L.S.D.) between averages determined at 5% level.

RESULTS AND DISCUSSION

The optimization of the HSSPME conditions

The type of SPME fibre coating used is crucial to sampling efficiency. Some useful and specific factors should be taken into consideration, including polarity, matrix, etc. Figure 1(A) shows desorption capacity of three different fiber coatings for extraction of oblate-peach volatile compounds. The results of the fiber screening confirmed that the DVB-CAR-PDMS fibers produced the best results for the compounds investigated. Of this, fiber had strong extraction capacity for esters, lactones, aldehydes, ketones, and acids. It seemed to be the best fiber for the analysis of volatile compounds in peach.

Extraction time is another important factor to sampling efficiency. In this experiment, the different extraction times (20, 30, 40, 50 and 60 min) were evaluated to obtain the optimized sampling efficiency (Figure 1B). According to Zhang et al. (2007), an increase in sampling extraction time increases the headspace concentration of the volatile compounds, favoring their extraction. However, as the absorption of analytes by the fiber coating is an exothermic process, the partition coefficient decreases by increasing extraction time, negatively

affecting the absorption of analytes. The results show that the best conditions to extract the volatile compounds of oblate-peach was extraction time close to 40 min.

Volatile compounds emanating from the peaches by HS-SPME

The aroma volatiles of oblate-peach pulp were identified according to the standard mass spectra of the National Institute of Standards and Technology (NIST) MS spectral library. Aroma volatiles were considered 'identified' when their mass spectral fit values were at the default value of 85 or above. When available, some aroma volatiles were further confirmed by comparing their retention times with standards. The variability of the retention times between the aroma volatiles and corresponding standards were within 0.05 min. Figure 2 shows the total ion chromatogram obtained for a pulp sample of oblate-peach before and after 4 weeks of cold storage (4±1°C) with the PDMS-CAR-DVB fibre at the optimal sampling conditions. The volatile compounds behaved differently during storage, a total of 58 volatiles included esters, lactones, alcohols, aldehydes, ketones, acids and hydrocarbons were identified in oblate-peach pulp, 53 prior to storage and 44 after 4 week's storage (Table 1). Thirteen of the pre-storage volatiles were not found after storage.

A total of 53 volatile compounds emanating from intact peaches just prior to storage were identified: 11 esters, 5 lactones, 17 alcohols, 7 aldehydes, 3 ketones, 1 acid and 7 hydrocarbons (Figure 3; Table 1). Based on the relative proportion of the primary classes, aldehydes (61.91%) comprised the major components, followed by alcohols (11.96%), esters (6.31%), lactone (2.49%), and ketones (0.42%). Peach volatiles previously identified, lactones, in

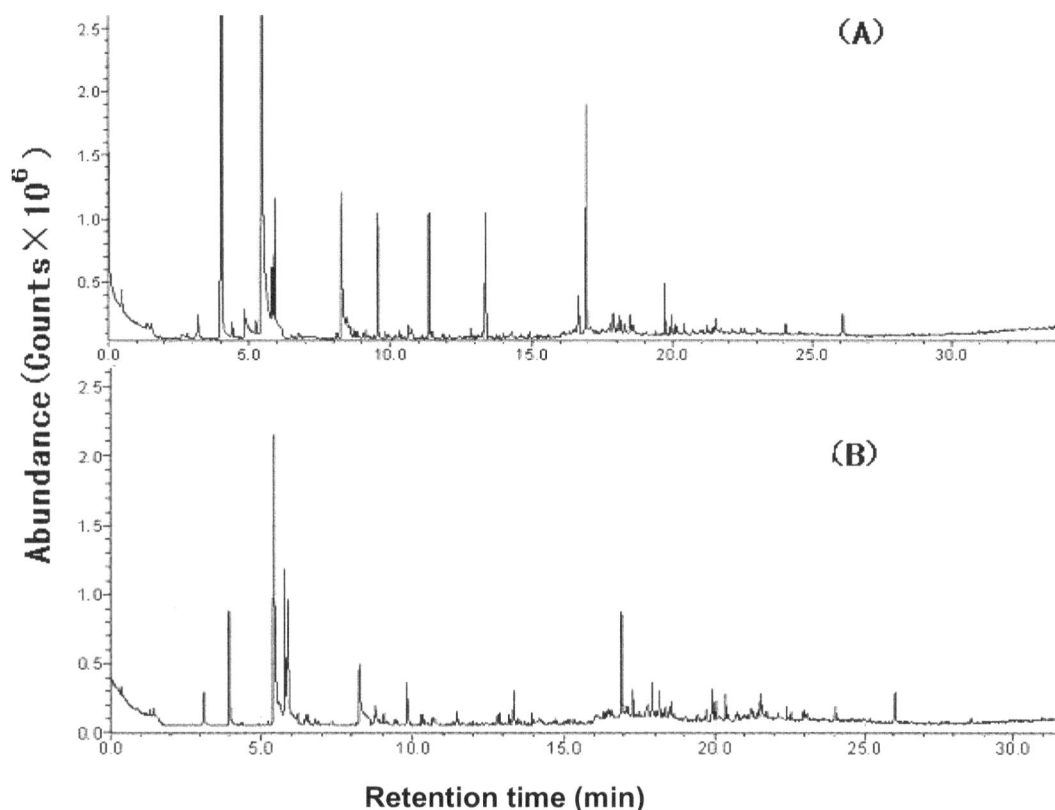

Figure 2. The aroma profile characteristics of oblate-peach pulp at before (A) and after (B) 4 weeks of cold storage (4±1°C) phases by HSSPME.

particular γ-C_8, C_{10}, δ-C10 and some unsaturated lactones have been reported as "character impact" compounds in peach aroma (Do et al., 1969). They act in association with other volatiles, such as C_6 aldehydes, C_6 alcohols and terpenoids, to produce the flavours specific to peach, and lactones contribute the "peachy" background whilst others contribute fruity and floral notes (Horvat and Chapman, 1990a).

In this study, δ-decalactone, which contributed only 0.18% of total volatiles (Table 1), has been identified as the major contributor to the overall aroma of peaches due to its low odor threshold and peach-like odor (Horvat and Chapman, 1990b). Pre-storage volatiles had significantly higher content of δ-decalactone than after storage. After 4 weeks of storage, the composition of volatile compounds varied quantitatively and qualitatively from the pre-storage composition (Table 1; Figure 2). The relative proportion of esters, alcohols, lactones, aldehydes, and ketones was 18.61, 23.48, 1.63, 35.62, and 0.9%, respectively. The relative proportion of alcohols and esters increased; however, the proportion of the esters of other classes decreased during storage. Volatile esters, which are formed by esterification of an alcohol and a carboxylic acid, play an important role in the characteristic aroma of many fruits (e.g., peach, pear, apple, apricots, and grapes); high content in esters

should give a pleasant flavour in peaches (Narain et al., 1990; Takeoka et al., 1992; Holland et al., 2005; Greger Schieberle, 2007; Franco et al., 2004). Octanyl acetate, heptyl acetate, and allyl methacrylate, present in the peach prior to storage, were no longer found after 4 weeks in storage. However, 1 ester and 4 alcohol compounds that were initially absent were formed during storage.

Volatile compounds emanating from the peaches by LLE and SDE

Conventional LLE and SDE methods were also used to sample the aroma volatiles from oblate-peach pulp (Figure 3). Table 2 shows the results of LLE and SDE sampling. Twenty-four aroma compounds from peach pulp were isolated by LLE and twenty-nine aroma volatiles were sampled by SDE. Among 5 volatile esters identified in the LLE and SDE procedures, ethyl acetate and ethyl 3-methylbut-2-enoate made up the main aroma profile. HSSPME found more esters than the conventional methods. Generally speaking, HSSPME recovered more volatiles than LLE or SDE methods. It is possible that the degradation of some unstable compounds during the LLE and SDE procedures

Table 1. Volatile compounds of oblate-peaches before and after 4 weeks of cold storage (4±1°C) by HS-SPME.

Volatile compounds	Retention time (min)	Aroma volatiles	Normalized amounts of aroma volatiles (%)	
			Pre-storage	Post-storage
Alcohols	0.474	Methanol	0.45	0.41
	3.101	Pentanol	0.63	0.6
	4.432	Phenethyl alcohol	0.3	ND
	5.774	trans-2-Hexen-1-ol	1.27	13.8
	6.544	Hexanol	ND	0.46
	8. 462	Geraniol	0.83	ND
	8.707	cis-3-Octen-1-ol	0.23	0.25
	9.033	5-Methyl -heptan-1-ol	ND	0.35
	9.114	5-Ethyl-heptan-2-ol	0.19	0.3
	10.631	trans-2-Nonen-1-ol	0.21	ND
	11.379	Linalool	2.93	ND
	12.830	Menthol	0.23	0.39
	13.352	Terpineol	3.01	1.4
	17.284	Tetradecan-4-ol	ND	0.67
	17.770	Nonane-1,9-diol	ND	0.46
	17.906	trans-2-Decene-1-ol	0.62	1.89
	18.279	2-Decanol	0.15	ND
	20.38	Dodecan-4-ol	0.24	0.61
	21.215	Decan-2-ol	0.14	0.37
	21.514	2- Butyl-octan -1-ol	0.38	0.99
	22.422	trans-2-Nonen-1-ol	0.15	0.53
Esters	5.876	Hexyl formate	3.62	10.76
	6.198	Heptyl formate	ND	0.58
	8.807	Vinyl hexanoate	0.16	0.51
	9.819	Octyl propionate	0.15	1.55
	13.191	Salicyateformate	0.11	0.38
	13.292	Linalyl isobutyrate	0.21	0.49
	13.723	Allyl methacrylate	0.07	ND
	19.922	Diethyl phthalate	0.6	1.55
	20.132	Heptyl acetate	0.08	ND
	20.703	Octanyl acetate	0.22	ND
	24.047	Dipropyl phthalate	0.3	0.81
	26.05	Dibutyl phthalate	0.79	1.98
Lactones	10.181	γ-Hexalactone	0.2	0.36
	14.92	γ-Heptalactone	0.15	ND
	18.492	γ-Dodecalactone	0.4	0.25
	18.542	δ-Decalactone	0.55	0.18
	19.707	γ-Undecalactone	1.19	0.84
Aldehydes	4.025	Hexanal	20.68	7.47
	4.873	Furfural	1.14	ND
	5.466	trans-2-Hexen-1-al	38.96	27.29
	8. 138	cis-2-Hepten-1-al	0.15	ND
	8. 575	Benzaldehyde	0.64	ND
	11.464	n-Nonanal	0.13	0.47
	18.337	Glutaraldehyde	0.21	0.39

Table 1. Contd.

Ketones	16.5	α-Ionone	0.11	ND
	18.633	2-Dodecanone	0.15	0.31
	20.071	3-Methylheptan-4-one	0.16	0.59
Acids	8. 271	Benzaldehyde	6.46	4.24
Hydrocarbons	16.642	β-Myrcene	0.92	ND
	18.007	β-Caryophyllene	5.28	0.3
	16.920	Decene	0.17	0.9
	18.143	Trideca ne	0.52	11.79
	21.575	Hexadecane	0.13	0.41
	22.545	1-Bromodecane	0.16	0.36
	22.992	Nonadecane	0.12	0.36
Others	9.547	Maltol	3.05	0.74
	13.991	Benzothiazine	0.1	0.66

[a] ND= not detected.

[b] Normalized amounts of aroma volatiles (%) = $\dfrac{\text{Peak area of an aroma volatile}}{\text{Total peak area of all aroma volatiles}}$.

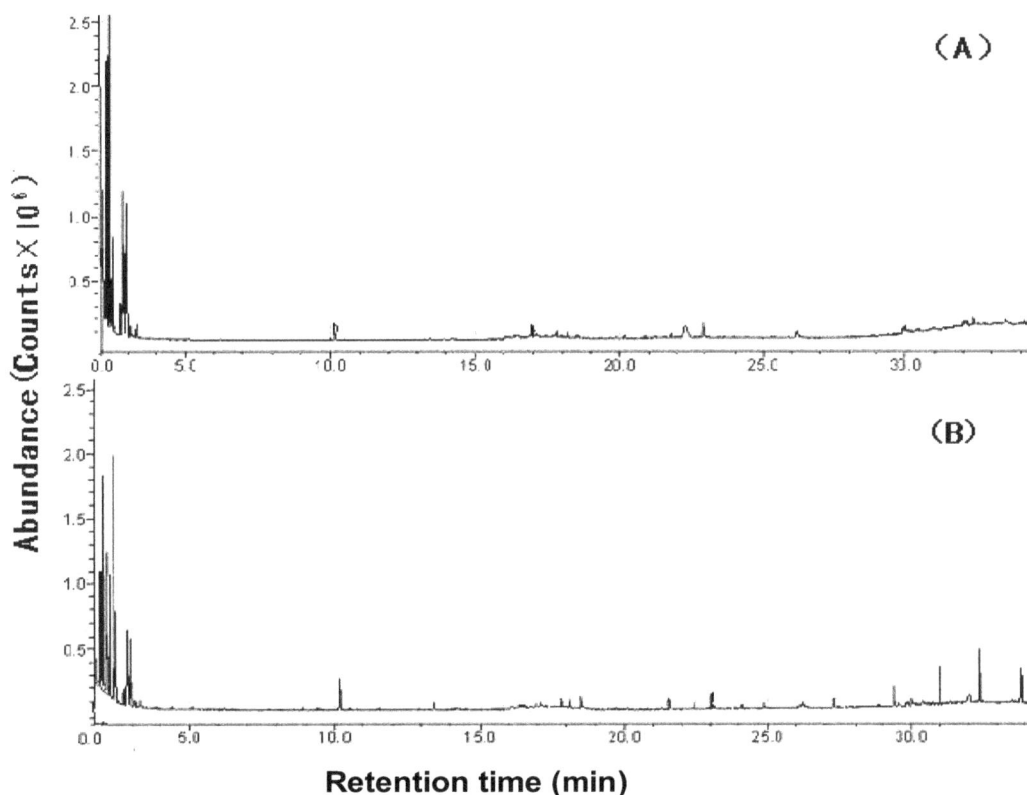

Figure 3. The aroma profile characteristics of oblate-peach pulp by LLE (A) and SDE (B).

produced some artifacts not found by the SPME method. The sample changed greatly during the LLE and SDE processes because of the complex extraction or distillation procedures.

Table 2. Volatile compounds of oblate-peach before and after 4 weeks of storage (4±1°C) by LLE and SDE.

Volatile compounds	Retention time (min)	Aroma volatiles	Normalized amounts of aroma volatiles (%)	
			LLE	SDE
Alcohols	2.864	Salicyl alcohol	9.6	7.54
	4.432	Phenethyl alcohol	ND	3.2
	17.284	Tetradecan-4-ol	ND	0.35
	17.906	trans-2-Decene-1-ol	0.21	0.53
	29.404	Nonadecanol	ND	2.29
	29.902	Oleyl Alcohol	0.25	0.47
Esters	2.363	Ethyl acetate	16.45	16.5
	2.465	Ethyl 3-methylbut-2-enoate	14.3	11.42
	3.111	Isoamyl hexanoate	0.67	0.51
	22.932	p-Methoxyphenyl acetate	0.59	ND
	33.483	Decyl butyrate	0.56	1
Lactones	10.181	γ-Hexalactone	0.85	0.2
	18.542	δ-Decalactone	ND	0.43
	18.092	γ-Octalactone	ND	0.53
Aldehydes	2.244	Isovaleraldehyde	17.75	11.53
Ketones	2.761	Heptane-4-one	1.95	1.43
Acids	26.195	Tetradecanoic acid	0.89	0.8
Hydrocarbons	2.046	Butane	8.58	5.21
	2.139	Pentane	8.89	5.47
	2.91	Hexane	12.57	8.91
	3.296	Heptane	0.82	0.63
	13.415	Nonane	0.18	0.58
	16.456	Undecane	0.34	1.69
	16.525	Dodecane	0.66	0.38
	21.575	Hexadecane	0.19	1.02
	22.375	Octadecane Tridecane	1.7	0.23
	22.992	Nonadecane	0.56	2.12
	29.942	Allylcyclohexane	1.37	ND
	31.028	Docosane	ND	3.47
	32.015	Tricosane	ND	1.87
	32.375	Tetracosane	ND	5.65
	33.817	Heptacosane	0.28	4.04

[a] ND= not detected.

[b] Normalized amounts of aroma volatiles (%) = $\dfrac{\text{Peak area of an aroma volatile}}{\text{Total peak area of all aroma volatiles}}$

On the other hand, LLE and SDE were suitable for the sampling of stable compounds, such as hydrocarbons, isovaleraldehyde, and isoamyl hexanoate, which were not found by HS-SPME. Although different sampling methods resulted in a different sampling efficiency for some typical aroma volatiles, based on the normalized amounts, the main aroma volatiles were the same compounds in both HS-SPME and conventional sampling methods, such as phenethyl alcohol, γ-hexalactone, 3-methylheptan-4-one, γ-undecalactone, trans-2-decene-1-ol and so on. Our results support the idea that the methods are complementary for sampling the aroma volatiles of oblate-peach pulp. Similar information was reported on the aroma of the durian pulp (Zhang et al.,

Table 3. Physical characteristics of the oblate-peaches before and after 4 week's storage.

Parameter	Storage time (weeks)				
	0	1	2	3	4
Fruit appearance	5.0±0.3[a]	4.7±0.2[b]	4.3±0.3[c]	4.1±0.2[c]	3.5±0.2[d]
SSC (%)	12.5±1.0[c]	12.8±1.2[b]	11.5±0.8[d]	14.7±0.6[a]	10.3±0.2[e]
Rot index (%)		0[d]	1.8±0.5[c]	14.11±0.6[b]	22. 63±2.4[a]
Weight loss (%)		0.25±0.08[d]	0.37±0.07[c]	0.49±0.11[b]	0.67±0.08[a]
Firmness (kg/m^2)	11.8±2.5[a]	9.2±0.2[b]	7.5±0.5[c]	6.8±3.3[d]	5.5±1.8[e]
Acidity (%)	0.430±0.012[a]	0.391±0.007[b]	0.266±0.007[c]	0.252±0.002[d]	0.232±0.003[e]

Different letter in a row indicate a significant difference ($p < 0.05$). Values = means of standard deviations (n = 3).

2007).

Physical characteristics of the oblate-peaches

Physical characteristics of oblate selection are presented in Table 3. Appearance is a major criterion for determining the acceptability of products. As shown in the table, visual quality scores decreased after 4 weeks to 3.5 (good and limit of marketability) for the sample. Table 3 also shows that slight differences of SSC existed over the storage time. Fruit weight loss is mainly associated with respiration and moisture evaporation through the skin (Sanz et al., 1997).

The thin skin of peach fruits makes them susceptible to rapid water loss, resulting in shrivelling and deterioration. In our study, the fruits demonstrated a gradual loss of weight during storage. Throughout storage, the loss of weight of control fruit was significantly greater than that of the treated fruit. At the end of storage, fruits showed 2.81% loss in weight. It was found that the peach fruits started to rot after 2 weeks storage.

Conclusions

A combination of sampling methods (HS-SPME, LLE and SDE) was developed to study the aroma profiles of oblate-peach pulp during storage, followed by GC–MS detection. Silica fiber coated with (DVB-CAR-PDMS) was found to be more efficient for collecting the SPME headspace volatile compounds, and the extraction time of 40 min was preferred in this study. The SPME headspace volatile constituents present in oblate-peach fruit before and after 4 weeks of cold storage (4±1°C) have been studied, as well as some physical characteristics such as weight, firmness, soluble solids content (SSC), and titratable acidity.

The volatile compounds behaved differently during storage, a total of 58 volatiles were identified, 53 prior to storage and 44 after 4 weeks storage, the content of lactones and esters, characteristic compounds of peach aroma, were much lower after 4 weeks of storage. Potential bio-markers were looked for, based on the aroma profile characteristics by common model strategy. The combination of HS-SPME and the conventional methods provided the most representative aroma information for oblate- peach pulp during storage.

ACKNOWLEDGEMENT

Our research was supported by the National "Twelfth Five-Year" Plan for Science and Technology Support: NO.2011BAD27B01-01-B.

REFERENCES

Aubert C, Baumann S, Arguel H (2005). Optimization of the Analysis of Flavor Volatile Compounds by Liquid–Liquid Microextraction (LLME). Application to the Aroma Analysis of Melons, Peaches, Grapes, Strawberries, and Tomatoes. J. Agric. Food Chem., 53: 8881-8895.

Aubert C, Gunata Z, Ambid C, Baumes R (2003). Changes in physicochemical characteristics and volatile constituents of yellow- and white-fleshed nectarines during maturation and artificial ripening. J. Agric. Food Chem., 51: 3083-3091.

Aubert C, Milhet C (2007). Distribution of the volatile compounds in the different parts of a white-fleshed peach (*Prunus persica* L. Batsch). Food Chem., 102:375-384

Baldry J, Dougan J, Howard GE (1972). Volatile flavouring constituents of Durian. Phytochemistry, 11: 081-2084.

Biale JB (1960). Respiration of fruits. In: Encyclopedia of Plant Physiology, Ruhland, W. (Ed.), Springer Verlag, New York, USA, pp. 536-592.

Do JY, Salunkhe DK, Olson LE (1969). Isolation, Identification and Comparison of the Volatiles of Peach Fruit as Related to Harvest Maturity and Artificial Ripening. J. Food Sci., 34: 618-621.

Franco M, Peinado RA, Medina M, Moreno J (2004). Off-vine grape drying effect on volatile compounds and aromatic series in must from Pedro Ximenez grape variety. J Agric Food Chem., 52: 3905-3910.

Greger V, Schieberle P (2007). Characterization of the key aroma compounds in apricots (*Prunus armeniaca*) by application of the molecular sensory science concept. J. Agric. Food Chem., 55: 5221-5228.

Hardenburg RE, Watada AE, Wang CY (1986). The commercial storage of fruits, vegetables, and florist and nursery stocks. In: USDA Department of Agriculture Handbook, U.S. Dept.of Agriculture, Agricultural Research Service, Washington,D.C., USA, p. 66.

Harker FR, Gunson FA, Jaeger SR (2003). The case for fruit quality: an interpretive review of consumer attitudes, and preferences for apples. Postharvest Biol. Technol., 28: 333-347.

Holland D, Larkov O, Bar-Yáakov I, Bar E, Zax A (2005). Developmental and varietal differences in volatile ester formation and acetyl-CoA: alcohol acetyl transferase activities in apple (*Malus domestica Borkh.*) fruit. J. Agric. Food Chem., 53: 7198-7203.

Horvat RJ, Chapman GW (1990a). Comparison of volatile compounds from peach fruit and leaves (cv. Monroe) during maturation. J. Agric. Food Chem., 38: 1442-1444.

Horvat RJ, Chapman GW, Robertson JA, Meredith FI, Scorza R, Callahan AM, Morgens P (1990b). Comparison of the volatile compounds from several commercial peach cultivars. J. Agric. Food Chem., 38: 234-237.

Kühn BF, Thybo AK (2001). The influence of sensory and physiochemical quality on Danish children's preferences for apples. Food. Qual. Prefer., 12: 543-550.

Moser R, Düvel D, Greve R (1980). Volatile constituents and fatty acid composition of lipids in *Durio zibethinus*. Phytochemistry, 19: 79-81.

Narain N, Hsieh TCY, Johnson CE (1990). Dynamic Headspace Concentration and Gas Chromatography of Volatile Flavor Components in Peach, J Food Sci., 55:1303-1307.

Peneau S, Hoehn E, Roth HR, Escher F, Nuessli J (2006). Importance and consumer perception of freshness of apples. Food. Qual. Prefer., 17: 9-19.

Sanz C, Olías JM, Pérez AG (1997). Aroma biochemistry of fruits and vegetables. In: Phytochemistry of Fruit and Vegetables. Tomás-Barberán, F.A., Robins, R.J. (Eds.), Oxford Science Publications, Oxford, UK, pp. 125-155.

Takeoka GR, Buttery RG, Flath RA (1992). Volatile constituents of Asian pear (*Pyrus serotina*). J Agric Food Chem., 40: 1925-1929.

Vermeir S, Hertog ML, Vankerschaver K, Swennen R, Nicolai BM, Lammertyn J (2009). Instrumental based flavour characterisation of banana fruit. Lwt-Food Sci. Technol., 42: 1647-1653.

Zhang ZM, Zeng DD, Li GK (2007). The study of the aroma profile characteristics of durian pulp during storage by the combination sampling method coupled with GC–MS. Flavour Frag. J., 22: 71-77.

24

Variation in germination and seed longevity of kenaf (*Hibiscus canabinus*) as affected by different maturity and harvesting stages

O. J. Olasoji[1]*, O. F. Owolade[1], R. A. Badmus[1], A. A. Olosunde[2] and O. J. Okoh[3]

[1]Institute of Agricultural Research and Training, Obafemi Awolowo University, Moor Plantation, Ibadan, Oyo State, Nigeria.
[2]National Centre for Genetic Resources and Biotechnology, Moor Plantation, Ibadan, Oyo State, Nigeria.
[3]Department of Plant Breeding and Seed Science, University of Agriculture, P. M. B. 2373, Markurdi, Benue State.

Laboratory experiments were carried out during November 2010 to October 2011 at the seed testing Laboratory of the Institute of Agricultural Research and Training, Obafemi Awolowo University, Moor Plantation, Ibadan, to determine the effects of harvesting stages on seed viability and longevity of kenaf. Kenaf seeds of four genotypes were harvested at four stages of seed maturity namely; (1) four weeks after flowering (4 WAF), (2) five weeks after flowering (5 WAF), (3) six weeks after flowering (6 WAF) and (4) at full maturity (FM). All the seeds were packed inside envelop and wrapped with polyethylene bags and kept at 10°C and 60% RH for a period of 12 months. The highest germination percentage average seed moisture content, average dry seed weight and the lower ion leakage concentration were recorded in all the kenaf genotypes when harvested at 5 WAF. Upon storage, after harvesting, the germination percent declined, while the highest viability was recorded by genotype A-60-282b at 4 months after storage, followed by genotype Ex shika at 8 and 12 months storage periods, respectively. The seed deterioration during storage could be minimized by proper harvest timing. Seeds harvested 5 WAF and stored at 10°C showed the highest seed viability. Therefore, mature seeds are recommended for harvested 5 WAF for enhanced storability.

Key words: Kenaf seed, harvesting stage, seed longevity, viability.

INTRODUCTION

Seed development is the period between fertilization and maximum fresh weight accumulation. Seed maturation begins at the end of seed development and continues till harvested (Mehta et al., 1993). The decision on when to harvest a seed crop is influenced amongst other things by the state of fruit at maturity. The stage of seed maturity determines the storage potential of the seeds if all other factors are kept constant (Fontes and Ohlrogge, 1972). Maturity and storability of seeds are important but the exact relationship between these variables is not elucidated. Harvesting stage influences the quality of seed in relation to germination, vigor, viability and also storability. Seeds harvested at right stage (physiological maturity) will be well developed, matured and possess maximum viability and vigor. Nkang and Umoh (1997) reported that germinability of cultivars harvested at different maturity periods differed significantly after six months of dry storage.

On the contrary, early harvesting prior to physiological maturity drastically lowers seed yield and quality on account of under developed and immature seeds. It is well documented that seeds harvested at an early stage of development are germinable but generally have poor vigor, indicating that germinability does not mean complete physiological development (Coolbear et al., 1997). Storability of seeds is mainly a genetic character and is influenced by pre-storage history of seed, seed maturation and environmental factors during pre-and

*Corresponding author. E-mail: juliusolasoji@yahoo.com.

Table 1. Mean value of moisture content (%) of kenaf at different harvesting stages.

Genotype	4 WAF	5 WAF	6 WAF	FM
A-60-282b	44.93	15.99	15.30	11.26
Ex- Shika	44.41	20.35	14.88	11.94
Ifeken-100	47.67	19.86	17.94	11.91
S-72-78-10	49.61	19.02	21.23	18.07
Mean	46.66	18.81	17.34	13.30
CV (%)	4.20	12.31	9.60	21.18
S.E.	0.49	0.73	0.69	0.74

WAF: Weeks after flowering; FM: full maturity.

post-harvest stages (Mahesha et al., 2001b). Early harvested seeds will be immature and poorly developed and as such, are poor storers compared to seed harvest at physiological maturity (Singh and Lachanna, 1995; Deshpande et al., 1991).

A high vigor seed lot has good storage potential and retains germination potential during storage, whereas low vigor seed lot show poor storage potential and may show a rapid decline in germination (Delouche and Baskin, 1973; Hampton, 1992). Research on kenaf seed storage indicated that when stored at 8% RH, kenaf seeds remained fully viable for 5.5 years when stored at either -10 or 10°C, and fully viable for 5.5 years when stored at -10,or 0°C and 12% RH (Toole et al., 1960). Meints and Smith (2003) reported that seed germination remained acceptable and was unaffected by storage for up to 4 years, when kenaf (Everglades 41) seeds were stored at 10°C and ambient relative humidity. We are willing to decipher whether kenaf seeds harvested at different developmental stages (physiologically immature, mature, and over matured) store well to give a reasonable high percent viability following sowing. In light of this, there was need to determine at what developmental stage kenaf should be harvested, so as to store well and promote better seed viability.

MATERIALS AND METHODS

The experiment was carried out in 2010 at the Institute of Agricultural Research and Training, Moor Plantation, Ibadan, Nigeria. Four kenaf lines were evaluated in the experiment. All lines were planted in a randomized complete block design with 3 replications. Four seeds per hill were planted at a spacing of 50 by 20 cm in a plot of 2.5 by 5 m with 6 rows per plot. The pre-emergence herbicide, Pendimethalin (500 EC) was applied at the rate of 1.7 kg ai/ha, using a knapsack sprayer. Two weeks after sowing, the plants were thinned to 2 plants per hill. NPK fertilizer was applied 2 days after weeding at the rate of 80:30:30. The insecticide lambda-cyhalothrin (2.5% EC) was applied at the concentration of 0.68 L/ha. At about 50% flowering, 200 flowers were tagged within the experimental rows. Seeds for the experiment were harvested at four different stages of development, namely; 4 weeks after flowering (WAF), 5 WAF (when the pods had turned brown), 6 WAF and lastly, at full seed maturity (FM).

Harvested pods were carefully shelled and the seeds extracted and bulked to make average samples for determination of moisture content, seed mass and germination percentage. Calculation of seed mass and moisture content were based on weight determined before and after oven drying seed samples at 105°C for 24 h. The harvested seeds were later packed inside envelopes and wrapped with polyethylene bag and kept at 10°C and 60% RH for a period of 12 months. The following tests were conducted on the stored seeds at 4 month interval for a period of 12 months.

Standard germination test

The test was performed according to International Seed Testing Association (ISTA, 2001). Fifty seeds of each genotype were germinated and replicated three times in 11 cm diameter Petri dishes inside a moistened filter paper with 5 ml of distilled water. The Petri dishes were arranged inside an incubator at 30°C in a completely randomized design (CRD). After seven days of germination, the proportion of germinated seed that produced normal seedlings was expressed as germination percentage.

Electrical conductivity

Fifty intact seeds from the three replicates were counted, weighed, and placed in a glass flask containing 100 ml of distilled water. The flasks were covered with aluminum foil to prevent contamination by flying insects and the flasks were gently shaken intermittently. Conductivity measurements were taken after 24 h at 25°C reference temperature using Mettler Toledo MC126 conductivity meter. All measurements were expressed as $\mu Scm^{-1}g^{-1}$ and the results were interpreted as suggested by Hampton and Tekrony (1995). Data were taken on seed viability and conductivity. The collected data were subjected to statistical analysis using SAS (2003) version 9.1 Software for test of significance and accurate inferences.

RESULTS AND DISCUSSION

Effect of maturity stages on dry seed weight and moisture content

The obtained results (Table 1) indicated that marked reduction in seed moisture content occurred in all the kenaf genotypes tested between 4 and 5 WAF. The average moisture content for the two harvesting stages are 46.66 and 18.81%, respectively, where 60% of the seed moisture content was lost during this period. The average lowest moisture (13.3%) was observed with the seeds that were collected at harvesting maturity. Mehta et al. (1993) reported that seeds of chickpea harvested at 29 days after anthesis (DAA) showed the highest moisture percentage, while seeds harvested at 45 DAA showed the lowest moisture percentage. Moisture WAF (Mahesha et al., 2001a). In all the genotypes, the content was the highest at 4 WAF and was lowest at 7 dry seed weight increased from 25.67 mg/seed at 4 WAF and reached the maximum of 26.49 mg/seed at about 5 WAF (Table 2). There was little change in dry weights of the seed harvested at later dates. The mass maturity or the end of grain filling period was attained at 5 WAF. At

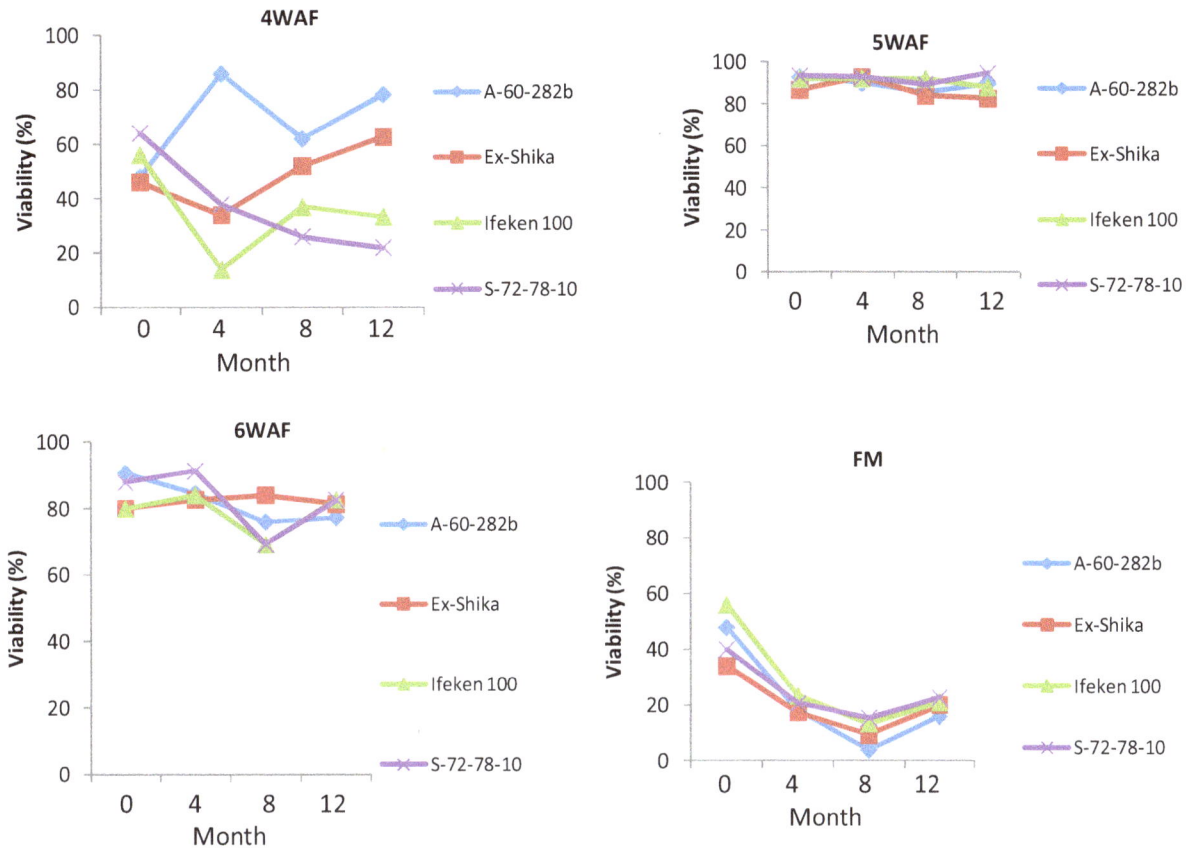

Figure 1. Percentage viability of kenaf seeds after dry storage.

Table 2. Mean value of seed mass (mg) of kenaf at different harvesting stages.

Genotype	4 WAF	5 WAF	6 WAF	FM
A-60-282b	25.87	27.32	25.06	24.20
Ex- Shika	25.26	25.52	25.26	24.06
Ifeken-100	25.92	27.00	26.52	25.60
S-72-78-10	25.66	26.12	25.72	24.20
Mean	25.67	26.49	25.64	24.52
CV (%)	6.65	5.72	4.43	4.27
S.E.	0.014	0.013	0.014	0.011

WAF: Weeks after flowering; FM: full maturity.

full maturity, the seed moisture content ranged between 15.99 and 20.35% among the fours kenaf genotypes with a mean of 18.81%.

Effects of maturity stages on seed germination after storage

Different longevity of seed storage as well as storage condition exert significant influence on seed viability (Nkang and Umoh, 1997). The results clearly pointed out declining trends in seed germination during storage at different harvesting stages. Overall seed germination was influenced by seed maturity stages. Germination performance was smaller for seeds collected at harvesting maturity (Figure 1). The reduction in germination performance of seeds harvested at harvesting maturity (HM) from the initial stage could be as the result of a decline of inherent germination due to unnecessary delay after physiological maturity before harvesting. Seed germination performance among different categories of seed maturity was higher in seeds harvested at 5 WAF (> 95%) followed by those harvested at 6 WAF and lowest in seeds harvested at usual harvesting maturity. Initial viability at 4 WAF (< 60%) was the same in all the tested kenaf genotypes. After some months in storage, the germination percent declined in 3 out of the 4 kenaf genotypes. Highest viability was recorded by genotype A-60-282b at 4 months after storage followed by genotype, Ex shika at 8 and 12 months storage periods, respectively. This agreed with the works of Nkang and Umoh (1997) that reported that the germinability of cultivars harvested at different maturity periods differed significantly after six months of storage for soybean. At 5 WAF, the result showed that the viability of all the kenaf genotypes throughout storage period were almost the same. This stage of harvest coincides with achievement

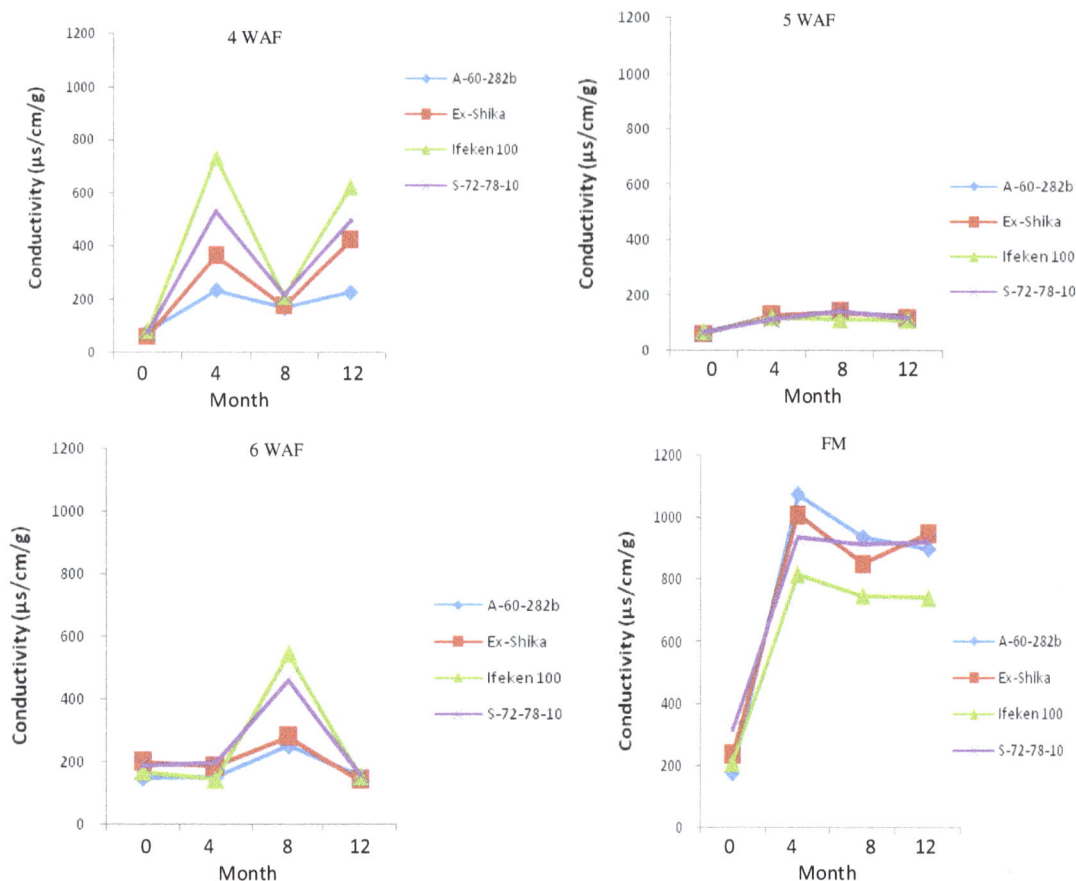

Figure 2. Conductivity (μs/cm/g) of kenaf seeds after dry storage.

of maximum dry weight of the seeds. Kumar et al. (2002) reported that seed yield and quality largely depends on the stage of maturity. As such, harvesting of seeds at the right stage of maturity is most important since harvesting either at early or late stage results in poor quality seeds. The viability was at the highest throughout the storage period because the seed had acquired maximum quality at harvest. At 6 WAF, Ex-shika maintained its viability percentage from the time of harvest to 12 months after storage suggesting that regardless of other factors, some genotypes store better than others. Our findings agreed with the earlier work of Meints and Smith (2003). Other genotypes had their viability potentials reduced 4 months after storage. At full maturity (FM), the initial viability had gone down even before storage. During storage, the viability decreased significantly. This could be as a result of unfavorable weather conditions the seeds were exposed to or experienced due to delayed harvesting after physiological maturity on the field.

Effects of maturity stages and storage on ion leachate

The mean conductivity of seeds harvested at 5 WAF and

stored at 10°C for 12 months tended to be lower than that of seed harvested at harvesting maturity (Figure 2). The conductivity of seeds harvested at 4 WAF was high in 3 of the kenaf genotypes with the exception of A-60-282b that showed moderate value. At 6 WAF, 2 of the genotypes showed increasing of ion leakage concentration 8 months after storage. The bulk conductivity of seeds harvested at FM showed huge increase in ion leakage concentration after 4 months of storage. The bulk conductivity test demonstrated the degree of the loss of solutes from the seeds, which reflects the extent of membrane deterioration resulting from seed aging (Roberts, 1986). Higher seed vigor was found at 5 WAF, which was also the stage at which lower ion leakage concentration was recorded. This means that the seeds at this stage of harvest had maintained their membrane integrity. The lower bulk conductivity values in seeds harvested at 5 WAF coincided with the higher values of standard germination (Figure 1).

Conclusion

The viability of kenaf seed in terms of standard germination and bulk conductivity was affected by the stage of seed maturity at harvest. The seed deterioration process could be minimized with proper harvest timing. Seed viability of seeds harvested at 5 WAF and stored at 10°C, stored better. Therefore, mature seeds of kenaf should be harvested at the right time, that is, 5 WAF are recommended for seed storage, to enhance viability and promote storability in the seeds

ACKNOWLEDGEMENT

This research was supported by a research grant from the Institute of Agricultural Research and Training, Obafemi Awolowo University, Moor Plantation, Ibadan, Nigeria.

REFERENCES

Coolbear P, Hill MJ, Pe W (1997). Maturation of grass and legume seed. In: Fairey DT, Hampton JG (eds.), Forage Seed Production, Vol. 1 Temperate Species. University Press, Cambridge. pp. 71-104.

Delouche JC, Baskin CC (1973). Accelerated ageing techniques for predicting the storability of seed lots. Seed Sci. Technol. 1:427-452.

Deshpande VK, Kulkarni GN, Kurdikeri MB (1991). Storability of maize as influenced by time of harvesting. Curr. Res. 20:205-207.

International Seed Testing Association, ISTA (2001). International Rules for Seed Testing. Seed Science Technololgy Vol. 29. Zurich, Switzerland.

Fontes LAN, Ohlrogge AJ (1972). Influence of seed size and population on yield and other characteristics of soybeans (Glycine max). Agron. J. 64:833-836

Hampton JG (1992). Prolonging seed quality. Proceeding of the 4th Australian Seed Research Conference. pp. 181-194.

Hampton JG, Tekrony DM (1995). Handbook of Vigour Test Methods. ISTA, Zurich.

Kumar V, Shahidhan SD, Kurdikeri MB, Channaveeraswami AS, Hosmani RM (2002). Influence of harvesting stages on seed yield and quality in paprika (Capsicum annuum L.). Seed Res. 30(1):99-103

Mahesha CR, Channaveeraswami AS, Kurdikeri MB, Shekhargouda M, Merwade MN (2001a). Seed maturation studies in sunflower genotypes. Seed Res., 29(1):95-97.

Mahesha CR, Channaveeraswami AS, Kurdikeri MB, Shekhargouda M, Merwade MN (2001b). Storability of sunflower seeds harvested at different maturity dates. Seed Res. 29(1):98-102.

Mehta CJ, Kuhad MS, Sheoran IS, Nandwal AS (1993). Studies on seed development and germination in chickpea cultivars. Seed Res. 21(2):89-91.

Meints DD, Smith CA (2003). Kenaf seed production. Proceedings of the America Kenaf Society. pp. 90-95.

Nkang A, Umoh EO (1997). Six month storability of five soybean cultivars as influenced by stage of harvest, storage temperature and relative humidity. Seed Sci. Technol. 25:93-99

Roberts EH (1986). Quantifying seed deterioration. In: McDonald MB (ed.). Physiology of seed deterioration. Crop Science Society of America, Madison, WI. pp. 101-123.

SAS Institute (2003). SAS/STAT User' guide, Version 9.1. SAS Institute, Inc. Cary NC USA.

Singh AR, Lachanna A (1995). Effect of dates of harvesting, drying and storage on seed quality of sorghum parental lines. Seed Res. 13:180-185.

Toole EH, Toole VK, Nelson EG (1960). Preservation of hemp and Kenaf Seeds. USDA Technical Bulletin, No. 1215. Washington, DC.

25

Methodology assessment on melting and texture properties of spread during ageing and impact of sample size on the representativeness of the results

Chitra Pothiraj*, Ruben Zuñiga, Helene Simonin, Sylvie Chevallier and Alain Le-Bail

ONIRIS - GEPEA – UMR CNRS 6144, BP 82225, F -44 322 Nantes Cedex 3, France.

A large number of methods and instruments have been used for measuring the melting and texture properties of margarine and spread. All these methods assume that margarine or spreads are isotropic materials. Depending on the scale of the sample, such statement is sometimes questionable in particular when using miniaturized samples. This paper gives an overview of the methods adopted and evaluates its suitability to analyze the melting and textural characteristics of spreads. Differential Scanning Calorimetry (DSC) was used to analyze the melting property of spread. Textural evaluation was carried out on spread with cone penetration, creep analysis and compression test using cylinder. DSC was found to be not reproducible due to the small size of the sample; larger sample are recommended. Creep analysis by DMA was found to be a sensitive method in detecting the differences in textural attributes of spread.

Key words: Emulsion, texture, calorimetry, spread, lipid.

INTRODUCTION

Butter, margarines and table spreads are water-in-oil emulsions; they differ by the fat to aqueous phase ratio as indicated in Table 1. The type and corresponding ratio of these products may differ in different countries. Structurally, margarine consists of a continuous liquid fat phase with fat globules, crystalline fats and aqueous phase dispersed in it (Juriaanse and Heertje, 1988). The low fat spreads cannot be easily formulated to be similar to butter. Poor or slow meltability in the mouth and slow flavor release are frequently encountered difficulties in developing low fat spread products. The rheology of low-fat spread is governed by emulsion characteristics such as the proportion of the aqueous phase and the size of the water droplets (Borwankar et al., 1992). The plastic character of fat products such as margarine, shortening and butter is the result of the presence of a three-dimensional network structure of fat crystals. The rheological properties of such products can be influenced

greatly by thermal and or mechanical treatments during processing (de Man, 1969). Spreadability of butter and margarine is an important aspect of the consumer acceptability of these products. The ratio of solid to liquid fat in a product is probably the most important factor determining hardness and spreadability (deMan et al., 1979). Storage temperature of the finished product is also important as a factor influencing the course of hardness changes and should, therefore, receive attention.

In the food industry, the texture of fat-containing products strongly depends on the microscopic, mesoscopic and finally macroscopic structure of the fat network formed within the finished product. The fat network provides firmness or solid-like behaviour to products such as margarine or spreads. Liquid fat surrounding the fat globule acts as a viscous fluid and flows on application of stress (Diener and Heldman, 1968). Margarine thus exhibits viscoelastic characteristics. For margarine, a very essential property in practical uses is its storage stability. During storage, the changes of physical properties are reflected by the changes in the crystals and the crystal network in the margarine (Zhang et al., 2005). The rheological characteristics of finished

---

*Corresponding author. E-mail: chitrarengaraj2004@yahoo.com.hk.

Table 1. Water-in-oil emulsions.

Type of emulsion	Country	Fat phase	Aqueous phase
Low fat spread.	US (patent 4071634).	30-50% fat phase.	Phosphatides, proteinaceous ingredients.
Low fat spread.	Europe (EP patent 0327288).	25-70% w/w of fat phase.	Non gelling starch hydrolysate.
Lactoprotein-free fat spread.	Canada (CA 2007770).	20-60% by weight.	0.1 to 5% by weight of gelatine or agar agar and 0.1 to 5% by weight of amylopectin rice starch.
Margarine.	Canada (CA 1271364).	Fat content 50-60%.	
Low-fat butter or margarine.	Canada (CA 2032337).	15 to 50% lipid.	40 to 60% moisture.
Butter.	France.	82% minimum.	18% maximum.
Butter.	UK.	80% minimum.	20% maximum.
Margarine.	Central Europe.	80% or less.	20% or more.
Margarine.	US.	80% or less.	20% or more.

margarine are expressed in terms such as consistency, texture, plasticity, hardness, structure and spreadability. Studies on the effect of storage conditions on the quality of retail margarine have a tendency to focus on the changes in the physical, chemical and rheological properties that occur during storage (Laia et al., 2000). The water droplet size distribution of fat spread is an important quality characteristic. The growth of microorganism is delayed when the water droplets in which they live are so small (<5 μm) that the amount of nutrients per droplet is insufficient (Van Dalen, 2002).

Product attributes such as spreadability, hardness and work softening are determined at least partly by the shape and size of the individual fat crystals and the way in which these fat crystals interact to form clusters, agglomerates and networks. For the fat blends cooled to 35°C, the surface area sizes ranges from 0.1 to 0.2 μm^2, approximately with an average of 0.34 μm^2. For samples cooled to 5°C, the surface area sizes ranges from 0.003 to 0.07 μm^2 with an average of 0.018 μm^2 stating that more rapid cooling leads to smaller crystals (Heertje and Leunis, 1997). Margarines and butter containing relatively large crystals (>5 μm) at high solid content are harder, more brittle and grainy than those containing small crystals. At low solid content, large crystals cannot incorporate as much liquid oil as small crystals and the product becomes oily (Chrysam, 1985). Large needle like crystals are usually beta crystals, while small ones are beta prime crystals (de Man et al., 1990). The crystal form of soft margarines were analysed by deMan et al. (1991a). The canola margarine containing only beta crystals had no surface sheen, appeared dull and crystals

were large. The crystals of other margarines were small containing beta and beta prime crystals.

Different studies were made on physical and textural properties of margarine and spread since the product development. Melting characteristics of these products are important for flavour release and consumer acceptance. DSC measurements were used to quantify the melting of fat crystals in these products. For melting characteristics of Vanaspathi, a hydrogenated vegetable fat commonly used in India as a substitute for butter, the samples were stabilised according to the IUPAC method (1987) and the thermograms were recorded by heating at the rate of 2°C/min from -5 to 60°C (Jeyarani and Reddy, 2005). The DSC analysis of shortenings and margarines were carried out with a model 900 du Pont Thermal Analyzer. Heating and cooling rates were 5°C/min (de Man et al., 1989). Thermal characteristics of butter were analysed using SETARAM Micro DSC-II type ultra-sensitive scanning calorimetry (Schaffer et al., 2001). The measurements were carried out in the temperature range of 0 to 50°C with the heating and cooling rate of 0.3°C/min. Consistencies of plastic fat products are closely related with their flow properties. One of the manifestations of viscoelastic material is that they undergo creep, that is, continue to deform under constant stress or load (Purkayastha et al., 1985). Relative studies have been reported to study viscoelastic properties of plastic fat products. de Man (1985) developed a creep analysis instrument to measure strain under constant stress as a function of time to determine the elastic and viscous components of butter and margarine. The selected force range from 4.9 to 19.6 N and cylindrical samples 2.3 cm diameter and 2.0 cm length were prepared by using stainless steel boring tube. The most common method used for evaluation of textural properties of margarine is the cone penetrometer (AOCS, 1974 method Cc 16-60). The cone is driven into the product by the force of gravity and the penetration depth is measured. Hayakawa and de Man (1982) suggested the term "hardness index" where the weight of the cone assembly is divided by the depth of penetration. Instron Universal Testing Machine (IUMM) and Ottawa Texture Measuring System (OTMS) were used in both the penetration and compression test to analyse the texture of 'stick margarine' (de Man et al., 1990). The texture of fat blends were analysed using a penetrometer PNR 10 equipped by a cone with the angle 40° (Unilever cone) and connected with a plunger of the total weight of 159 g (Piska et al., 2006). The aim of the present study is to assess the different techniques used for the evaluation of physical and textural properties of spread. A slight modification was made with the established reference method for melting profile and texture and the technique was assessed to determine the melting behaviour, uniformity in fat distribution and texture in the spread produced from the local company.

Results obtained from this study will be helpful in optimising the techniques used for the characteristics of spread for future studies. A secondary objective is to assess the impact of the size of the sample on the measured property and finally to decide if a sample can be representative for the selected method.

MATERIALS AND METHODS

The sample was obtained from the local company soon after production and was immediately stored at 4°C to analyse the melting and texture changes during its ageing for 4 days from production day.

Water content measurement

Water content of spread was determined using the AOAC method 925.10 (AOAC, 1996). Water content values were average of three measurements.

Thermal measurement

The thermal behaviour of spread was determined by differential scanning calorimetry (DSC Q100, TA Instruments – Waters, France). The equipment was calibrated with indium (m.p. = 156.61°C and ΔH = 28.54 J/g), water and sapphire and an empty pan was used as reference for these calorimetric measurements. About 15 to 20 mg of spread stored at 4°C was sampled from the tub of margarine and was installed in the DSC pan in a walk-in-cold chamber. Then the pan was wrapped in a thermal insulation and was quickly installed in the DSC oven. This procedure allows a better control of the cold chain of spread and permit to reduce heat up that may have a strong impact on fat crystals that have crystallised during storage. Sample was cooled down at -5°C and was equilibrated for 3 min in situ in the DSC oven. This short cool

down was made to have a noise less calorimetric signal starting at around 5°C, which represents the lower temperature of the temperature range of interest when studying lipids in the case of eatable spreads and butter (roughly temperature of a domestic refrigerator. This rapid cooling was applied to all samples and it was assumed that it had a minimal impact on the solid content of the sample. A heating rate of 3°C/min was then used until reaching a final temperature of 50°C, followed by an isothermal plateau at 50°C for 10 min. Nitrogen gas flow of 50 ml/min was used to avoid any water condensation in the calorimeter head. The TA instruments software was used to record and analyse the thermograms.

Since the repeatability of the measured melting enthalpy was not very good, larger pans, containing 60 mg were used for the DSC tests. Melting enthalpy was determined by averaging three replicates. Calorimetric measurements method used in this study is slightly different from that used by Borwankar et al. (1992) with Perkin-Elmer DSC-7 where the samples are cooled from 4.4 to 0°C at -40°C/min, and then the heating scan was begun immediately upon reaching 0°C with heating rate at 10°C/min. In another method the samples were chilled with liquid nitrogen to -50°C before measuring the melting profiles (de Man et al., 1979). In AOCS official method Cj 1-94 (AOCS, 1995) of fats and oils, the samples (7 ± 0.2 mg) were heated rapidly and held at 80°C for 10 min, cooled to –60°C at 10°C/min and held for 30 min and then heated to 80°C at 5°C/min.

Texture measurement – Large deformations

Different techniques were tested and compared to find out methods suitable for the characterisation of the texture of spread during ageing. Texture profile analysis (TPA) was performed with a Universal Testing Machine Lloyd LR 5K (AMETEK SAS, France) equipped with a force capacity of 50 N. NEXYGEN MT Data Analysis software was used to analyse the data. A conical probe (60°) was driven at 1 mm/s to a depth of 19 mm twice. A force-time curve is obtained with this TPA test and hardness is defined as the peak force during the first compression cycle (Bourne, 1978). The samples were tested in their original containers which was a plastic pot (250 ml). The pot was installed on a solid plastic holder frame which was machined to perfectly match the geometric profile of the bottom of the spot. Such a system permits to prevent any artifact due to incorrect contact between the pot and the seat of the texture analyzer. Three replicates were done during 1, 2 and 3 days of storage. Two compression tests were performed with a Texture Analyzer LFRA (Brookfield, United States) equipped with a capacity of 10 N. LFRA software was used to analyse the data.

A first test was performed with a conical probe (40°) at 1 mm/s down to 10 mm in the sample. A second test was performed with a cylindrical probe (diameter 12.7 mm; height 35 mm) at 1 mm/s to a depth of 10 mm. The penetration force (in gram) was reported as hardness. For both tests, the samples were tested in their original containers and the test was replicated three times at ambient temperature.

Texture measurement - Small deformations

Dynamic mechanical analysis (DMA) offers a potentially interesting alternative to conventional texture tests such as those performed so far. Indeed, the sample installed in a DMA is much smaller (typically in our case a cylinder of 13 mm diameter and 8 mm height) and a much better control of the ambient temperature of the sample can be obtained with a stability of 0.1°C. Liquid nitrogen was used to refrigerate the equipment, inducing more expansive experimental conditions. The measurement of the force and of the displacement is also done with a much higher sensitivity and accuracy than with

Table 2. DSC analyses of different spots from the same tub of spread.

Spot	Melting program – I peak		Melting program – II peak		Melting program (I+II) peak
	T (°C)	Enthalpy (J/g)	T (°C)	Enthalpy (J/g)	Enthalpy(J/g)
1	16.95	1.90	24.15	2.72	4.62
2	13.71	0.55	21.44	3.10	3.65
3	12.93	1.03	22.13	3.99	5.01
4	14.12	0.49	21.66	2.93	3.42
5	13.99	0.26	22.02	3.93	4.19
6	13.92	0.21	22.19	3.20	3.41
7	13.77	0.14	20.90	3.68	3.82
8	14.77	0.88	23.66	4.28	5.16
Average	14.27±1.12	0.68±0.55	22.27±1.03	3.48±0.53	4.16±0.68

the Lloyd machine. With our equipment (TA Q800 – Waters Instruments, France), the force resolution was 0.0001 N and the strain resolution was 1 nm. Specific tests have been carried out and have been compared to texture tests. Sample preparation was done in cold chamber to avoid melting of the margarine. The sample is then insulated and immediately transferred to DMA with settings at DMA creep with preload force of 0.001 N (preload force is the force necessary to have contact between the sample and movable clamp).

The sample was equilibrated at 5°C for 10 min and stress of 0.002 MPa is applied on the sample for 10 min and recovery is observed for 20 min to determine permanent deformation. The test was conducted for 3 days with 2 replications.

RESULTS AND DISCUSSION

Uniformity analysis

Since some heterogeneous results were preliminary obtained on samples of small size, the uniformity of the spread composition within a same container (commercial container of 250 ml) was assessed. 8 samples within a single container stored at 4°C for 3 months were picked up at different locations to represent the overall volume of spread. Thermal and water content analysis were performed for each sample. The results of the uniformity of fat distribution by DSC analysis of 8 samples taken from the same tub of spread are shown in Table 2 while the thermograms are illustrated in Figures 1 and 2. Upon melting, two peaks are observed. The thermogram shows a small first peak at temperature between 12 and 14°C and a second broader peak at 20 to 24°C. The melting profile is not uniform for the 8 samples taken for the test. There is difference in the evolution of first and second peak and the melting enthalpy values. There are small differences in the temperature of first and second peaks but significant differences in the melting enthalpy values. The first peak occurs at 14.27 ± 1.12°C and the second peak at 22.27 ± 1.03°C. The melting enthalpy for the first peaks is 0.68 ± 0.55 J/g and 3.48 ± 0.53 J/g for the second peak. There was no significant difference in the water content of the 8 samples; the water content was

between 39.02 and 39.23% for the 8 different samples.

Ageing analysis

Mean values of melting temperature and enthalpy of spread as function of storage is displayed in Table 3. There is a slight decrease in the peak temperature and an increase in the enthalpy of spread during ageing from production to 4 days. After 4 weeks of storage there is an increase in the melting temperature and decrease in the enthalpy of spread. The fats in the soft diet margarine and stick products melted below 40°C (Borwankar et al., 1992). DSC analyses of some North American shortening showed that at temperature around 25 to 30°C substantial amounts of solids melted in these mixtures (de Man et al., 1991b).

Texture profile analysis

The texture of fat-structured food products is strongly influenced by the structure and mechanical properties of their underlying fat crystal networks (Marangoni, 2002). In texture measurements it is important not to disturb the integrity of the crystal network when sampling the material (de Man et al., 1990). In this study, all the texture measurements were made with the original containers as supplied by the company. By 'texture profile analysis', a force time curve is plotted and analyses of the force-time curve led to the extraction of different texture parameters namely, hardness, cohesiveness, adhesiveness, springiness, gumminess and chewiness (Bourne, 1978). Textural attributes of margarine analysed during the study using TPA test (Lloyd) is displayed in Table 4. Mean value for hardness for the margarine for day 1, 2 and 3 were 15.05, 15.31 and 15.38 N, respectively. There is not much difference in the texture during the post crystallisation period. The increase in hardness is less pronounced, as the crystallisation of margarine is essentially complete during the production process and

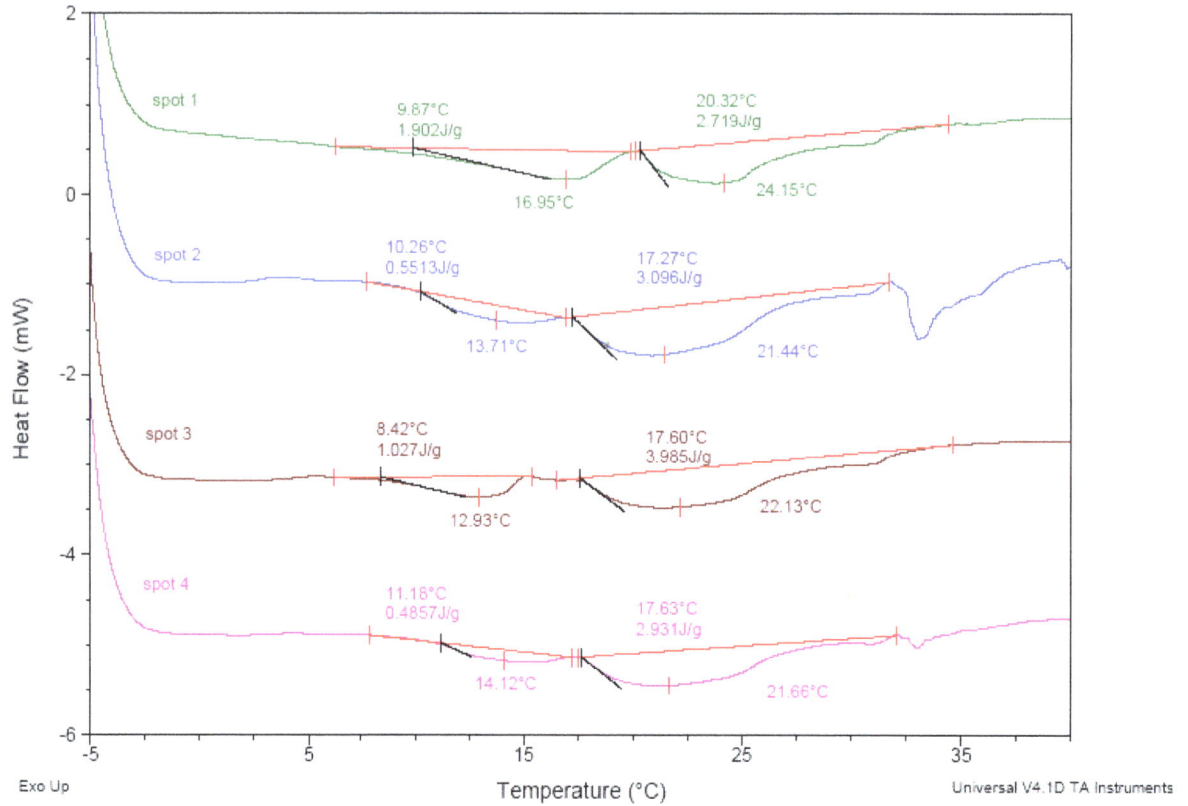

Figure 1. DSC thermogram of spot 1 to 4 taken from the same tub of spread; a relatively large dispersion of the melting enthalpy (in J/g) and of the peaks temperatures is observed even though the sample was taken from the same tub. The sample mass for the DSC was 15 to 20 mg.

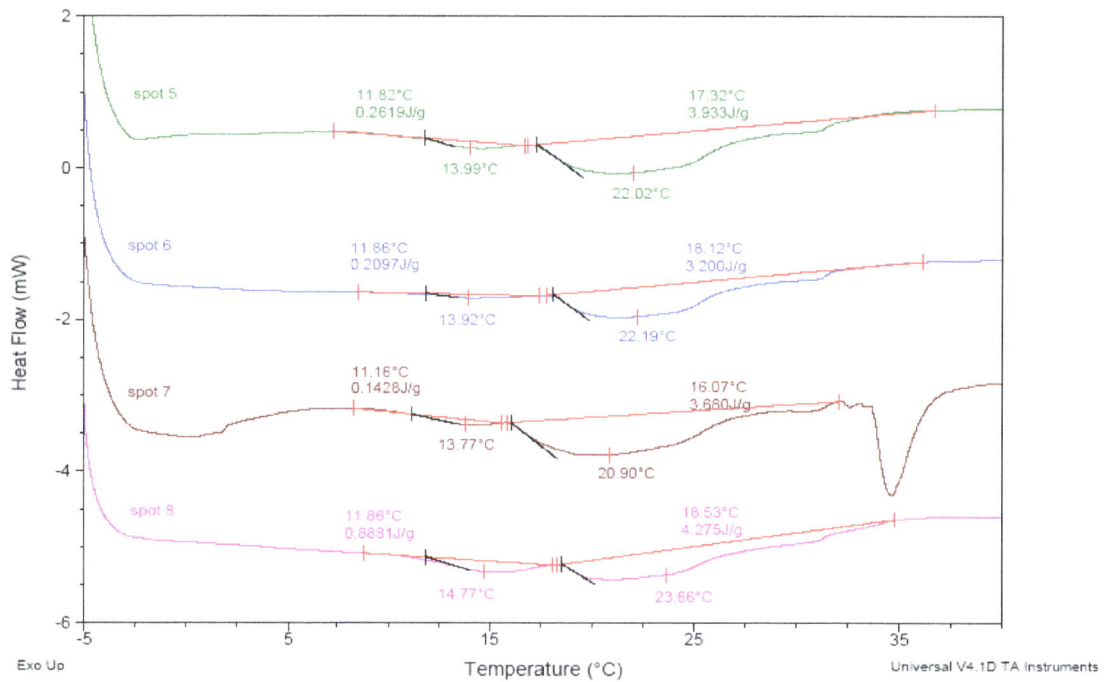

Figure 2. DSC thermogram of spot 5 to 8 taken from the same tub of spread; a relatively large dispersion of the melting enthalpy (in J/g) and of the peaks temperatures is observed even though the sample was taken from the same tub. The sample mass for the DSC was 15 to 20 mg.

Table 3. Mean value of melting temperature and enthalpy on the DSC thermograms of spread during storage.

Storage period	Melting temperature (°C)	Melting enthalpy (J/g)
1 day	23.10	5.45
2 days	22.24±0.01	7.15±0.04
4 days	21.94±0.00	8.50±0.03
After 4 weeks	24.75±0.02	6.76±0.21

Table 4.   Texture profile analysis of spread evaluated using Lyold testing equipment.

Days	Hardness (N)	Adhesiveness (J)	Cohesiveness	Elasticity	Gumminess (N)	Chewiness (N)
1	15.05±0.66	0.012±0.003	0.301±0.035	0.321±0.027	4.53±0.54	1.46±0.27
2	15.31±1.08	0.015±0.002	0.337±0.035	0.325±0.021	5.13±0.27	1.67±0.19
3	15.38±0.32	0.015±0.001	0.351±0.015	0.345±0.026	5.53±0.28	1.86±0.22

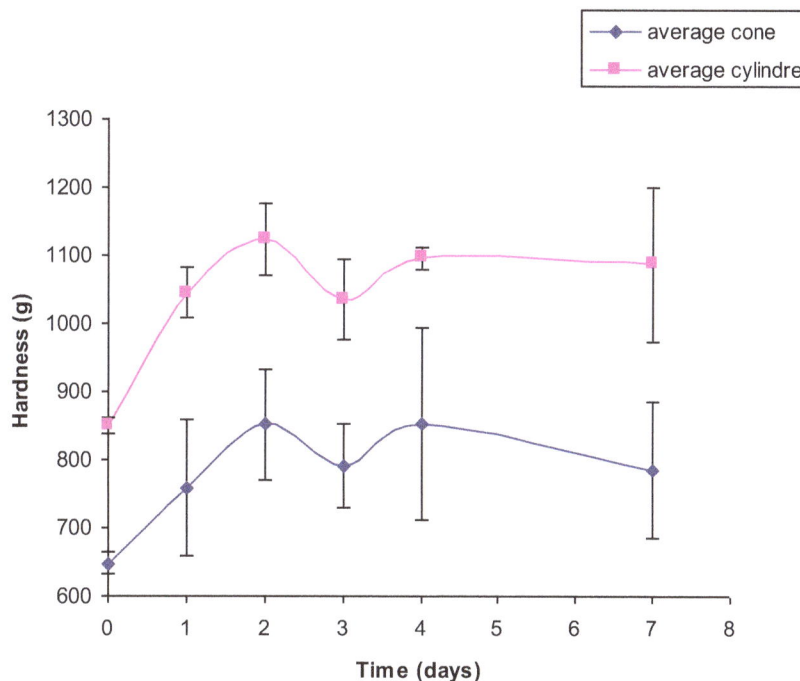

Figure 3. Texture of spread analysed using cone and cylinder during ageing (LFRA).

the crystal particles sets in rigid structure texture within 9 to 12 h of production. In studies by de Man et al. (1990 and 1991), mean value for hardness, penetration and compression for soy-bean stick margarines at 10°C were 15.6, 11.0 and 2.5 N and 8.5, 6.5 and 1.56 N for soybean soft margarines at 5°C while those for canola stick margarines were 17.1, 16.0 and 5.3 N at 10°C and for canola soft margarines were 7.3, 4.7 and 1.13 N at 5°C, respectively.

The changes in the hardness of margarine determined using cone and cylinder using LFRA texture analyser during post crystallisation period are displayed in Figure 3.

An increase in hardness is found until the first 2 days during ageing and thereafter a wide variation in texture was observed in both techniques used for analysing the texture. The results of the two techniques used for determining the texture shows the same trend in hardness of margarine during ageing. The texture of shortenings and margarines were analysed by cone penetrometer and compression (de Man et al., 1989). The margarines, although similar in solid fat content exhibited different textural properties and hardness values showed the same trend as Instron results. In this study, the texture analysis was done at 20°C whereas the

Table 5. Creep analysis of spread using DMA (mean value).

Storage period (days)	Initial deformation (%)	Recovery (%)	Permanent deformation (%)
1	1.69±0.34	0.58±0.16	1.12
2	1.86	0.81	1.05
3	3.50±0.61	1.01±0.00	2.49

Table 6. Representation of sample for DSC and texture by different authors.

Reference	Sample size	Measurement	Comments
de Man et al. (1979)	10-16 mg	DSC	The fats of the soft margarine have different thermograms with melting area in the 10-30°C range.
	25 x 16 mm	Penetration test – texture	Compression test is a sensitive method. Correlation of values within the textural method was significant.
de Man et al. (1991)			
	Cylindrical sample 20 x 20 mm	Compression test	
AOCS Official method (2000)	7±0.200 mg	DSC of fats and oil	DSC melting properties of fats and oils.
Laia et al. (2000)	Cone angle at 40, penetration depth in 0.1 mm	Hardness-cone penetrometer	ANOVA analyses showed no significant difference in the hardness of all samples during storage.
Schaffer et al. (2001)	700-900 mg	DSC	Melting of butter fat.
Campos et al. (2002)	5-10 mg	DSC	Melting behaviour of milk fat and lard.
Campos et al. (2002)	2 x 3.4 cm cone angle at 50.3	Hardness-cone penetrometer	Samples crystallised rapidly have higher values than those crystallised slowly.
Jeyarani and Yella (2005)	15 mg	DSC	Melting characteristics of Vanaspathi.

sample was initially stored at 4°C. In the case of TPA tests (cone), the cone was at 20°C which may induce an artefact due to the heat transfer between the cone and the spread. This could partially explain the higher variability observed in the TPA tests.

Creep analysis

Creep analysis is done at small deformations and involves none or minimal structure breakdown. The creep curves of a plastic fat represent the relationship of strain and time at constant stress (de Man et al., 1985). The results of the deformation during ageing of spread for 3 days are shown in Table 5. From the results it was observed that permanent deformation increases with storage period and so it implies that the product becomes less elastic with time of storage period. This study permitted to point out the importance of the size of the sample and the difference which may be observed when using different methodology to assess the texture and the melting enthalpy of lipids in the case of eatable emulsions (spread, butter). The production of spread and butter is

made under the shearing combined to refrigeration (scrapped heat exchanger). The solid content (crystallised lipids) is also known to change during storage. Therefore, cluster of recrystallized lipids may be present in the emulsion, which looks uniform and isotropic, but which is relatively non isotropic. This being said, it is obvious that the size of the sample and that the methodology used may induce some artefact. This is especially the case with modern equipment for which the trend is often to miniaturize the sample.

Crystallisation and melting in water-in-oil emulsions is a complex phenomenon affected by many factors such as emulsion droplet size, droplet-droplet interaction, polymorphism, effects of cooling rate and subsequent temperature variation (Campos et al., 2002). The size of the sample for various measurements represented by different authors in their studies is given in Table 6. Different authors have used different sample size for DSC and texture measurements and from Table 6 it is difficult to assume which sample size can be taken as representative for particular analysis. There should be a representative sample size which is suitable for each specific analysis. From our DSC tests on the uniformity of

Table 7. Recommendation on the sample size for DSC and texture (TPA test).

Size - scale	10 cm	1 cm or 1 cm/1 g	0.1 cm or 10 to 100 mg
Adapted	Texture - TPA	Melting enthalpy DSC	-
Fair		Texture - DMA	Melting enthalpy - DSC
Risky		Texture – TPA	Melting enthalpy - DSC
Not concerned	DSC texture – DMA	-	Texture – TPA; Texture - DMA

spread composition and melting behaviour of spread during ageing, recommendation can be made to take the sample of 60 to 65 mg (or more) from the centre of the tub of margarine and also from the same sample for each analysis. Additional recommendations for DSC and texture analysis are discussed in Table 7.

Conclusions

Results of the different methods adopted to assess melting behaviour and texture of spread shows its suitability to analyse shows that sample preparation method should be carefully considered for DSC and DMA as it can influence on the variability on the experimental results. Since the crystallisation of lipids in an emulsion is not isotropic, it is desirable to use large pans for the DSC tests containing about 65 mg of the product or more. Compression test using cylinder with LFRA texture analyser can be used to produce reproducible results at room temperature. The texture of spread depends on solid content of fat, crystal network, number of crystals as well as their size, morphology and polymorphism. Hence, future work is being carried out to analyse the above mentioned factors and to relate these structural characteristics to the rheological properties of spread.

REFERENCES

AOAC (1996). Method 925.10 (Air Oven Method) for moisture in flour. In: P. Cunniff, Editor, official Methods of Analysis of AOAC International, AOAC International, Maryland.

AOCS (1995). Method Cc 16-60, Cone Penetration Method. AOCS Method Cj 1-94, DSC melting properties of fats and oils. In: Official Methods and Recommended practices of the American Oil Chemists' Society, 16th ED. AOCS Press, Champaign, IL.

Borwankar RP, Frye LA, Blaurock AE, Sasevich FJ (1992). Rheological characterization of melting of margarines and tablespreads. J. Food Engr., 16, 55-74.

Bourne MC (1978). Texture profile analysis. Food Technology., July, 62-66.

Campos R, Narine SS, Marangoni AG (2002). Effect of cooling rate on the structure and mechanical properties of milk fat and lard. Food Research International., 35, 971-981.

Chrysam M (1985). Table spreads and shortenings.In: T.H. Applewhite (Ed.), Bailet's Industrial Oil and Fat Products, Wiley-Interscience, New York,pp. 41-126.

de Man JM, Dobbs JE, Sherman P (1979). Spreadability of butter and margarine. Food Texture and Rheology, London: Academic Press, 43-54.

deMan JM (1969). Effect of mechanical treatment on the hardness of margarine and butter. J. Texture Stud., 1, 109-113.

deMan JM (1985). Viscoelastic properties of plastic fat products. J. Am.Oil Chem.Soc. 62, 1672-1675.

deMan JM, Gupta S, Kloek M, Timbers GE. (1985). Viscoelastic properties of plastic fat products. J. Am.Oil Chem.Soc.,62 (12), 1672-1675.

deMan L, deMan JM, Blackman B. (1989). Physical and textural evaluation of some shortenings and margarines. J. Am.Oil Chem.Soc.,66 (1), 128-132.

deMan L, DeMan JM, Blackman B (1991). Physical and Textural characteristics of some North American shortenings. J. Am.Oil Chem.Soc., 68 (2), 63-69.

deMan L, Postmus E, deMan JM (1990). Textural and physical properties of North American stick margarines. J. Am.Oil Chem.Soc., 67 (5), 323-328.

deMan L, Shen CF, deMan JM (1991a). Composition, physical and textural characteristics of soft (tub) margarines. J. Am.Oil Chem.Soc., 68 (2), 70-73.

Diener RG, Heldman DR. (1968). Methods of determining rheological properties of butter. Trans. ASAE., 11(4), 444-447.

Hayakawa M, DeMan J. (1982). Interpretation of cone penetrometry consistency measurements of fats. J Texture Stud., 13(2), 201- 210.

Heertje I, Leunis M (1997). Measurement of shape and size of fat crystals by electron microscopy. Food Sci. Technol., 30, 141-146.

IUPAC (1987). Standard methods for the analysis of oils Fats and Derivatives. Blackwell Scientific Publications, 7th edition.

Jeyarani T, Yella Reddy S (2005). Physicochemical evaluation of vanaspati marketed in India. J. Food Lipids., 12, 232-242.

Juriaanse AC, Heertje I (1988). Microstructure of shortenings, margarine and butter: A review. Food Microstructure., 7, 181-188.

Laia OM, Ghazali HM, France Cho, Chong CL. (2000). Physical and textural properties of an experimental table margarine prepared from lipase-catalysed transesterified palm stearin:palm kernel olein mixture during storage. Food Chemistry., 71, 173-179.

Marangoni AG (2002). Special issue of FRI – crystallisation, structure and functionality of fats. Food Research International., 35 (10), 907-908.

Piska I, Marketa Zarubova, Tomas Louzecky, Hisham Karami, Vladimir F (2006). Properties and crystallisation of fat blends. J. Food Engr., 77, 433-438.

Purkayastha S, Peleg M, Johnson EA, Normand MD (1985). A computer aided characterisation of the compressive creep behaviour of potato and cheddar cheese. J. Food Sci., 50, 45-50.

Schaffer B, Szakaly S, Lorinczy D, Schaffer B (2001). Melting properties of butter fat and the consistency of butter. J. Thermal Analysis and Calorimetry., 64, 659-669.

Van-Dalen G (2002). Determination of the water droplet size distribution of fat spreads using confocal scanning laser microscopy. J. Microscopy., 208(2), 116-133.

Zhang H, Jacobsen C, Adler-Nissen J (2005). Storage stability study of margarines produced from enzymatically interesterified fats compared to margarines produced by conventional methods. I. Physical properties. Euro. J. Lipid Sci. Technol., 107(7-8), 530 - 539.

26

Microbiological and physicochemical characterization of shea butter sold on Benin markets

Fernande G. Honfo[1,3]*, Kerstin Hell[2], Noël Akissoé[1], Anita Linnemann[3] and Ousmane Coulibaly[4]

[1]Faculty of Agronomic Sciences, University of Abomey-Calavi, 01 BP 526, Cotonou, Benin Republic.
[2]International Potato Center (CIP), Cotonou, Bénin Republic.
[3]Department of Agrotechnology and Food Sciences, Wageningen University, Wageningen, Netherlands.
[4]International Institute of Tropical Agriculture (IITA-Cotonou), Cotonou, Bénin Republic.

Shea butter, a fat from the nuts of shea tree, is of great nutritional and commercial value for local communities of Africa. The sanitary and physicochemical qualities of shea butter sold in Benin markets are unknown. This study assesses the quality characteristics of 54 samples of shea butter collected from eight markets in Benin, West Africa. Total germs, yeasts and mould varied with markets. Moisture content ranged between 2.5 to 6.2%. Iodine index was around 49 mgI$_2$/100 g, but acid index (4.1 to 6.0 mgKOH/g), peroxide value (9.4 to 11.8 meq O$_2$/kg) and saponification values (186.4 to 193.7 mgKOH/g) showed high variability both within and between samples of different markets. Quality characteristics were poorer for butter collected in the main urban markets (Cotonou, Bohicon and Malanville), due mainly to poor storage conditions. Shea butter could be stored in a clean package before sale to preserve its beneficial qualities.

Key words: Shea butter, microbial status, acid index, peroxide index, iodine index, saponification value, Benin market.

INTRODUCTION

Shea butter, a vegetable fat extracted from the kernels of the fruit of *Vitellaria paradoxa* Gaertner, Sapotaceae, is an ancient African commodity that still plays an important role in village life (Hall et al., 1996; Kengue and Ndo, 2003; Elias and Carney, 2004; Honfo et al., 2011). Shea tree is the main indigenous oil-producing wild plant spontaneously growing in Africa, and native of dry savannah zones from Senegal to Uganda. In *V. paradoxa* producing countries, such as Benin, shea butter is generally extracted by traditional processing that involves roasting, churning and boiling in the fruit-producing areas and then marketed in village or urban markets (Agbahungba and Depommier, 1989). However, the traditional extraction process in African countries implies unequal water qualities, at worst leading to increased

oxidized material observable by high peroxide values of the resulting fat (Di Vincenzo et al., 2005). Locally, shea butter is widely used as cooking oil, for producing soap, and used in pharmacological and cosmetic products. Shea butter has an increasing international demand by cosmetic and pharmaceutical industries, and it is also used as a cocoa butter additive in chocolate manufacture (Elias and Carney, 2004; CNUCED, 2006).

Shea butter is composed of triglycerides and fatty acids including oleic acid (60 to 70%); stearic acid (15 to 25%); linolenic acid (5 to 15%); palmitic acid (2 to 6%); linoleic acid (<1%) and an unsaponifiable content (3 to 15%) (CENUCED, 2006). Due to their high content of unsaturated fatty acids (49 to 63%), shea nuts are susceptible to deterioration (Maranz et al., 2004; Di Vincenzo et al., 2005).

Shea butter is the main edible oil for the communities of northern Benin; it provides food oil for more than 80% of the population in this zone, it is therefore the most important source of fatty acids and glycerol in the diet

*Corresponding author. E-mail: fernandehonfo@gmail.com.

Figure 1. Map of Benin showing the shea butter samples provenances.

(Agbahungba and Depommier, 1989). Its daily consumption in this zone is estimated at 26.3 g per person (Honfo et al., 2010), and its consumption is increasing due to the high cost of imported oils. Because of the dietary importance of shea butter, it is important to analyze the quality of the butter, and assess improvements opportunities if it is necessary. The objective for this study was to assess the major microbial and physicochemical characteristics of the shea butter sold in the different urban markets in Benin.

MATERIALS AND METHODS

Sample collection

Samples of shea butter were collected in the main urban markets of eight regions of Benin (Bohicon, Cotonou, Djougou, Kandi, Malanville, Natitingou, Parakou, and Tanguieta) (Figure 1). Most of the sample collection locations are the production zones except Bohicon and Cotonou. A total of 54 samples were collected, packed in aluminum foil, wrapped in polyethylene bags and numbered according to their origins (market, town and area). They were stored in an icebox containing ice cubes and transported to the laboratory, where they were kept at 4°C until analysis. For each parameter, sample determinations were made in triplicate.

Microbiological characteristics

Aerobic mesophilics bacteria (on plate count agar at 30°C for 72 h), total coliforms (on violet red bile agar at 30°C for 24 h), faecal or thermotolerant coliforms (on Violet Red Bile Agar at 44°C for 24 h), yeasts and moulds (on malt extract agar at 25°C for 72 h) were determined in each shea butter sample according to the methods described by Megnanou et al. (2007) and AOAC (2002).

Physicochemical characteristics

Moisture content was determined according to AOAC (2002). Colour measurements were performed using the chromameter (Minolta (CR210b). Results were expressed as L* (brightness), b* (yellowness), and ΔE (difference of color from the white ceramic standard). Acid, peroxide, iodine and saponification values were determined according to the methods of the Beninese shea butter characterization standards using NB ISO 660 (2006), NB ISO 3960 (2006), NB ISO 3961 (2006), and NB ISO 3657 (2006) respectively.

Statistical analysis

The tests of conformity by Student's t-test were performed to compare the microbiological counts of butter with the international standards. The analysis of variance (Proc GLM) was used to compare the different parameters measured on the shea butter among the provenances and SNK (Student–Newman–Keuls) was used to classify these parameters using SAS 9.1 software. Correlations were also established between variables.

RESULTS

Microbial characteristics

Samples of shea butter had various microbial load with

Table 1. Microbiological characteristics of marketed shea butter from Benin.

Provenance	Germs identified (CFU/g)		
	Aerobic mesophilic bacteria	Yeast and mould	Total coliforms
Bohicon(n = 5)	$3.810^5\pm410^{3a}$	$1.810^3\pm1.910^{2a}$	79 ± 42^a
Cotonou (n = 4)	$4.710^5\pm4.210^{3a}$	$1.510^3\pm1.710^{2ab}$	56 ± 32^{abc}
Djougou (n = 8)	$10^5\pm210^{2a}$	$3.610^2\pm29^{bc}$	23 ± 13^{bc}
Kandi (n = 5)	$1.610^4\pm10^{2a}$	$3.710^2\pm35^{bc}$	24 ± 13^{bc}
Malanville (n = 7)	$9.110^5\pm1.810^{3a}$	$10^3\pm10^{2abc}$	64 ± 34^{ab}
Natitingou (n = 6)	$2.810^5\pm6.510^{3a}$	$2.510^2\pm10^{2c}$	15 ± 11^c
Parakou (n = 10)	$2.510^5\pm2.310^{2a}$	$8.910^2\pm85^{abc}$	13 ± 10^c
Tanguieta (n = 9)	$5.510^4\pm9.810^{2a}$	$3.410^2\pm85^{bc}$	11 ± 5^c
Norms*	10^{4a}	10^c	$25b^c$

*: Codex Alimentarius, 1992; NBF 01-005, 2006; For each parameter (in column), mean ± standard deviation with the same letter are not significantly different.

Table 2. Color and moisture content of marketed shea butter from Benin.

Provenance	Moisture content	Color characteristics		
		Luminance (L*)	Yellow saturation index (b*)	Color difference (ΔE)
Bohicon (n = 5)	6.06 ± 0.7^a	62.43 ± 2.6^e	19.77 ± 6.1^{bc}	44.15 ± 1.5^a
Cotonou (n = 4)	6.24 ± 1.7^a	68.53 ± 1.7^{bc}	19.81 ± 1.6^{bc}	38.74 ± 1.9^b
Djougou (n = 8)	6.17 ± 1.9^a	70.29 ± 2.5^a	22.12 ± 3.6^{ab}	42.13 ± 2.3^a
Kandi (n = 5)	5.67 ± 1.6^a	67.36 ± 1.1^{cd}	25.55 ± 5.9^a	43.45 ± 3.9^a
Malanville (n = 7)	6.13 ± 1.7^a	65.69 ± 2.6^d	22.09 ± 5.7^{ab}	42.69 ± 3.7^a
Natitingou (n = 6)	2.48 ± 1.7^b	69.69 ± 1.1^{abc}	21.88 ± 3.3^{ab}	39.26 ± 2.8^b
Parakou (n = 10)	3.64 ± 1.4^b	69.53 ± 2.6^{abc}	21.75 ± 3.1^{ab}	39.44 ± 2.4^b
Tanguieta (n = 9)	3.81 ± 1.1^b	71.24 ± 3.1^a	15.93 ± 2.8^c	34.69 ± 3.7^c

For each parameter (in column), mean ± standard deviation with the same letter are not significantly different.

high count of aerobic mesophilic germs (4.2 to 5.70 $\log_{10}$ CFU/g), yeasts and moulds (2.4 to 3.3 $\log_{10}$ CFU/g) and total coliforms (1.0 to 1.9 $\log_{10}$ CFU/g) (Table 1). Great variation were observed between and within the geographical locations for the number of aerobic mesophilic bacteria (CV = 142 to 225%), but the differences were not statistically significant. The number of aerobic mesophilic bacteria in all collected samples was close to the international standard of 4 $\log_{10}$ CFU/g (Codex Alimentarius 1992; NBF 01-005 2006). However, the highest microbial count was found in sample collected in Cotonou and Bohicon markets, which are the biggest markets in the Southern and Central regions respectively (Table 1). Differences between locations for the number of yeasts and moulds (p = 0.0014) and total coliforms (p = 0.0056) were observed, with samples collected at the Bohicon (Central Benin) market giving significantly higher values of yeasts and moulds compared with the samples from other markets. In addition, samples from some locations gave higher numbers of yeasts and moulds than the international standards value of 1.0 $\log_{10}$ CFU/g for yeasts and moulds. Total coliforms count for all sample were closed to the standard of 1.4 $\log_{10}$ CFU/g,

but the higher value were observed in the sample collected at Bohicon. No faecal coliforms were detected.

Physicochemical characteristics

The moisture content ranged between 2.5 to 6.2% (Table 2), and significant differences between locations were found (p = 0.0001) with two distinctive groups: one with a moisture content ranging between 2 and 4% (from Natitingou, Parakou and Tanguieta) and the other with a higher moisture content (from Cotonou, Djougou, Malanville, Kandi and Bohicon).

Mean values for the color parameters of the shea butter samples are presented in Table 2. The highest L* value was observed for the samples collected in Tanguieta while the lowest L* value was from a sample from Bohicon. Reversely, low b* values and ΔE were observed in Tanguieta and the highest in Bohicon. Significant differences were observed between the shea butter provenance for L* value (p = 0.0001), b* value (p = 0.0001) and ΔE (p = 0.0001) in spite of great variability within locations.

Table 3. Biochemical characteristics of marketed shea butter from Benin.

Provenance	Acid index (mgKOH/g)	Peroxyde index (meqO$_2$/kg)	Iodine index (mgI$_2$/100 g)	Saponification index (mgKOH/g)
Bohicon (n = 5)	5.47±1.4[ab]	11.54±2.4[a]	49.37±1.1[a]	193.74±6.0[a]
Cotonou (n = 4)	5.70±1.2[a]	11.16±1.2[ab]	49.80±1.2[a]	187.19±1.5[b]
Djougou (n = 8)	4.88±1.4[ab]	9.78±1.1[ab]	50.26±1.2[a]	188.43±2.4[b]
Kandi (n = 5)	5.06±1.3[ab]	10.70±2.5[ab]	48.30±1.3[a]	190.74±5.1[ab]
Malanville (n = 7)	6.01±1.2[a]	11.76±2.3[a]	48.93±1.5[a]	190.39±4.3[ab]
Natitingou (n = 6)	4.51±1.2[ab]	9.41±0.8[b]	49.46±1.5[a]	188.59±3.4[b]
Parakou (n = 10)	4.90±1.2[ab]	10.91±1.6[ab]	49.07±1.4[a]	186.43±2.6[b]
Tanguieta (n = 9)	4.10±1.6[b]	9.45±1.9[b]	49.64±1.7[a]	186.69±3.5[b]

For each parameter (in column). mean ± standard deviation with the same letter are not significantly different.

Mean acid values ranged from 4.1 to 6.0 mgKOH/g, and mean peroxide values from 9.4 to 11.8 meq O$_2$/kg (Table 3). Samples collected in the markets of Tanguieta and Natitingou (Northern Benin) seemed to have low acid and peroxide values.

The mean values of the iodine index were around 49 mgI$_2$/100 g for the samples from all locations. No significant differences for this index were observed between the different sample provenances. However, relatively higher values were found in the samples collected in Djougou (Table 3). The mean of the saponification values varied from 186.4 to 193.7 mgKOH/g, with significant differences between locations for this index ($p = 0.0001$) (Table 3).

DISCUSSION

Traditional shea butter sold on the urban markets of Benin presented a great variability in microbial and physicochemical characteristics. Total coliform counts of most of the shea butter samples were close to the microbiological international standard for edible fat (Table 1) (Codex Alimentarius, 1992). However, difference was observed between samples provenances; the highest values of the total coliforms yeasts and moulds observed in samples collected in the three biggest markets (Bohicon, Cotonou and Malanville) could be explained basically by the suboptimal transportation conditions and storage materials used by traders or by the exposure to the atmospheric air during sale, engine exhaust and dust, since these elements could be vectors of microbial germs (Roquebert, 1997; Pfohl-Leszkowicz, 2000). Indeed, most of the butter is stored for 6 months before sale in major urban markets (Honfo et al., 2011). Two locations (Bohicon and Cotonou) are outside the shea butter production zone and during transportation traders watered the butter to maintain its humidity. This practice could lead to favourable environmental conditions for the development of microorganisms, such as yeasts and moulds, as showed by the high value in these two markets. In addition, these locations also have a climate with a high relative humidity (65 to 85%), which conditions could also influence the sanitary quality of the butter.

The moisture content, acid, and peroxide values of the butters sold in Benin were lower than those of shea butter from Côte d'Ivoire (Megnanou et al., 2007), while the iodine values of shea butter in Benin were higher. This variability could be explained by the diversity of butters sold which come from many locations, but also by the different traditional processes used for their extraction (Dieffenbacher et al., 2000, Kapseu et al., 2005; Honfo et al., 2011). The traditional process is generally without control of the unit operations (Kapseu et al., 2005) and could result in the presence of low quality shea butter in the market.

Irrespective of locations, the moisture content in all samples was higher than the international standard (0.05 to 2%) for non refined shea butters (NBF 01-005, 2006), but lower than values found in shea butter (10.2%) processed by the Bangoua method in Cameroon (Kapseu et al., 2005). The high moisture content of shea butter can activate lipase, which can potentially catalyze the hydrolysis of triglycerides leading to rapid deterioration of shea butter (Mittal and Paul, 1997). The hydrolysis of the triglycerides leads to high levels of acidity in the butter; this is corroborated by the positive and significant correlation between moisture content and acid index ($r = 0.437$, $p = 0.0001$). A high moisture content can also promote the growth of microorganisms and in this study high levels of total germs were positively correlated with moisture content ($r = 0.306$; $p = 0.0032$).

The high level of the acid index in samples collected in Malanville, Cotonou, and Bohicon, might be related to the hydrolysis of triglycerides that occurred during the storage of butter. In addition, the high number of yeasts and moulds found on the samples in these locations could increase the enzymatic hydrolysis since some of these microorganisms would have the capacity to secret

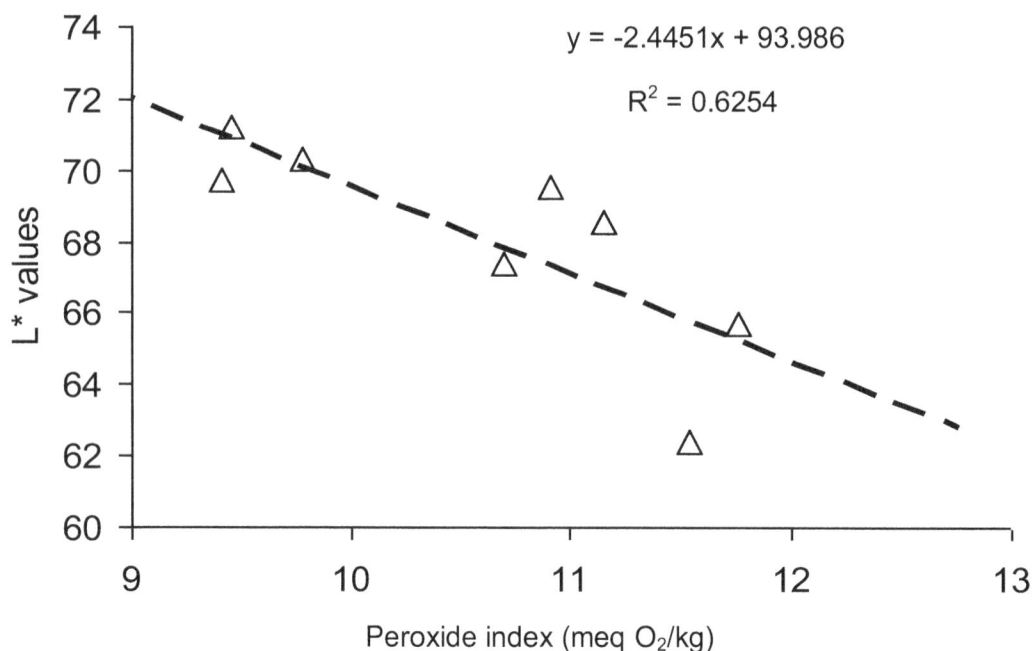

$$y = -2.4451x + 93.986$$
$$R^2 = 0.6254$$

Figure 2. Relationship between luminance (l*) and peroxide index of shea butter.

lipase, responsible for enzymatic hydrolysis in lipids (Hultin, 1994). This result is in agreement with the positive correlation between the acid value and the number of yeast and moulds ($r = 0.306; p = 0.0032$).

All samples of shea butter collected had peroxide values close to the maximal value of 10 meq02/kg tolerated by the cosmetic and pharmaceutical industry (Codex Alimentarius, 1992; NBF 01-005, 2006). Furthermore, the negative correlation between L* value and peroxide index ($r = -0.79; p = 0.0013$) (Figure 2) is consistent with the previous work of Akissoe et al. (2003), who observed the increase of brown index (100-L*) with the increase in peroxidase activity in yam cultivars (*D. rotundata* spp). Then, any deterioration factor (enzymic, hydrolysis) which increases peroxide values, can result in the low luminance of the shea butter. In addition, a positive correlation was observed between the peroxide index and the moisture content ($r = 0.509; p = 0.0001$). The relatively high acid and peroxide values in marketed shea butter of Benin could be associated with the lack of quality control in the traditional process; in particular, these high values could be explained by the long drying period for nuts and prolonged roasting of kernels (Kapseu et al., 2005; Womeni et al., 2006). Exposure of shea butter to sun and air assumedly causes hydrolysis of glycerides and the oxidation of the unsaturated fatty acids (Schreckenberg, 2004), potentially resulting in high acid and peroxide indices as observed in the presented data.

The value of the iodine index, which is a measure for the level of unsaturation of oils, was lower than the norms (58-72 mgI2/100g) (NBF 01-005, 2006). The oxidation of unsaturated fatty acids could be responsible for the relatively low iodine values of the marketed shea butter (Dieffenbacher et al., 2000).

Conclusions

The quality of shea butters sold in urban markets of Benin varied widely in terms of microbiological quality (yeasts and moulds, total coliforms) and physicochemical characteristics (color parameters, acid, peroxide and saponification values). This variation is probably associated with the transportation and storage conditions since the butter is often stored in makeshift packages for up to six months before being sold. It is recommended to transport and store shea butter in a clean package, safe from heat (solar), air and dust to preserve its beneficial qualities.

ACKNOWLEDGEMENTS

This research was supported by the International Institute of Tropical Agriculture and "Institut National de Recherches Agricoles du Bénin (INRAB)", through STDF 48 project, financed by FAO. Many thanks to all people involved in the laboratory analysis.

REFERENCES

Agbahungba G, Depommier D (1989). Aspects du parc à karités-nérés (Vitellaria paradoxa Gaertn. f. parkia biglobosa Jacq. Benth.) dans le sud du Borgou (Bénin). BFT, 222: 41-54.

Akissoe N, Hounhouigan JD, Mestres C, Nago MC (2003). How parboiling and drying affect the colour and functional characteristics of yam (Dioscorea cayenensis-rotundata) flours. Food Chem., 82: 257-264.

AOAC (2002). Official Methods of Analysis.16th edn, Association of Official Analytical Chemists, Washington DC.

CNUCED (2006). Le karité: production, consommation et marché, (Accessed 23/03/2009) http://www.unctad.org/infocomm/francais/karité/marché.htm.

Codex ALimentarius (1992). Programme mixte FAO/OMS sur les normes alimentaires, FAO, Rome.

Di Vincenzo D, Maranz S, Serraiocco A, Vito R, Wiesman Z, Bianchi G (2005). Regional variation in shea butter lipid and triterpene composition in four African countries. J. Agric. Food Chem., 53: 7473-7479.

Dieffenbacher A, Buxtorf P, Derungs R, Friedli R, Zurcher K (2000). Graisses comestibles, huiles comestibles et graisses émulsionnées. In Neukom, Zimmermann (eds.), Graisses comestibles, huiles comestibles et graisses émulsionnées, Société des chimistes analystes suisses, Berne, pp. 1-249.

Elias M, Carney J (2004). La filière féminine du karité: productrice burkinabée,«éco-consommatrices» occidentales et commerce équitable. Cah. Géogr. Québec, 48: 71-88.

Hall J, Aebischer D, Tomlinson H, Osei-Amaning E, Hindle J (1996). Vitellaria paradoxa: a monograph, School of Agricultural and Forest Sciences, University of Wales, Banghor, 105 pp.

Honfo FG, Akissoe N, Linnemann A, Soumanou M, van Boekel MAJS (2011). Nutritional composition of shea products and chemical properties of shea butter: A review. Crit. Rev. Food Nutr. (accepted).

Honfo FG, Hell K, Akissoe N, Dossa RAM, Hounhouigan JD (2010). Diversity and nutritional value of foods consumed by children in two agro-ecological zones of Benin. Afr. J. Food Sci., 4: 184-191

Hultin HO (1994). Oxidation of lipids in seafoods. In Shahidi F, Botta JR (eds.) Seafoods: Chemistry, Processing Technology and Quality. Blackie Academic and Professional, New York. pp. 49-74.

Kapseu C, Womeni H, Ndjouenkeu R Tchouanguep MF, Parmentier M (2005). Influence des méthodes de traitement des amandes sur la qualité du beurre de karité. Proc. Biol. Alim., 3: 1-18.

Kengue J, Ndo E (2003). Les fruitiers forestiers comestibles du Cameroun: aspects agronomique. In Matig E, Ndoye J, Kengue J, Awano A (eds.) Les fruitiers forestiers comestibles du Cameroun: aspects agronomique, IPGRI Regional Office for West and Central Africa, Cotonou, Benin, pp. 51-82.

Maranz S, Wiesman Z, Bisgaard J, Bianchi G (2004). Germplasm resources of Vitellaria paradoxa based on variations in fat composition across the species distribution range. Agrofor. Syst., 60: 71-76.

Megnanou R, Nianke S, Diopoh J (2007). Physicochemical and microbiological characteristics of optimized and traditional shea butters from Côte d'Ivoire. Afr. J. Biochem. Res., 1: 41-47.

Mittal G, Paul S (1997). Regulating the use of degraded oil/fat in deep fat/oil food frying. Crit. Rev. Food Sci. Nutr., 37: 635-662.

NB ISO 660 (2006). Normes Béninoises pour les corps gras d'origine animale et végétale: détermination de l'indice d'acide et de l'acidité, CEBENOR, Cotonou, Benin, 13 pp.

NB ISO 3657 (2006). Normes Béninoises pour les corps gras d'origine animale et végétale : détermination de l'indice de saponification. CEBENOR, Cotonou, Benin, 9 pp.

NB ISO 3960 (2006). Normes Béninoises pour les corps gras d'origine animale et végétale: détermination de l'indice de peroxyde. CEBENOR, Cotonou, Benin, 9 pp.

NB ISO 3961 (2006). Normes Béninoises pour les corps gras d'origine animale et végétale: détermination de l'indice d'iode. CEBENOR, Cotonou, Benin, 8 pp.

NBF 01-005 (2006). Normes Burkinabées pour le beurre de karité non raffiné, Ouagadougou, Burkina Faso, 21 pp.

Pfohl-Leszkowicz A (2000). Ecologie des moisissures et des mycotoxines : situation en France. Cah. Nutr. Diet., 35: 379-388.

Roquebert M (1997). Les moisissures: nature, biologie et contamination. (Accessed 06/04/2009) http://www.culture.gouv.fr/culture/conservation/fr/cours/roqueber.htm.

Schreckenberg K (2004). The contribution of shea butter (Vitellaria paradoxa CF Gaertner) to local livelihoods in Benin. In Sunderland T, Ndoye O (eds.), Forest Products, Livelihoods and ConservationIndonesia, pp. 91-113.

Womeni H, Ndjouenkeu R, Kapseu C, Parmentier M, Fanni J (2006). Application du procédé séchage-friture aux amandes de karité: influence sur les indices chimiques de qualité et les propriétés de fusion du beurre. OCL, 13: 297-302.

27

Storability and quality indices of palm oil in different packaging containers in Nigeria

Ego U. Okonkwo*, Kayode A. Arowora, Bukola A. Ogundele, Mike A. Omodara and Sunday S. Afolayan

Nigerian Stored Products Research Institute Headquarters, Km. 3 Asa Dam Road, P. M. B. 1489, Ilorin, 240001, Kwara State, Nigeria.

Palm oil was obtained from Linkjon commercial processing oil mill with food grade equipment at Umunze in Anambra State. The samples were stored at ambient conditions for one year in five different containers (transparent plastic, opaque blue plastic, opaque white plastic, green bottle and transparent bottle). Quality indices were determined for moisture content, free fatty acid (FFA), acid value and saponification value at bimonthly interval using standard methods. Moisture content varied between 1.0 and 1.5% in all treatments during the 12 months storage period. The initial value for FFA was 4.18% and this increased to 11.85% for palm oil stored in green bottle and lowest with 10.16% in palm oil stored in opaque blue plastic container at the end of storage. Saponification values increased from initial 191 to 250 among the treatments within the storage period. The increases observed in FFA, acid and saponification values could be attributed to moisture absorption from storage environment and oxidation. Although the values of the quality parameters varied with the type of storage containers used, there were no significant differences (P≤0.05) among treatments.

Key words: Palm oil, quality indices, storability, containers.

INTRODUCTION

Palm oil is an essential part of diet of man and animals and also plays leading roles in some manufacturing industries. The oil is unique having approximately 50% saturated fats and 40% unsaturated fats (Arowora and Fafunso, 1999). The distinctive colour of the oil is due to fat soluble carotenoids that are also responsible for the high vitamin content (Kruger et al., 2007). Most of the crude palm oil for domestic consumption and industrial purposes is processed in mills without food grade equipment. The process involved does not meet International standards for food quality and safety. The quality of the oil therefore varies depending on the processing method and different packaging materials (Okonkwo, 2011). One of the most important quality parameters in edible oil refining industry is low content of

free fatty acid (FFA) and oxidative products (Kusum et al., 2011). Therefore, the importance of quality and safe palm oil low in FFA content in human nutrition for healthy life cannot be over emphasized (Ghot, 2007).

This work determined the storability and quality indices of palm oil produced from food grade equipment at Linkjon agro-processing oil mill using different packaging containers.

MATERIALS AND METHODS

The palm oil was obtained from Linkjon agro-based oil mill, Eziagu, Orumba South LGA in Anambra State. Red palm oil was dispensed into five different storage containers. The transparent plastic, opaque white plastic and opaque blue plastic were 750 ml container of 2 mm thickness and aperture of 25 mm with threaded seal cap, while the green bottle and transparent bottle (control) were 750 ml glass bottles of 5 mm thickness and aperture of 23 mm with threaded seal cap. Triplicate samples were used for treatments and control, arranged on laboratory table at completely randomized

*Corresponding author. E-mail: egoulu@yahoo.com.

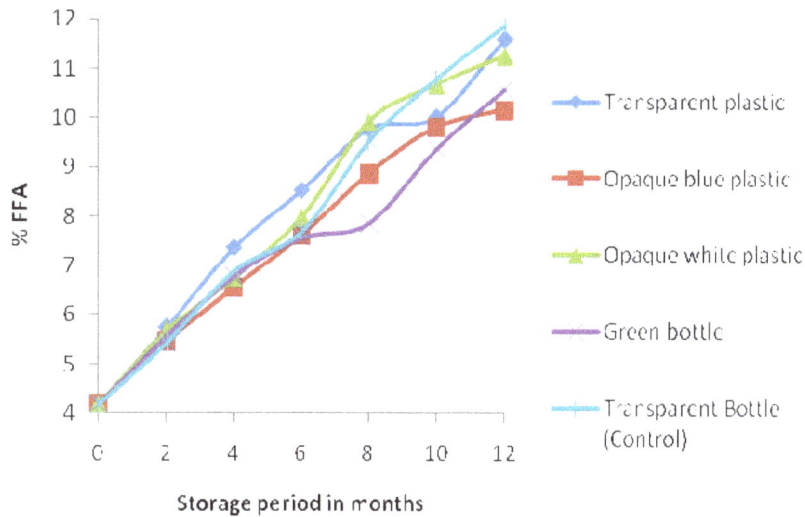

Figure 1. Effect of storage on FFA values.

design (CRD) at ambient conditions. The quality indices (moisture content, free fatty acid (FFA), acid value, and saponification value) for the palm oil were determined at initial and bimonthly interval for the 12 months storage period using the AOAC (2000) methods of analysis. The data obtained were subjected to ANOVA and differences between means tested.

RESULTS AND DISCUSSION

The results of the quality indices of palm oil during 12 months storage in the different containers are as shown in Figures 1 to 3. The initial FFA content of the batch of palm oil used for the experiment was 4.18%. There were increases among treatments in FFA (10.16 to 11.85%) at the end of storage period which did not differ significantly (P≤0.05) in all the storage containers (opaque blue plastic, green bottle, opaque white plastic, transparent plastic and transparent bottle respectively). The quality of the palm fruits used in processing palm oil is an important factor in the storage and quality of the oil. Deteriorated fruits would yield oil high in FFA and other quality indices of the palm oil. Storage containers are also important in keeping quality of palm oil in storage.

The initial FFA value of 4.18% was high which was reflected in the initial moisture content of 1.0 and 1.4% after 12 months; and the trend increased throughout the storage period. There were no significant differences (P≤0.05) among treatments in FFA values obtained (Figure 1). These findings are in accordance with Frank et al. (2011) who reported FFA value of 4.71% at initial and 10.26% in semi-mechanized extraction of palm oil after 10 weeks of storage. This is contrary to Abulude et al. (2007) who obtained FFA values (0.72-1.02 and 0.6-1.14%) for *Jatropha curcas* (Physic nut) and *Helianthus annuus* (Sunflower), respectively for oils stored at ambient conditions for 4 months in polythene, glass, metal and plastic bottles. High acid values are usually

indicative of spoilage or high moisture which enables the enzyme lipase to convert the triglycerides to free fatty acids. FFA content is the most used index for determining the quality of palm oil and must not exceed 5% (expressed as palmitic acid) according to Codex Alimentarius/FAO/WHO (2005). Figure 2 showed that the initial acid value was 9.15 and increased to between 22.27 and 25.95 among treatments at the end of 12 months storage. This was contrary to the work of Abulude et al. (2007) who obtained (13.00-50.00 and 36.00-59.00) for *Jatropha curcas* and *Helianthus annuus* after 4 months storage in different containers. Figure 3 showed that the initial saponification value (mg KOH/g oil) was 191.12 at initial and increased to 223.45 in palm oil stored in transparent plastic compared to 229.61 to 249.97 in other storage containers at the end of 12 months.

Generally, it was observed in this study that the FFA, acid value (AV) and saponification value (SV) increased in all treatments during the 12 months storage period which is in accordance with previous studies (Abulude et al., 2007; Frank et al., 2011). The increase in the quality indices as indicators of reduction in quality of the palm oil may be attributed to the initial quality of the palm fruits used for processing of the oil. When this occurs there is the likelihood of microorganisms affecting the oils, which in turn may lead to spoilage. Another factor is absorption of moisture from the laboratory environment and oxidation of the red palm oil, since Linkjon oil mill is food grade mill (Okonkwo, 2011).

Conclusion

The opaque bottles reduced light absorption by the stored palm oil which normally leads to oxidation of the product and increase in free fatty acids and rancidity.

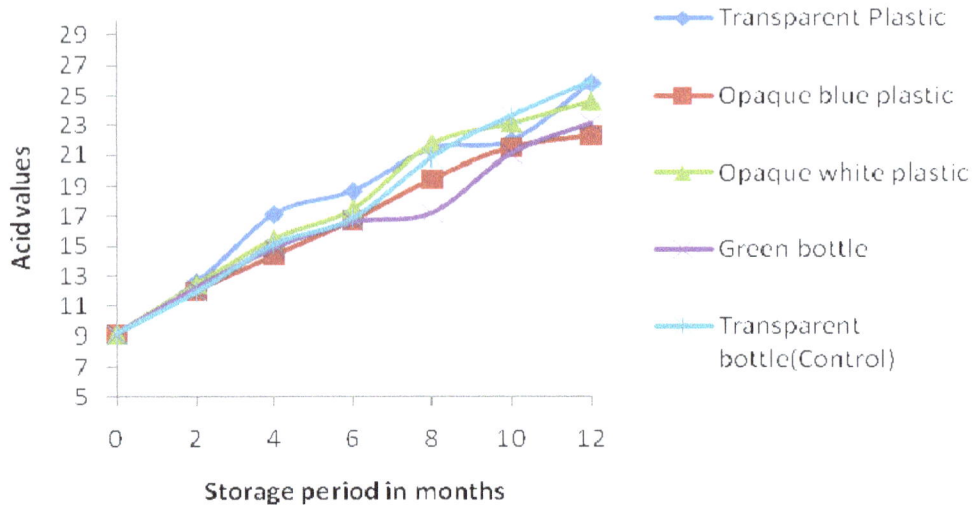

Figure 2. Effect of storage period on acid value.

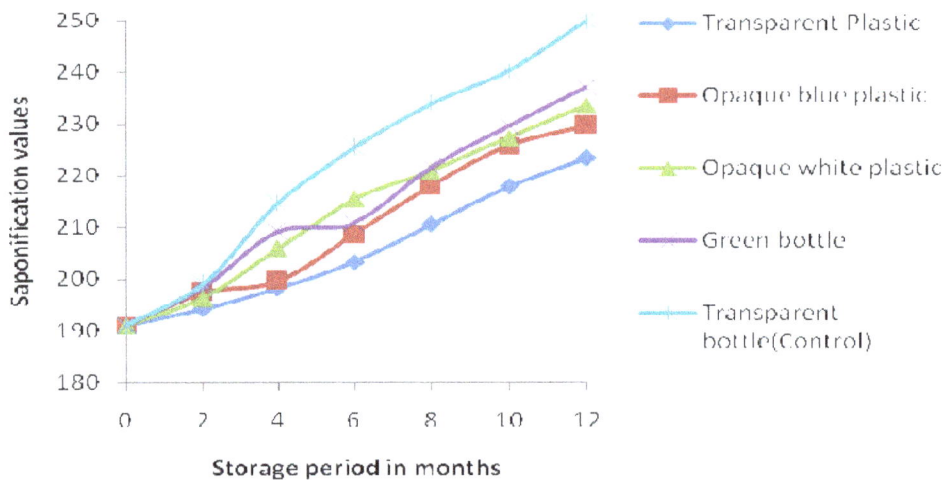

Figure 3. Effect of storage period on saponification value.

REFERENCES

Abulude FO, Ogunkoya MO, Ogunleye RF (2007). Storage properties of oils of two Nigerian oil seeds *Jatropha c urcas* (Physic nut) and *Helianthus annuus* (Sunflower). Am. J. Food. Technol. 2:207-211.

AOAC (2000). Association of Official Analytical Chemists. Official Methods of Analysis, 17th Edition, Washington D.C.

Arowora KA, Fafunso M (1999). Effect of plant lipids formulated diet on the histopathology of rat liver. West Afr. J. Biol. Appl. Chem. 44:1-7.

Codex A limentarius Commission (2005). Recommended international standards for edible arachis oil. Food and Agriculture Organization of the United Nations, World Health Organization, Geneva, Switzerland.

Frank NEG, Albert MME, Laverdure DEE, Paul K (2011). Assessment of the quality of crude palm oil from smallholders in Cameroon. J. Stored Prod. Postharvest Res. 2(3):52-58.

Ghot M (2007). Recent Trends in Rice Bran Oil Processing. J. Am. Oil Chemists' Soc. 84(4):315-324.

Kruger MJ, Engelbretcht AM, Esterhuyse J, Du Toit EF, Van Rooyen J (2007). Diet red palm oil reduces ischaemia-reperfusion injury in rats fed a hypercholesterolemic diet. British J. Nutr. 97(4):653-660.

Kusum R, Bommayya H, Fayaz Pasha P, Ramachandran HD (2011). Palm oil and rice bran oil: Current status and future prospects. Int. J. Plant Physiol. Biochem. 3(8):125-132.

Okonkwo EU (2011). Hazard analysis and critical control points in palm oil processing in Anambra State, Nigeria. Afr. J. Agric. Res. 6(2):244-247. Available online at http://www.academicjournals.org/AJAR.

28

Microwave-vacuum drying effect on drying kinetics, lycopene and ascorbic acid content of tomato slices

ABANO Ernest Ekow[1,2], M. A. Haile[1]*, OWUSU John[1,4] and ENGMANN Felix Narku[1,3]

[1]School of Food and Biological Engineering, Jiangsu University, 301 Xuefu Road, Zhenjiang 212013, China.
[2]Department of Agricultural Engineering, University of Cape Coast, Cape Coast, Ghana.
[3]Department of Hotel, Catering and Institutional Management, Kumasi Polytechnic, Kumasi, Ghana.
[4]Department of Hospitality, School of Applied Science and Technology, Koforidua Polytechnic, Koforidua, Ghana.

This study investigated the microwave-vacuum drying on the drying kinetics and quality attributes of dried tomatoes such as lycopene and ascorbic acid contents. Among the thirteen thin layer drying models that were used to fit the experimental data, the Midilli et al gave the highest correlation coefficient, lowest residual sum of squares, root mean square error, and reduced chi-square, thus indicating that the model of Middilli et al best described the microwave-vacuum drying of tomato slices. The highest ascorbic acid retention occurred in the samples dried at 200 W and 0.06 MPa, with a significant decrease (p<0.05) from an initial mean value of 2.74 ± 0.29 to 1.87 ± 0.13 mg/g dry matter, representing a decrease of about 32% in relation to the fresh tomato. On the other hand, the lycopene content of the dried tomatoes significantly (p<0.05) increased from 2.96 mg/100 g dry matter to a maximum value of 25.44 mg/100 g dry matter after microwave-vacuum drying at 700 W and 0.04 MPa.

Key words: Tomato slices, drying models, lycopene, ascorbic acid, microwave-vacuum.

INTRODUCTION

Tomato (*Lycopersicon esculentum*) is the second most important vegetable after potato due to its wide spread cultivation (Celma et al., 2009) and health benefit particularly in fighting prostate cancer, cervical cancer, cancer of the stomach and rectum as well as pharynx and oesophageal cancers (FAO, 2010; HSPH, 2010). Drying is an important process in the processing of agricultural materials. It does not only help address post-harvest challenges facing the agricultural sector but reduces the bulk volume of materials to an appreciable one. Burning of dried whole strawberries was found when microwave power of 600W was applied (Venkatachalapathy and Raghavan, 2000). Microwave drying has been found to produce poorer quality of dried mushrooms products (Walde et al., 2006).

In Poland, microwave drying of grains for propagation was found not to germinate at all. To help address some of the problems in microwave drying, microwave-vacuum drying has gained attention in the drying of fruits and vegetables. The idea to apply microwave-vacuum was to speed up drying process, to increase mass transfer by an increased pressure gradient between inner and outer layers, and to maintain drying process at low temperature (Pere and Rodier, 2002; Therdthai and Zhou, 2009). It is noted that in food drying processes, process engineers need simple, accurate, and analytical models with verified experimental data to carry out optimized unit operations (McMinn, 2006). The contribution of modeling in this respect is to help understand the heat and mass transfer phenomena, and computer simulations for designing new processes and improving existing ones (Kardum et al., 2001). It is a well known fact that thin-layer equations have found wide application in experimental convective air drying data, but its application to microwave vacuum

---
*Corresponding author. E-mail: mhl@ujs.edu.cn.

drying is more limited. Previous work done support the argument that the Page, Lewis models could not adequately describe the experimental drying curves of microwave assisted drying of carrot (Prabhanjan et al., 1995), microwave vacuum drying of mint leaves (Therdthai and Zhou, 2009), microwave -vacuum drying of lactose powder (McMinn, 2006).

Various drying models have been developed for solar drying of organic tomatoes (Sacilik et al., 2006), microwave drying of carrot (Prabhanjan et al., 1995), okra (Dadali et al., 2007), microwave vacuum drying of mint leaves (Therdthai and Zhou, 2009), lactose powder (McMinn, 2006), ultrasonic drying of red bell pepper and apple slices (Schössler et al., 2012), and infrared drying of apple slices (Togrul, 2005). At present, there are limited models that adequately represent the drying of tomato slices in a microwave-vacuum dryer. Therefore, this study was undertaken to investigate the thin-layer drying kinetics of tomato slices in a microwave-vacuum dryer and to simulate the experimental data to the mathematical models available in scientific literature. In addition, the quality attributes of dried tomatoes such as lycopene and ascorbic acid contents were investigated.

MATERIALS AND METHODS

Sample preparation

The fresh tomatoes used in this study were tomatoes from the same cultivar procured from the Zhenjiang local Market, China. Selection was based on visual assessment of uniform colour and geometry. The initial moisture content of the tomatoes was determined at 105°C for 24 h. The tomato samples were washed under running tap water and stored in a refrigerator at a temperature of 4°C in order to slow down the physiological and chemical changes (Maskan, 2001; Karaaslan and Tuncer, 2008). Prior to drying, the individual tomatoes were cut into slices of thickness 7 mm using a cutting machine (SS-250, SEP Machinery Company Ltd, Guangzhou, China).

Combined microwave-vacuum drying

The microwave-vacuum experiment was conducted in a 4 × 3 factorial design with four microwave powers (200, 300, 500 and 700 W) in three treatments combinations with vacuum pump pressures (0.04, 0.05 and 0.06 MPa). All observations were replicated three times. A laboratory scale microwave-vacuum dryer (NJZ07-1, Jiangsu University, Zhenjiang, China) with maximum power output of 1000 W at 2500 MHz was used for this experiment.

The dimensions of the microwave cavity were 40 cm by 28 cm by 33 cm. The microwave was fitted with rotating table of diameter 230 mm, which is the load density of the dryer. The microwave dryer is combined with vacuum pump, which has maximum pressure of 0.1 MPa. In this machine the microwave dryer can only function when the pressure is at least 0.04 MPa. The schematic diagram of the combined microwave vacuum dryer is shown in Figure 1. Microwave powers of 200, 300, 500 and 700 W at vacuum pressures of 0.04, 0.05 and 0.06 MPa were used for this experiment at constant sample loading density of 100 g. The sample was put in thin layer in a bowl with one slice not touching the other and placed on the Teflon turn table in the microwave

vacuum cavity. Depending on the drying condition, the moisture loss was recorded every 1, 2 and 4 min until constant mass was observed for the determination of the drying curves (Karaaslan and Tuncer, 2008). After the set time, the sample was taken out of the drying chamber and weighed with an electronic balance (accuracy of 0.001 g) within 10 s (Karaaslan and Tuncer, 2008).

Ascorbic acid analysis

The ascorbic acid in the fresh and dried samples was determined volumetrically according to the method described by Sadasivam and Balasubraminan (1987). A dye solution was prepared by dissolving 52 mg of 2, 6–dichlorophenol indophenols in small volume of distilled water containing 42 mg of sodium bicarbonate and made up to 200 ml. The stock standard solution of concentration (1 mg/ml) was made with 100 mg standard ascorbic acid in 100 ml of 4% oxalic acid solution in a flask. Ten millilitre of the stock solution was taken and diluted to 100 ml with oxalic acid solution. To 5 ml of the working standard solution was added 10 ml of oxalic acid in a 100 ml conical flask and titrated against the dye (V_1 ml). The appearance of the pink colour, which persists for few minutes, indicates the end point and the dye consumed is equivalent to the amount of ascorbic acid. Ascorbic acid was extracted with 50 ml oxalic acid from 5 g fresh and 0.5 g grounded dried tomatoes and centrifuged at 4000 rpm for 10 min. To 5 ml of the supernatant was added 10 ml of oxalic acid and the mixture titrated against the dye (V_2 ml). The titration was done in triplicate. The ascorbic acid (AA) content in mg/g was calculated following Equation 1:

$$AA = \frac{0.5}{V_1} \times \frac{V_2}{5\ ml} \times \frac{50\ ml}{\text{Sample weight}} \tag{1}$$

Lycopene content analysis

The lycopene in the fresh and dried tomato samples was extracted in acetone and then taken up in petroleum ether following the protocol of Ranganna (Ranganna, 1976). The pigment was repeatedly extracted from 5 g of fresh tomato pulp and 0.5 g of dried tomatoes with acetone and assisted with ultrasound until the residue was colourless. The acetone extract was transferred into a separating funnel containing 20 ml petroleum ether and mixed gently. Then, 20 ml of 5% sodium sulphate solution was added and the separating funnel shaken gently. Another 20 ml of petroleum ether was added to make up for any evaporated petroleum ether. The coloured pigment noticed in the upper petroleum ether phase was separated and the lower phase re-extracted with additional petroleum ether until colourless. The petroleum ether was washed with a little distilled water and poured into a brown bottle containing 10 g of anhydrous sodium sulphate and kept for at least 30 min. The petroleum ether was decanted into a 100 ml volumetric flask through a cotton wool in a funnel. The sodium sulphate slurry was washed with petroleum ether until it was colourless and transferred into the volumetric flask. It was topped up to the mark with petroleum ether and the absorbance measured in a spectrophotometer at 503 nm with petroleum ether as blank. The lycopene content (mg/100 g sample) was calculated using Equation 2.

$$L_c = \frac{31.206 \times Abs_{503nm}}{w_t} \tag{2}$$

where L_c is the lycopene content (mg/100 g), Abs is the absorbance, and w_t is the weight of the sample (g).

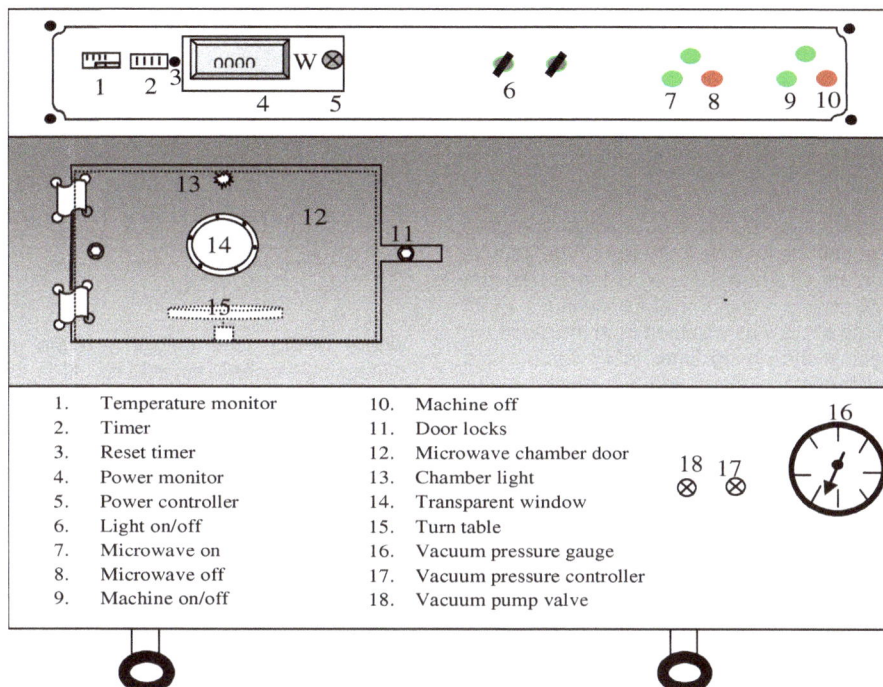

Figure 1. Schematic diagram of the microwave vacuum equipment.

Table 1. Mathematical models that was applied to the experimental data.

Model name	Model expression	Reference
Lewis	$MR = \exp(-kt)$	Lewis (1921)
Page	$MR = \exp(-kt^n)$	Page (1949)
Modified Page	$MR = \exp[-(kt)^n]$	Overhults et al. (1973)
Henderson and Pabis	$MR = a\exp(kt)$	Henderson and Pabis (1961)
Modified Henderson and Pabis	$MR = a\exp(-kt) + b\exp(-gt) + c\exp(-ht)$	McMinn (2006)
Logarithmic	$MR = a\exp(-kt) + c$	McMinn (2006)
Two-term	$MR = a\exp(-k_0 t) + b\exp(-k_1 t)$	Sacilik et al. (2006)
Two-term exponential	$MR = a\exp(-kt) + (1-a)\exp(-kat)$	Sharaf-Eldeen et al. (1980)
Diffusion approach	$MR = a\exp(-kt) + (1-a)\exp(-kbt)$	Kassem (1998)
Verma et al.	$MR = a\exp(-kt) + (1-a)\exp(-gt)$	Verma et al. (1985)
Wang and Singh	$MR = 1 + at + bt^2$	Demir et al. (2007)
Midilli et al.	$MR = a\exp(-kt^n) + bt$	Midilli et al. (2002)
Parabolic	$MR = a + bt + ct^2$	Sharma and Prasad (2004)

Modeling of the experimental data

This experimental moisture ratio data were fitted to 13 thin-layer drying models presented in Table 1 to describe the microwave vacuum drying kinetics of tomato slices. The Non-linear regression analysis was performed with statistical package program SPSS 16.0 (SPSS, 2007). Regression and correlation analysis are very useful tools in modeling the drying behaviour of biological materials (Erbay and Icier, 2008). Four primary criteria used to assess the accuracy of fit to the models were: the correlation coefficient (R^2), residual sum of squares (RSS), the root mean square error (RMSE), and the reduced chi square (χ^2). In non-linear regression modeling, designing the fitting procedure to achieve a minimum RSS is very crucial. The highest R^2, and the lowest values of RSS, χ^2 and RMSE values were used to predict the goodness of fit. Several authors have used these criteria to select the best model for drying mistletoe (Kose and Erenturk, 2010), onion slices (Mota et al., 2010), aromatic plants (Akpinar, 2006), olive leaves (Erbay and Icier, 2008), okra (Doymaz, 2005), thyme (Doymaz, 2010), lactose powder (McMinn, 2006), and aloe vera (Vega et al., 2007).

Calculation of moisture diffusion coefficient and activation energy

The moisture ratio values for the various drying conditions were generated according to the simplified diffusion equation given by Equation 3.

$$MR = \frac{M - M_e}{M_0 - M_e} = \frac{8}{\pi^2}\exp\left(\frac{-\pi^2 D_{eff} t}{4L^2}\right) \tag{3}$$

where, MR is the moisture ratio, D_{eff} is the effective moisture diffusivity (m^2/s), and L is half the thickness of slice of the sample (m), M is the moisture content at any time, t, M_e is the equilibrium moisture content, and M_0 is the initial moisture content. From Equation 3, D_{eff} of the tomato slices was obtained from the slope (K) of the graph of $lnMR$ against the drying time. $lnMR$ versus time results in a straight line with negative slope and K is related to D_{eff} by equation 4.

$$K = \frac{\pi^2 D_{eff}}{4L^2} \tag{4}$$

In a standard microwave drying process, the internal temperature of the sample is not a measurable variable. The tailored form of the Arrhenius type equation is used to illustrate the relationship between the diffusivity coefficient and the ratio of the microwave power output to sample thickness instead of the temperature, for the calculation of the activation energy. The equation as suggested by Dadali et al. (2007) is of the form:

$$D_{eff} = D_0 \exp\left[\frac{-E_a q}{P}\right] \tag{5}$$

where D_o is the constant in the Arrhenius equation (m^2s^{-1}), E_a is the activation energy (W/mm), P is the microwave power (W), and q is the sample thickness (mm). Equation 5 can be rearranged into the form:

$$In\left(D_{eff}\right) = In\left(D_0\right) - \frac{E_a q}{P} \tag{6}$$

The activation energy for moisture diffusion was obtained from the slope of the graph of $In\left(D_{eff}\right)$ against q/P. The drying rate of tomato slices was calculated using Equation 7 (Doymaz, 2010), where M_{t+dt} is the moisture content (kg water per kg dry matter) at t +dt, and t is the drying time (min).

$$DR = \frac{M_{t+dt} - M_t}{dt} \tag{7}$$

Statistical analysis

One-way analysis of variance (ANOVA) was carried out with SPSS 16.0 to determine the main influence of microwave-vacuum on the parameters studied. The Fishers least significance difference (LSD) was used to compare differences at the 95% probability level. Where significant differences exist, the Duncan Multiple range test was employed to separate the means.

The different statistical evaluation equations (8, 9, 10 and 11) were used to describe the goodness of fit of the dried tomato slices to the drying models applied.

$$R^2 = \frac{N\sum_{i=1}^{N}MR_{pred,i}MR_{expt,i} - \sum_{i=1}^{N}MR_{pred,i}\sum_{i=1}^{N}MR_{expt\,x,i}}{\sqrt{\left(N\sum_{i=1}^{N}MR_{pred,i}^2 - \left(\sum_{i=1}^{N}MR_{pred,i}\right)^2\right)\left(N\sum_{i=1}^{N}MR_{expt,i}^2 - \left(\sum_{i=1}^{N}MR_{expt,i}\right)^2\right)}} \tag{8}$$

$$RSS = \frac{1}{N}\sum_{i=1}^{N}\left(MR_{expt} - MR_{pred}\right)^2 \tag{9}$$

$$RMSE = \sqrt{\frac{1}{N}\sum_{i=1}^{N}\left(MR_{expt,i} - MR_{pred,i}\right)^2} \tag{10}$$

$$\chi^2 = \frac{\sum_{i=1}^{N}\left(MR_{expt,i} - MR_{pred,i}\right)^2}{N - z} \tag{11}$$

where $MR_{expt,i}$ and $MR_{pred,i}$ are the experimental and predicted dimensionless MR respectively, N is the number of observations, and z is the number of model constants. The drying rate constants and coefficients of the model equations were determined with nonlinear regression of SPSS 16.0 (SPSS, 2007), and the goodness of fit of the curves was determined with correlation analysis.

RESULTS AND DISCUSSION

Effect of microwave power and vacuum pressure on drying kinetics

Figures 2 and 3 show the variation of moisture ratio versus drying time for the various microwave powers of 200, 300, 500 and 700 W and vacuum pressures of 0.04, 0.05 and 0.06 MPa. The initial average moisture content of the tomatoes was 24.71 kg water/kg dry matter, which reduced to 0.15 kg water/kg dry matter after drying. It is clear how drying followed a falling rate period and the increase in microwave power accelerated the drying process. As microwave power increased, moisture removal also increased and ultimately resulted in the reduction in drying time. Drying time reduced from 84 to 14 min as the microwave power increased from 200 to 700 W. This means that there was significant savings in time as microwave power increased. The results agree with what was reported by Contreras et al. (2008), Bai-Ngew et al. (2011), Figiel, (2009), Karaaslan and Tuncer (2008) for microwave drying of apple and strawberry, durian chips, beetroot, and spinach respectively.

The increase in vacuum pressure enhanced drying rate as shown in Figure 3. This was expected because as vacuum pressure increases there is an accelerated removal of moisture build-up in the microwave chamber, which consequently enhanced the drying process. The drying time was fitted to a linear model (that is, β_0+β_1 P_0+β_2 P_r), where β_0, β_1, β_2, are estimated coefficient of the linear model, P_o is microwave power, and P_r is vacuum pressure. The analysis of variance for the main or linear effect of microwave power and vacuum pressure indicated that microwave power was the significant model term (Table 2). Variations in the estimated coefficients show that there were different relative importance of the microwave power and vacuum pressure. From the model result shown in Table 2, it could be seen that the effect of microwave power was higher than vacuum pressure.

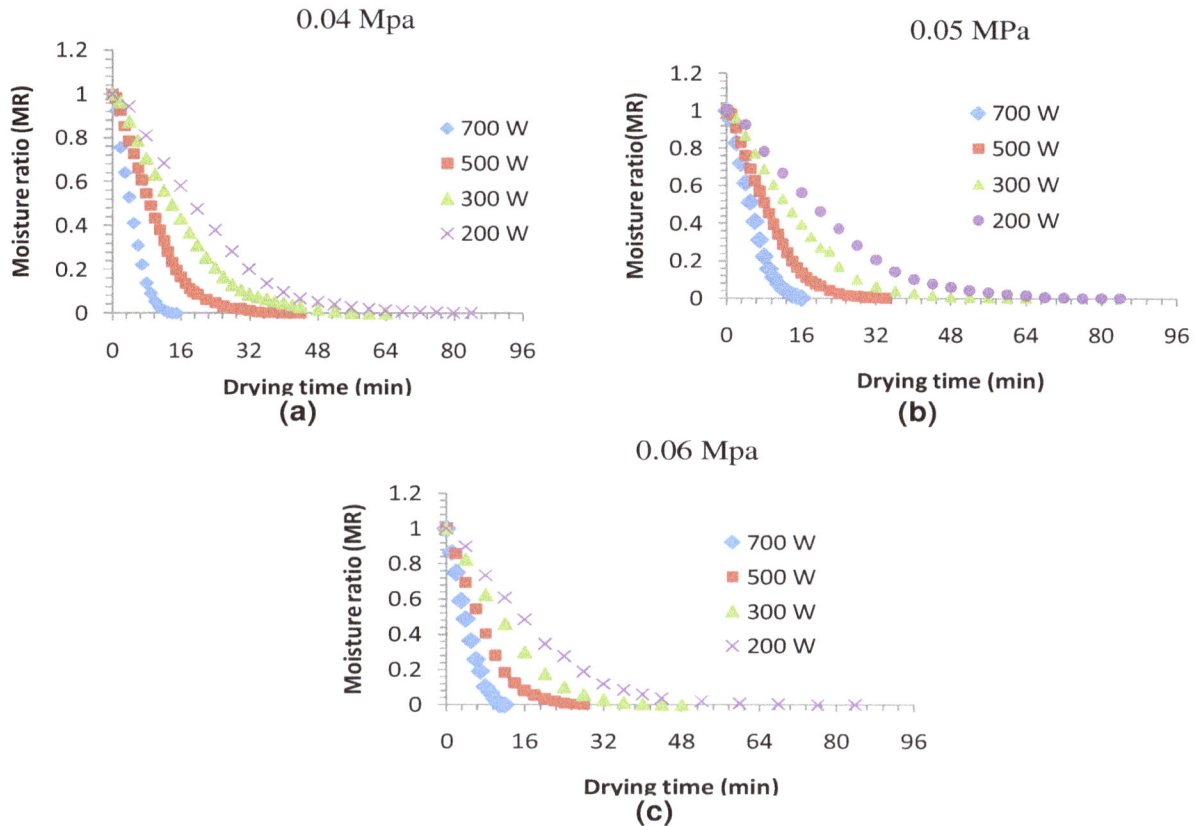

Figure 2. Variation of moisture ratio versus drying time at various microwave powers at a vacuum pressures of 0.04, 0.05, and 0.06 MPa.

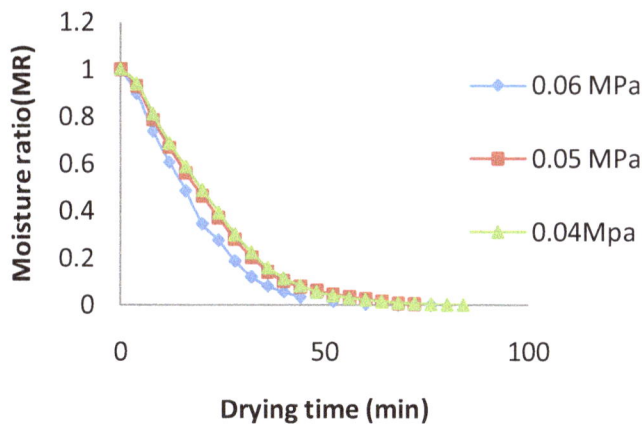

Figure 3. Variation of moisture ratio versus drying time for the various vacuum pressures at 200 W microwave powers.

The drying rates against moisture content for the various drying conditions are illustrated in Figures 4 and 5. It was observed from the Figure 4 that two types of drying rates period were observed; the constant rate and the falling rate period. The two drying rates period were mostly observed in the various microwave powers at 0.04 and 0.05 MPa. However, there was no constant rate period

observed for the samples dried at 700 W and 0.06 MPa. The constant rate period was noticed in the sample moisture content range of 17.96 to 5.98, 14.08 to 7.17, and 18.19 to 8.57 kg water/kg dry matter for the 0.04, 0.05 and 0.06 MPa vacuum pressures respectively. The drying rate trend indicated for 700 W and 0.06 MPa show that at higher microwave power and vacuum pressure, the drying rate may cause excessive and unexpected burning if care is not taken. This trend is further explained by the drying rate and moisture content plot for 500W and various vacuum pressures of 0.04, 0.05 and 0.06 MPa displayed in Figure 5.

Effect of drying parameters on coefficients of effective moisture diffusion

The variation of Ln(MR) against drying time plot for the various microwave-vacuum drying conditions is shown in Figure 6. The slopes of the straight line generated by the plot of Ln(MR) against drying time were used to determine the effective moisture diffusion values. The plot gave a straight line with high correlation coefficients ranging between 0.9069 and 0.9825. The effective moisture diffusion values ranged between 6.74×10^{-9} and 41.71×10^{-9} m^2/s.

Table 2. Analysis of variance for main effect of microwave power and vacuum pressure on drying time.

Source	Coeficient estimate	Sum of squares	df	Mean square	F Value	p-value Prob > F
intercept	47.75					
Model		8238.75	5	1647.75	50.06	< 0.0001*
P	36.25	8030.25	3	2676.75	81.32	< 0.0001*
Pr	4.0	208.50	2	104.25	3.17	0.1151**
R^2	0.9766					

*Significant ; ** not significant at $p<0.05$.

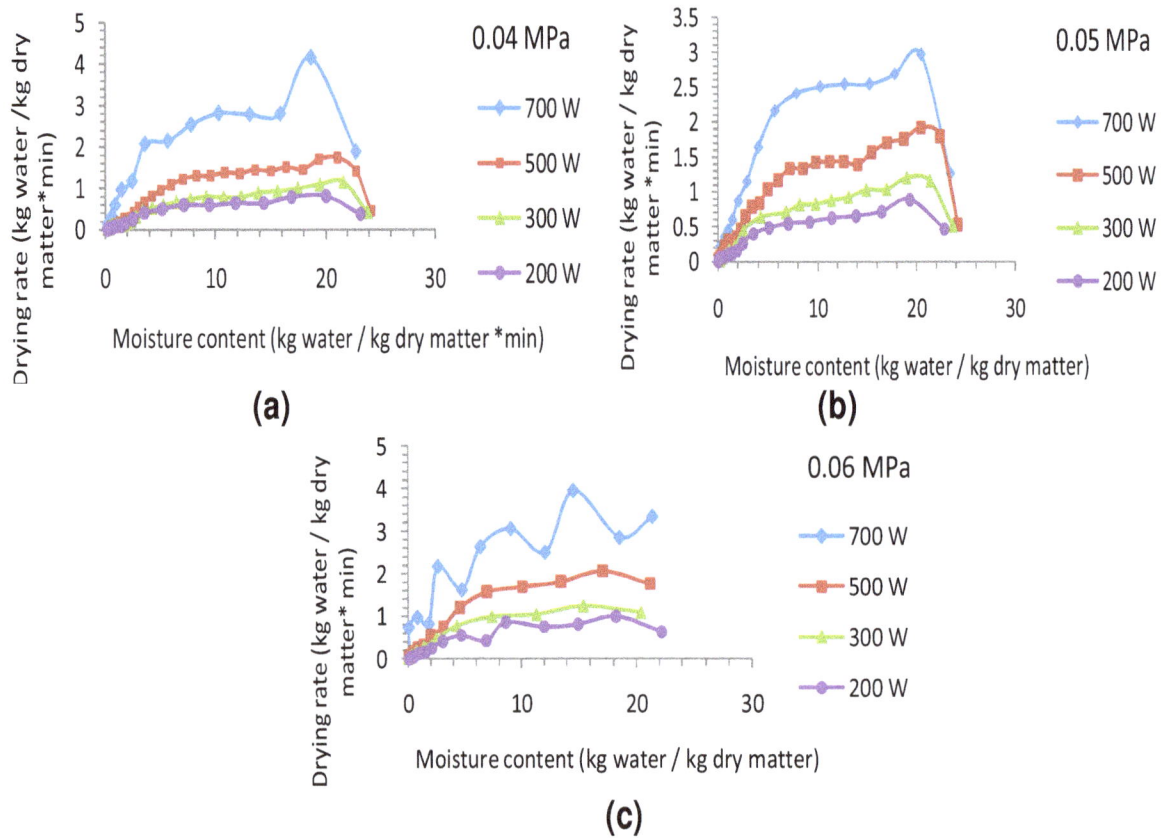

Figure 4. Drying rate against moisture content for various microwave power levels at a vacuum pressures of (a) 0.04 MPa, (b) 0.05 MPa, and (c) 0.06 MPa.

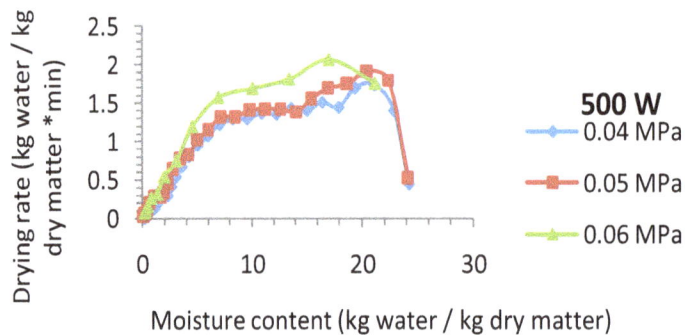

Figure 5. Drying rate against moisture content various microwave vacuum pressures at a microwave power of 500 W.

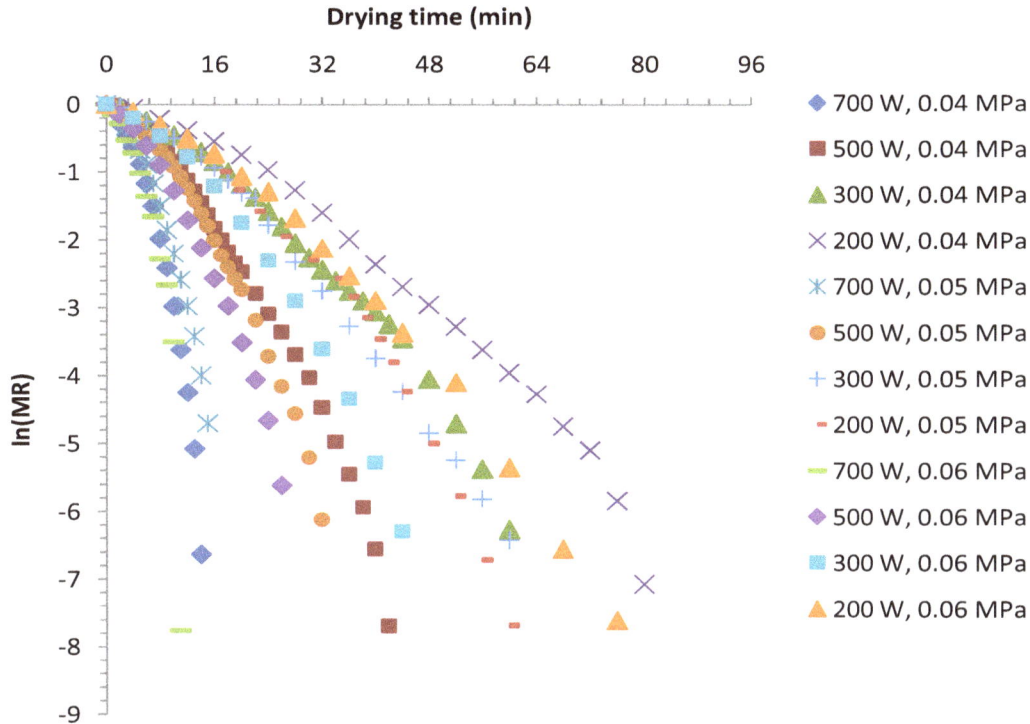

Figure 6. Variation of ln(MR) against drying time plots for the various drying condition.

The variation of microwave power and vacuum pressures on the effective moisture diffusion coefficients are displayed in Figures 7 and 8 respectively. It can be observed from the Figure 6 that increases in microwave power corresponding increased exponentially the effective moisture diffusion coefficients. A similar exponential increase with increasing vacuum pressure was observed for the diffusivity values within the microwave power range of 200 and 500 W. However, a fluctuation in effective moisture diffusivity which is best described by a second order polynomial was realized for the 700 W microwave powers at various vacuum pressures. The D_{eff} values recorded in this study lie within the general range of 10^{-12} to 10^{-8} m^2/s for drying agricultural materials (Doymaz, 2010). The effect of the microwave power was however higher than the vacuum pressure (Figures 7 and 8 and Table 2).

The Arrhenius type microwave power dependence activation energy was calculated using equation (10) and illustrated in Figure 9 showing the microwave power dependence on the activation energy at various vacuum pressures with their statistical coefficient of correlation. It could be observed from the Figure 9 that the activation energy values recorded at the vacuum pressure of 0.04 MPa was the highest, followed by the 0.06 MPa, and the 0.05 MPa drying condition.. The results gave activation energy in the range of 47.8 and 58.94 W/mm for the three vacuum pressures used in this study. The results agree with microwave vacuum drying of mint leaves by

Therdthai and Zhou (2009). The activation energy recorded in this study was higher than the activation energy of 36.40 W/mm reported (Darvishi et al., 2012) for dying of carrot slices, 23.83 W/g for sweet pomegranates (Minaei et al., 2012) and lower than 554 W/100 g for okra (Dadali et al., 2007).

Effect of microwave-vacuum on ascorbic acid

Table 3 presents the ascorbic acid content of the fresh and dried tomatoes at various microwave-vacuum conditions. The ascorbic acid content decreased significantly from an initial mean value of 2.74 ± 0.29 mg/g dry matter of the fresh tomatoes to a least value of 0.91 ± 0.04 mg/g dry matter after drying at 500 W and 0.06 MPa. This represents a maximum decrease of about 66% in ascorbic acid. On the other hand, the samples dried at 200 W and 0.06 MPa had the least reduction to 1.87 ± 0.13 mg/g dry matter, representing a decrease of about 32% in relation to the fresh tomato. The reduction of ascorbic acid content observed during microwave-vacuum drying may be due to the destruction of vitamin C by the electromagnetic waves of the microwave power as the samples were dried. This is because thermal damage and irreversible oxidative reactions are the two main causes of ascorbic acid degradation during drying as a result long drying times. In this study, the drying times were relatively short compared with hot air drying.

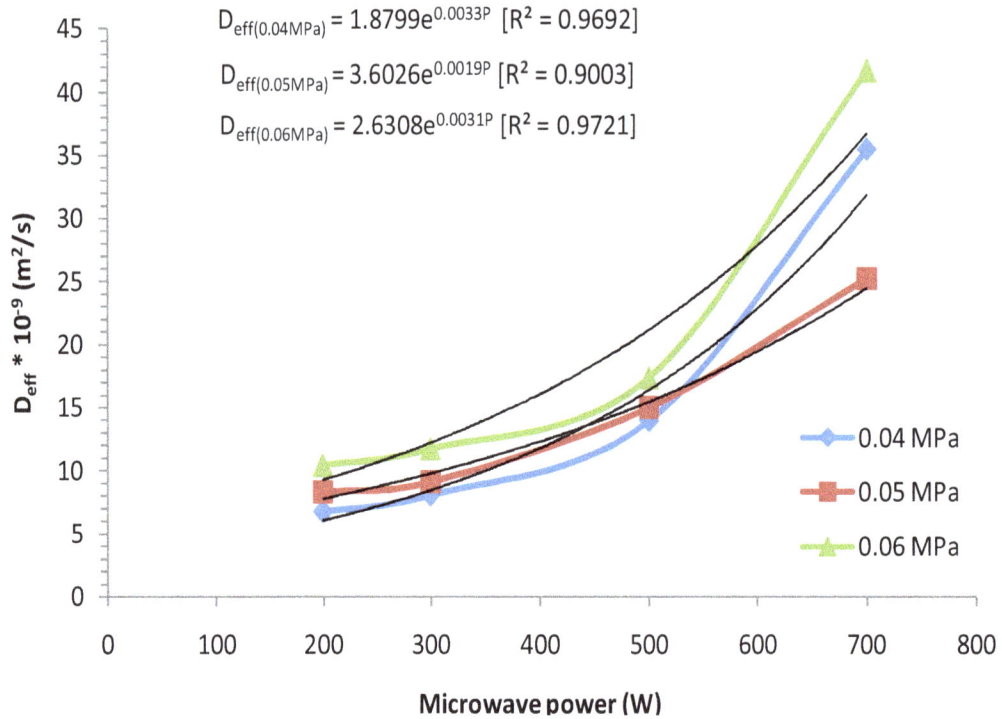

$$D_{eff(0.04MPa)} = 1.8799e^{0.0033P} \ [R^2 = 0.9692]$$

$$D_{eff(0.05MPa)} = 3.6026e^{0.0019P} \ [R^2 = 0.9003]$$

$$D_{eff(0.06MPa)} = 2.6308e^{0.0031P} \ [R^2 = 0.9721]$$

Figure 7. Effect of microwave power on the on the effective moisture diffusion coefficient.

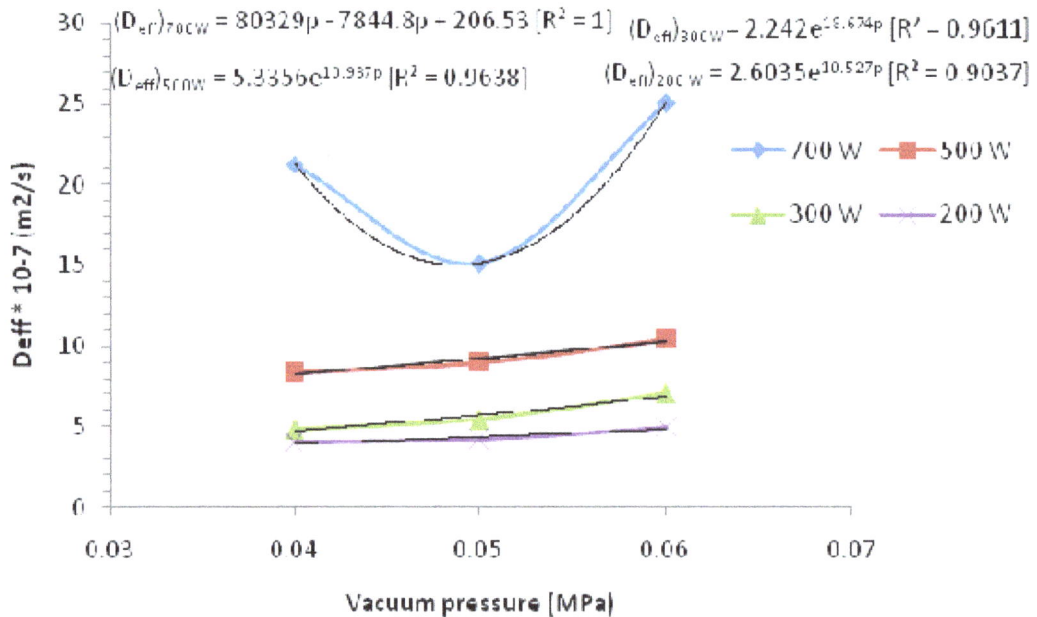

$$(D_{eff})_{70CW} = 80329p - 7844.8p - 206.53 \ [R^2 = 1]$$ $$(D_{eff})_{30CW} = 2.242e^{-6.674P} \ [R' - 0.9611]$$

$$(D_{eff})_{50W} = 5.3356e^{10.937p} \ [R^2 = 0.9638]$$ $$(D_{eff})_{20CW} = 2.6035e^{10.527p} \ [R^2 = 0.9037]$$

Figure 8. Effect of vacuum pressure on the effective moisture diffusion coeeficient.

Therefore, the degradation in AA content may be thermal damage resulting from the microwave energy as it impinges on the tomato surface. Alibas Ozkan et al. (2007) found a similar decrease in ascorbic acid content of dried spinach in the microwave power range of 350 to 1000 W. In their studies, there was a reduction in ascorbic acid content from 50.18 ± 1.36 to 23.30 ± 1.93 mg/100 g. A similar trend was observed for microwave cooking treatment of broccoli in a study by Zhang and Hamauzu (2004).

The reduction agrees with results obtained by Zheng and Lu (2011) for microwave pretreatment on the kinetics

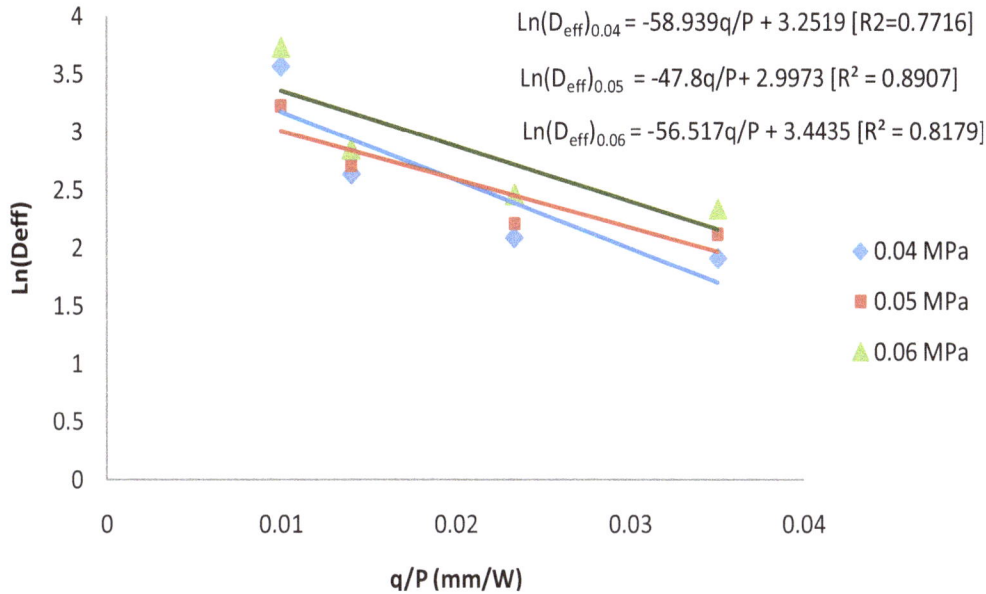

Figure 9. Ln(Deff) against q/P plot at various vacuum pressures.

Table 3. Ascorbic acid content (mg/g dry matter) of the fresh and microwave-vacuum dried samples

Drying condition	0.04 MPa	0.05 MPa	0.06 MPa
Fresh	2.74 ± 0.29^{c}	2.74 ± 0.29^{c}	2.74 ± 0.29^{c}
200 W	1.32 ± 0.08^{b1}	1.45 ± 0.12^{ab1}	1.87 ± 0.13^{b2}
300 W	1.01 ± 0.13^{a1}	1.27 ± 0.09^{a2}	1.84 ± 0.09^{b3}
500 W	1.19 ± 0.04^{ab2}	1.69 ± 0.12^{b3}	0.91 ± 0.04^{a1}
700 W	1.24 ± 0.13^{ab1}	$1.56\pm0.13b^{ab2}$	1.86 ± 0.13^{b3}

Mean values within a column ($^{a-c}$) and in a row ($^{1-3}$) bearing the same letters are not significantly different at the 0.05 level *Values are means ±standard deviation.

Table 4. Lycopene content (m g/100g dry matter) of the fresh and microwave-vacuum dried samples

Drying condition	0.04 MPa	0.05 MPa	0.06 MPa
Fresh	2.960 ± 0.002^{a}	2.960 ± 0.002^{a}	2.960 ± 0.002^{a}
200 W	9.278 ± 0.095^{b1}	12.253 ± 0.157^{d3}	11.359 ± 0.108^{e2}
300 W	17.018 ± 3.536^{c2}	7.843 ± 0.036^{b1}	9.278 ± 0.095^{c1}
500 W	9.695 ± 0.157^{b2}	11.088 ± 0.095^{c3}	8.217 ± 0.072^{b1}
700 W	25.44 ± 0.343^{d3}	13.939 ± 0.157^{e2}	10.776 ± 0.036^{d1}

Mean values within a column ($^{a-e}$) and in a row ($^{1-3}$) bearing the same letters are not significantly different at p=0.05 *Values are means ±standard deviation

of ascorbic acid in different parts of green asparagus. Marfil et al. (2008) investigated ascorbic acid content in osmotic drying of tomato havles at 50, 60 and 70°C and observed an ascorbic acid reduction from 4.00 ± 0.30 mg/g to 2.19 ± 0.24 mg/g dry matter, representing 35% reduction. In a study of microwave vacuum drying of carrot slices and microwave drying of potatoes, Lin et al. (1998) and Khraisheh et al. (2004) reported a decrease

of 21 and 25% ascorbic acid content respectively.

Effect of microwave-vacuum on lycopene content

To examine the influence of microwave-vacuum on the lycopene content, the dried tomato slices lycopene content were compared with that of the fresh (Table 4).

Table 5. Results of the fitting of the experimental data to the drying models, N=27.

Model name	Model constants	R^2	RSS	RMSE	$\chi^2 \times 10^{-3}$
Lewis	k:0.101	0.9585	0.105	0.0623	4.038
Page	k:0.033, n:1.465	0.9988	0.003	0.0105	0.12
Modified Page	k:0.033, n:1.465	0.9988	0.003	0.0105	0.12
Henderson and Pabis	k:0.120, a: 1.188	0.9842	0.040	0.0385	1.60
Modified Henderson and Pabis	k:0.120, a: 0.396, b: 0.397	0.9842	0.040	0.0385	1.904
Logarithmic	k:0.099, a: 1.224, c:-0.082	0.9921	0.020	0.0272	0.833
Two-term	a:96.816, b:-95.685, k_0:0.064, k_1: 0.064	0.9941	0.015	0.0236	0.652
Two-term exponential	k:0.157, a:1.984	0.9964	0.009	0.0182	0.36
Diffusion approach	k:0.100, a: 1.00, b: -20.262	0.9586	0.105	0.0623	4.375
Verma et al.	k:0.095, a: 1.00, g:-2.00	0.9546	0.115	0.0652	0.68
Wang and Singh	a:-0.074, b: 0.001	0.9933	0.017	0.0251	0.0869
Midilli et al.	k:0.031, a: 0.989, n:1.482, b: 0.000	0.9992	0.002	0.0086	0.0869
Parabolic	a: 1.047, b: -0.079, c: 0.001	0.9961	0.01	0.0192	0.4167

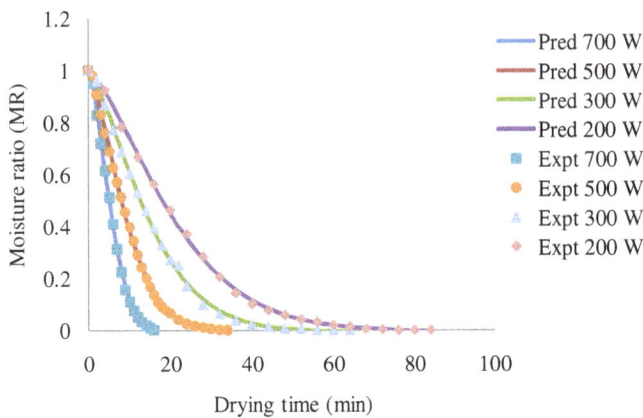

Figure 10. Fitting of microwave-vacuum drying of experimental and simulated data to the Midilli et al. model.

The lycopene levels of the dried tomatoes significantly (p<0.05) increased from an initial value of 2.96 mg/100 g dry matter to a maximum value of 25.44 mg/100 g dry matter after microwave-vacuum drying. It has been reported that in tomatoes, lycopene content degradation depends on many factors including processing temperature, that is, the heat energy from the microwave power. Lycopene of the fresh tomato can isomerizes from its trans-form into the cis-lycopene as a result of thermal treatment or degrade into colorless form (Sharma and Le Maguer, 1996).

Many studies have recently indicated that common heat treatments of tomato products do not result in a shift in the distribution of cis-lycopene isomers (Goula et al., 2006). Thus, the increase in lycopene content reported in this study from the microwave heat treatment may be due to a progressive conversion from the all-trans lycopene to a less strongly colored, less intensely absorbing cis form than an actual degradation of lycopene. The values of the lycopene content obtained after microwave-vacuum were relatively lower than the 82.90 mg/100 g reported by

Takeoka et al. (2001) for tomato paste and 65.28 mg/100 g dry matter reported by Abano et al. (2011) for hot air drying of tomato slices at 80 °C. The lower values recorded in this study might be due to the electromagnetic waves in the microwave-vacuum, which might have negatively affected the lycopene in the tomatoes.

Modeling of the drying curves

The dimensionless moisture ratio against drying time for the experimental data of the 0.05 MPa vacuum pressures at various microwave powers was fitted to thirteen thin-layer drying models available in scientific literature. The results of such fitting obtained with SPSS 16.0 software are displayed in Table 5, which show the values of the estimated constants with their corresponding statistical R^2, χ^2, and RMSE values characterizing each fitting. From the results obtained, it was evident that the experimental data fitted to the models used in this study. The correlation coefficients obtained are in the range of 0.9546 to 0.9992. This means that the thirteen models could satisfactorily describe the microwave-vacuum drying of tomato slices. The relatively high values of correlation coefficients, low reduced chi-square, and low root mean square errors indicate a good predicting capacity for the microwave-vacuum drying conditions tested over the entire duration of the drying process. Among the thirteen thin-layer drying models tested, Midilli et al. model obtained the highest R^2 values and the lowest χ^2, and RMSE values in the microwave-vacuum drying conditions studied. Figure 10 displays the fitting of the experimental and simulated points to the Midilli et al. model. It can be seen from Figure 11 that the experimental data are closely bounded to the simulated data for Midilli et al. model around logarithmic curves. The consistency of the middilli model for the microwave vacuum drying of tomato slices is evident in Figure 12, showing the predicted and simulated data points. Figure

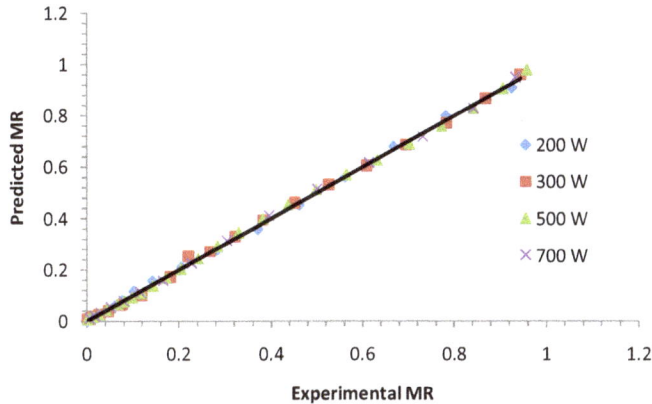

Figure 11. Predicted and experimental moisture ratio for 0.05 MPa vacuum pressures.

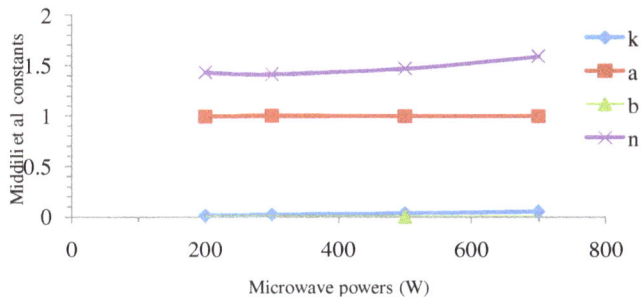

Figure 12. Variation of drying constants against microwave powers for the Midilli *et al.* model.

12 shows the variation of drying constants against microwave powers in the Midilli et al model. It can be verified from Figure 12 that the drying rate constant k indicating drying bahaviour for the models tested increased with increasing microwave power in the Midilli et al. model. The increase in drying rate constant with increasing microwave power indicates an enhancement of drying potential. Additionally, drying rate constant increased exponentially with increasing vacuum pressure. Multiple non-linear regression analysis for this exponential relationship between drying rate constant k, microwave power (Po), and vacuum pressure (Pr), for the Midilli et al. model was:

$$k = 6.8 \times 10^{-3} \exp^{0.002 P_0} [R^2 = 0.9724]$$

$$k = 2.3 \times 10^{-3} \exp^{31.43 P_r} [R^2 = 0.9999]$$

Conclusion

The effect of four microwave powers and three vacuum pressures on drying time lycopene content and ascorbic acid content was studied. The drying time reduced from 84 to 14 min as the microwave power increased from 200

to 700 W. It was shown that increase in microwave power and vacuum pressure enhanced the drying rate and significantly reduced the drying time of tomato slices. The effect of microwave power on the drying time was however higher than vacuum pressure. It was observed that the lycopene levels of the tomatoes increased significantly after drying whereas the ascorbic acid content reduced. Among the thirteen thin-layers drying models that were fitted to the experimental data, the Midilli et al. model showed the best fit although all the models could satisfactorily describe the microwave-vacuum drying kinetics of tomato slices. The drying rate constant in the Midilli et al model increased exponentially with microwave power and vacuum pressures.

ACKNOWLEDGEMENT

The authors acknowledge the financial support of the Priority Academic Program Development (PAPD) for Jiangsu Higher Education Institutions.

REFERENCES

Abano EE, Ma H, Qu W (2011). Influence of Air Temperature on the Drying Kinetics and Quality of Tomato Slices. J. Food Process Technol, 2(5):1-9.

Akpinar EK. (2006). Mathematical modeling of thin layer drying process under open sun of some aromatic plants. J. Food Eng, 77:864-870.

Alibas Ozkan I, Akbudak B, Akbudak N (2007). Microwave drying characteristics of spinach. J. Food Eng. 78:577–583

Bai-Ngew S, Therdthai N, Dhamvithee P (2011). Characterization of microwave vacuum-dried durian chips. J. Food Eng. 104:114-122

Celma AR, Cuadros F, López-Rodríguez F (2009). Characterisation of industrial tomato by-products from infrared drying process. J. Food. Bioprod. Proc. 87:282-291.

Contreras C, Martin ME, Martínez-Navarrete N, Chiralt A (2008). Influence of microwave application on drying kinetics, and optical mechanical properties of apple and strawberry. J. Food Eng. 88:55-64.

Dadali G, Dilek Kilic A, Ozbek B (2007). Estimatioin of Effective Moisture Diffusivity of Okra for Microwave Drying, Drying Technol, 25:1445-1450.

Demir V, Gunhan T, Yagcioglu AK (2007). Mathematical modeling of convection drying of green table olives. Biosyst. Eng. *98*:47–53.

Darvishi H, Banakar A, Zarein M (2012). Mathematical modeling and thin layer drying kinetics of carrot slices. *Global* J. Fronteir Sci. Math. Deci. Sci. 5(1):57-64.

Doymaz I (2005). Drying characteristics and kinetics of okra. J. Food Eng, 69:275-279.

Doymaz I (2010). Drying of thyme (*Thymus Vulgaris L.*) and Selection of a suitable thin-layer drying model. J. Food Proc and Preserv., 2010. Wiley Periodical. Available: http://www.wiley.com/jfpp/10.1111/j.1745-4549.2010.00488.x

Erbay Z, Icier F (2008). Thin-layer drying behaviors of Olive leaves (*Olea Europaea L.*). J. Food Proc. Eng. 33:287-308.

Figiel A (2009). Drying kinetics and quality of vacuum-microwave dehydrated garlic cloves and slices. J. Food Eng. 94(1):98-104.

Food and Agriculture Organization (2010). Water Development and Management Unit. Available: http:// fao.org/nr/water. Accessed on [2011-11-12]

Goula AM, Adamopoulos KG, Chatzitakis PC, Nikas VA (2006). Prediction of lycopene degradation during a drying process of tomato pulp. J. Food Eng. *74*:37-46.

Harvard School of Public Health, HSPH (2010). Harvard University

Press. Assessed on [2011-10-12]

Henderson SM, Pabis S (1961). Grain drying theory I: temperature effect on drying coefficient. J. Agric. Res. Eng. 6:169–174.

Karaaslan SN, Tuncer IK (2008). Development of a drying model for combined microwave-fan-assisted convection drying of spinach. J. Biosys. Eng, 100:44-52.

Kardum JP, Sander A, Skansi D (2001). Comparison of convective, vacuum, and microwave drying chlorpropamide. Dry Technol, 19(1):167–183.

Kassem AS (1998). Comparative studies on thin layer drying models for wheat. 13th international congress on agricultural engineering, 6, 2–6 February, Morocco.

Khraisheh MAM, McMinn WAM, Magee TRA (2004). Quality and structural changes in starchy foods during microwave and convective drying. Food Res. Int. 37(5):497–503

Kose B, Erenturk S (2010). Drying Characteristics of mistletoe (Viscum album L.) in convective and UV combined convective type dryers. Industr. Crops Prod. 32. 394-399.

Lewis WK (1921). The rate of drying of solid materials. J. Ind. Eng. 13:427-443.

Lin TM, Durance TD, Scaman CH (1998). Characterization of vacuum microwave, air and freeze dried carrot slices. Food Res. Int. 31(2):111–117

Marfil PHM, Santos EM, Telis VRN (2008). Ascorbic acid degradation kinetics in tomatoes at different drying conditions. Food Sci. Technol. 41:1642-1647.

Maskan M (2001). Drying, shrinkage, and rehydration characteristics of kiwi fruits hot air and microwave drying. J. Food Eng. 48 (2):177-182.

McMinn WAM (2006). Thin-layer modeling of the convective, microwave, microwave-convective and microwave-vacuum drying of lactose powder. J. Food Eng. 72:113–123.

Midilli A, Kucuk H, Yapar Z (2002). A new model for single layer drying. Drying Technol. 20(7):1503–1513.

Minaei S, Motevali A, Ahmadi E, Azizi MH (2012). Mathematical models of drying pomegranate arils in vacuum and microwave dryers. J. Agric. Sci. Technol. 14:311-325.

Mota CL, Luciano C, Dias A, Barroca MJ, Gine RPF (2010). Convective drying of onion: Kinetics and nutrional evaluation. Food. Bioprod. Proc. 88:115-123

Overhults DD, White GM, Hamilton ME, Ross IJ (1973). Drying soybeans with heated air. Transactions of the ASAE 16:195–200.

Page GE (1949). Factors influencing the maximum rate of air drying shelled corn in thin layers. MS Thesis, Purdue University, West Lafayette, IN.

Pere C, Rodier E (2002). Microwave vacuum drying of porous media: experimental study and qualitative considerations of internal transfers. Chem. Eng. Proc. 41:427–436.

Prabhanjan DG, Ramaswamy HS, Raghavan GSV (1995). Microwave-assisted convective air drying of thin layer carrots. J. Food Eng. 25:283–293.

Ranganna S (1976). In: Manual of Analysis of Fruits and Vegetable Products, McGraw Hill, New Delhi. p. 77.

Sacilik K, Keskin R, Elicin AK (2006). Mathematical modeling of solar tunnel drying of thin layer organic tomato. J. Food Eng. 73:231–238.

Sadasivam S, Balasubraminan T (1987). In: Practical Manual in Biochemistry,Tamil Nadu Agricultural University Coimbatore. p. 14.

Schössler K, Jäger H, Knorr D (2012). Effect of continuous and intermittent ultrasound on drying time and effective diffusivity during convective drying of apple and red bell pepper. J. Food Eng. 108:103-110.

Sharaf-Eldeen O, Blaisdell YI, Spagna G (1980). A model for ear corn drying. Trans. ASAE, 23: 1261–1271.

Sharma GP, Prasad S (2004). Effective moisture diffusivity of garlic cloves undergoing microwave-convective drying. J. Food Eng, 65: 609-617.

Sharma SK, Le Maguer M (1996). Kinetics of lycopene degradation in tomato pulp solids under different processing and storage conditions. Food Res. Intern. 29(3-4):309-315.

SPSS 16 (2007). SPSS 16.0 for Windows SPSS Inc., Chicago, IL.

Takeoka GR, Dao L, Flessa S, Gillespie DM, Jewell WT, Huebner B, Bertow D, Susan EE (2001). Processing Effects on Lycopene Content and Antioxidant Activity of Tomatoes. J. Agric. Food Chem. 49:3713-3717.

Therdthai N, Zhou W (2009). Characterization of microwave vacuum drying and hot air drying of mint leaves (Mentha cordifolia Opiz ex Fresen). J. Food Eng. 91(3):482-489.

Togrul H (2005). Simple modeling of infrared drying of fresh apple slices. J. Food Eng. 71:311–323

Vega A, Uribe E, Lemus R, Miranda M (2007). Hot air drying characteristics of aloe vera (Aloe barbadensis Miller) and influence of temperature on kinetic parameters. Lebensm.-Wiss. Technol. 40:1698-1707.

Venkatachalapathy K, Raghavan G SV (2000). Microwave drying of whole, sliced and pureed strawberries. Agri. Eng. J. 9(I):29-39.

Verma LR, Bucklin RA, Endan JB, Wratten FT (1985). Effects of drying air parameters on rice drying models. Trans. ASAE 28:296–301.

Walde SG, Velu V, Jyothirmayi T, Math RG (2006). Effects of pretreatments and drying methods on dehydration of mushroom. J. Food Eng. 74:108–115.

Zhang D, Hamauzu Y (2004). Phenolics, ascorbic acid, carotenoids and antioxidant activity of broccoli and their changes during conventional and microwave cooking. Food Chem. 88:503–509.

Zheng H, Lu H (2011). Effect of microwave pretreatment on the kinetic of ascorbic acid degradation and peroxidase inactivation in different parts of asparagus (Asparagus officinalis L.) during water blanching. Food Chem. 128:1087-1093.

29

The impact of cooking on the proximate composition and anti-nutritional factors of water yam (*Dioscorea alata*)

Ezeocha V. C.[1]* and Ojimelukwe P. C.[2]

[1]Post Harvest Technology Programme, National Root Crops Research Institute, Umudike, Abia State, Nigeria.
[2]Food Science and Technology Department, Michael Okpara University of Agriculture, Umudike, Abia State, Nigeria.

Raw and boiled water yam tubers (*Dioscorea alata*) were analyzed for proximate contents such as ash, crude protein, carbohydrate, crude fibre, crude lipid, energy and anti-nutrients using standard procedures and methods. The crude protein contents (10.27%), ash (2.93%) and lipid (0.15%) were significantly ($p < 0.05$) lowered in the boiled tubers while the carbohydrate (76.57%) significantly ($p < 0.05$) increased in the boiled tubers. The antinutrients; alkaloids (2.77%), saponins (2.71%), flavonoids (1.38%) and tannins (0.21%) significantly ($p < 0.05$) reduced in the boiled tubers. It was concluded that boiling had both positive and negative effect on water yam. A cooking time of between 30 and 60 min at 100°C was recommended for *D. alata*.

Key words: Water yam, boiling, proximate content, anti-nutrients.

INTRODUCTION

Yam is one of the staple foods in Nigeria and other tropical African countries. The yams are members of the genus *Dioscorea* in the section, Enantiophyllum. *Dioscorea* is the largest genus of the family Dioscoreaceae, containing between three and six hundred species (Vernier et al., 1998). Yam is grown and cultivated for its energy-rich tuber. Only a few species of yams are cultivated as food crops. The most important species of *Dioscorea* include *D. rotundata, D. alata, D. cayenensis, D. dumetorum, D. esculenta* and *D. bulbifera. Dioscorea alata*, called "water yam", "winged yam" and "purple yam", is the species most widely spread throughout the world and in Africa is second only to white yam in popularity (Mignouna et al., 2004). Yam is an excellent source of starch, which provides calorific energy (Coursey, 1973). It also provides protein three times more superior than the one of cassava and sweet potato (Bourret-Cortedellas, 1973). Yams are consumed as staple food. Apart from food, yams are also sources of pharmaceutical compounds like saponins and sapogenins, which are precursors of cortisone (a hormone from the adrenal gland used medically in the treatment of arthritis and some allergies) (Higdon et al., 2001). Water yam (*Dioscorea alata*) is the most economically important yam species which serve as a staple food for millions of people in tropical and sub-tropical countries (Coursey, 1967; Hahn, 1995). *D. alata* is a crop with potential for increased consumer demand due to its low sugar content necessary for diabetic patients (Udensi *et al.*, 2010). Yams generally have high moisture content; the dry matter is composed mainly of starch, vitamins, sugars and minerals. Nutrient content varies with species and cooking procedures. Yams may also contain small quantities of polyphenolic compounds (e.g. tannins), alkaloids (e.g. dioscorine), steroid derivatives (e.g. diosgenin), calcium oxalate crystals and phytic acid. *D. alata* cultivars possess a higher content of protein, vitamin C and low lipids than *D.* cayenensis, *D. escunlenta, D. rotundata* and *D. trifida* (Muzac-Tucker et al., 1993). Root crops are not easily digested in their

*Corresponding author. E-mail: avezeocha@yahoo.com.

Table 1. Proximate composition of raw and boiled *D. alata*.

Sample	Ash (%)	Lipids (%)	Moisture (%)	Fibre (%)	Protein (%)	CHO (%)	Energy (kcal/100 g)
RWY	2.93[a]	1.15[a]	6.79[a]	2.31[a]	10.27[a]	76.57[d]	357.65[c]
WYB30	2.70[b]	0.84[b]	4.85[b]	1.69[b]	9.12[b]	80.79[c]	367.36[ab]
WYB60	2.48[c]	0.15[c]	4.74[c]	1.52[c]	8.11[c]	83.02[b]	363.94[b]
WYB90	2.05[d]	0.09[c]	4.11[d]	1.37[d]	7.10[d]	85.20[a]	370.01[a]

Means with different letters on the same column are significantly different (P<0.05). RWY = Raw *D. alata*, WYB30 = *D. alata* boiled for 30 min (at 100 ℃), WYB60 = *D. alata* boiled for 60 min, WYB90 = *D. alata* boiled for 90 min.

natural state and should be cooked before they are eaten. Cooking improves their digestibility, promotes palatability and improves their keeping quality as well as making the roots safer to eat. However, cooking may affect the nutritional composition and phytoconstituents in food. The objective of this study is to evaluate the nutritional and phytochemical composition of yams and how they are affected by cooking duration.

MATERIALS AND METHODS

Water yam (*D. alata*) was obtained from the yam program of National Root Crop Research Institute, Umudike. The tubers were washed and divided into 2 portions of 1 kg each. One portion was peeled, washed and chipped with a chipping machine (locally fabricated). The second portion was washed, peeled and cooked by boiling in distilled water at 100 ℃ for 30, 60 and 90 min; and then chipped with a chipping machine (locally fabricated). The raw and boiled chips were spread thinly on a dark nylon and sun dried (at 40 ℃ for 48 h). The dried yam chips were then milled into powder using a Thomas Wiley mill (model ED-5) and stored in air tight containers before analysis. The AOAC (1990) method was used to determine the proximate composition of the yam flour. The method of Obadoni and Ochuko (2001) was adopted for the determination of the alkaloid, saponin, phenol and tannin composition while the method of Boham and Kocipai (1994) was used in the flavonoid content determination of the yam flour. The analysis was done in triplicates and data collected were analysed by analysis of variance (ANOVA) to indicate the significant differences between mean values of the different results using a SAS system 2008 version. Significant level was established at *P* <0.05.

RESULTS

The proximate composition of raw and boiled water yams are shown in Table 1. Protein content of the raw tuber was 10.27%, significant differences (P<0.05) were observed between the crude protein content of the raw and boiled tubers. The crude protein contents reduced significantly (p<0.05) with increase in the boiling time. Boiling effected a 30.83% reduction in the crude protein of *D. alata* after boiling for 90 min. Boiling *D. alata* for 90 min significantly reduced the ash content from 2.93% to 2.05% resulting in 29.91% loss in the ash content. Boiling for 30 and 60 min reduced the lipid significantly (P<0.05) from 1.15 to 0.15% but further boiling for 90 min did not affect the lipid significantly. The crude fibre content

reduced significantly with boiling from 2.31 to 1.37%. On the other hand the carbohydrate composition of *D. alata* increased significantly (p<0.05) with boiling time from 76.57 to 85.20%.

The level of phytochemicals in *D. alata* is shown in Table 2. *D. alata* had 2.71% of saponin; the saponin levels in the boiled samples were reduced significantly when compared with the raw sample. The levels of alkaloids in the tubers ranged from 2.765% when raw to 0.60% when boiled for 90 min. The levels were reduced significantly (p < 0.05) by boiling when compared to the raw samples. The concentration of flavonoids in raw *D. alata* was 1.375%, boiling for 30 min significantly reduced it to 0.93%, boiling for 60 min further reduced it significantly to 0.65%, however further boiling for 90 min did not significantly affect the flavonoid content. Raw *D. alata* had phenol concentration of 1.91%; the variation in phenol content cannot be solely attributed to heat application. *D. alata* had 0.21% tannin content, which reduced with boiling. A loss of 14% tannin content was observed in *D. alata*, after boiling for 90 min.

DISCUSSION

Protein content of raw *D. alata* was 10.27%, a similar result was obtained by Lebot *et al.* (2005). The value obtained for the raw *D. alata* were however, comparably higher than reported values of 5.15% for white yam and 3.64% for sweet potato (Alaise and Linden, 1999). Significant differences (P<0.05) were observed between the crude protein content of the raw and boiled tubers. This reduction may be as a result of the loss of free amino acids which took place through leaching. Ash content is a reflection of the mineral status, even though contamination can indicate a high concentration in a sample. Ash content of the raw tubers were 2.93%, a comparable result has been reported for *D. alata* tubers (Lebot *et al.* 2005). The significant (p<0.05) reduction in ash content of the tubers with increased boiling period is in agreement with the results of Onu and Okongwo (2006), who recorded decrease in ash content of pigeon pea seeds from 5.50% (raw seeds) to 4.00% (30 min boiled seeds). These losses could be as a result of leaching of the minerals into the boiling water. The observed decrease in ash content after cooking implies

Table 2. Effect of boiling on the phytochemical composition of *D. alata*.

Sample	Alkaloid (%)	Flavonoid (%)	Saponin (%)	Tannin (%)	Phenol (%)
RWY	2.77[a]	1.38[a]	2.71[a]	0.21[a]	1.91[a]
WYB30	1.91[b]	0.93[b]	1.37[b]	0.16[b]	1.24[c]
WYB60	0.61[c]	0.65[c]	1.23[c]	0.14[c]	1.66[b]
WYB90	0.60[d]	0.61[c]	1.03[d]	0.18[b]	1.21[d]

Means with different letters on the same column are significantly different (P<0.05); RWY = Raw *D. alata*, WYB30 = *D. alata* boiled for 30 min (at 100 °C), WYB60 = *D. alata* boiled for 60 min, WYB90 = *D. alata* boiled for 90 min.

that the potential ability of these tubers to supply essential minerals has been reduced. This is in accordance with the observation of Onyeike and Oguike (2008) on boiled groundnut (*Arachis hypogaea*) seeds. According to the authors, this may be due to water absorption during boiling leading to dilution and hence, low amount of ash.

Lipids are a distinct and diverse set of small molecules consisting of eight general compound classes including: Fatty acyls, glycerolipids, glycerophospholipids, sphingophospholipids, sterol lipids, prenol lipids, saccharolipids and polyketides. The lipid content (1.15%) was quite reasonable as all root crops contain very low lipid content (Ekpeyong, 1984). The lipid content of the raw sample was comparably higher than that of white yam, 0.56% but lower than that of sweet potato, 0.95% (Alaise and Linden, 1999). Boiling reduced the lipid content of the *D. alata*. However the percentage level of reduction is dependent on the duration of boiling as observed in Table 1. The fibre content of raw water yam (2.305%) is comparably higher than that of polished rice, 0.2% and sweet potato, 0.17% (Alaise and Linden, 1999). Decrease in Crude fibre with increase in duration of cooking agrees with the report of Akinmutimi (2007), who worked on mucuna species. The carbohydrate contents of raw water yam (76.570%) agrees with the work of Onyenuga (1968), which reported that the dry matter of most root and tuber crops is made up of about 60 to 90% carbohydrate. The carbohydrate values are comparable to that obtained by Longe (1986) for white yam (78%), water yam (75.65%) but lower than the carbohydrate value for sweet potato (82.55%). The increase in carbohydrate content with boiling observed in this work may have been due to the fact that carbohydrates may have absorb water to bulk up via cross-linking reaction probably induced by heat generated by boiling. This may increase the stability of the carbohydrates thereby enhancing resistance to further heat (Nzewi and Egbuonu, 2011).

Boiling resulted in reduction of all the phytochemicals analysed in this study, which is in agreement with Farris and Singhu (1990) and Balogun *et al.* (2001) reports, which state that most anti-nutritional factors in food can be reduced by proper application of heat. The reduction increased as boiling period increased. This trend may be due to higher ability of hydrolyzing the anti-nutritional factors as boiling period increased. The determination of the anti-nutritional substances was of interest because of their toxicity in yams, negative effects on mineral bioavailability and their pharmacological effect. These metabolites occur in varying concentrations in yam tubers.

Saponin is an antinutritional factor whose toxicological effects should be balanced with its benefits. Some of the general properties of saponins include formation of foams in aqueous solution, hemolytic activity and cholesterol binding properties and bitterness (Sopido *et al.*, 2000). Saponins natural tendency to ward off microbes makes them good candidates for treating fungal and yeast infections. These compounds serve as natural antibiotics, which help the body to fight infections and microbial invasion (Sopido et al., 2000). The availability of alkaloids in the tubers of *D. alata* indicates that yam tubers cannot be eaten raw. Most alkaloids are known for their pharmacological effects rather than their toxicity. However when alkaloids occur in high levels in foods, they cause gastro-intestinal upset and neurological disorders (Okaka et al., 1992).

Flavonoid concentrations were significantly (p<0.05) affected by boiling, this confirms the work of Mc Williams (1979) which reported that flavonoids are destroyed by heat processing methods like drying, roasting and boiling. Flavonoids are widely distributed group of polyphenolic compounds, characterized by a common benzopyrone ring structure that has been reported to act as antioxidants in various biological systems. The biological functions of flavonoids apart from its antioxidant properties include protection against allergies, inflamemation, free radicals, platelet aggregation, microbes, ulcers, viruses and tumors (Okwu, 2004; Okwu and Omodamiro, 2005). Flavonoids reduce cancer by interfering with estrogen synthetease, an enzyme that binds estrogen to receptors in several organs (Farquer, 1996; Okwu and Omodamiro, 2005).

In some species of yam tubers, browning reactions occur when the tissues are injured and exposed to air. This type of browning is due to the oxidation of phenolic constituents, especially o-hydroxy or trihydroxy phenolics, by a phenol oxidase present in the tissue (Martin and Rubeste, 1976). Phenol content of 1.91% was identified in *D. alata*; the presence of phenols indicates that Dioscorea species could act as anti-inflammatory, anticlotting, antioxidant, immune enhancers and hormone modulators (Okwu and Omodamiro, 2005). There was no regular pattern on how cooking affected the phenol content of *D. alata*.

Tannin affects the nutritive value of food products by forming complex with protein (both substrate and enzyme) thereby inhibiting digestion and absorption (Osuntogun et al., 1989). They also bind iron, making it unavailable (Aletor and Adeogun, 1995) and other evidence suggests that condensed tannins may cleave DNA in the presence of copper ions (Shirahata *et al.*, 1998). Tannin content of 0.21% was identified in *D. alata*, a similar report was given by Akin-Idowu *et al.* (2009). The reduction of tannin concentration of boiled Dioscorea varieties is expected, since earlier report indicated that processing methods such as soaking, boiling and fermentation lowered the tannin contents of the foods (Jude *et al.*, 2009). The decrease in the levels of tannin during cooking may be due to the thermal degradation and denaturation of the tannin as well as the formation of insoluble complexes (Kataria et al., 1989). The bitter principle of the wild *D. dumetorum* may be due to the presence of tannins in them. The trace quantities of tannin available in yam tubers act as a repellant against rots in yams.

Conclusion

From the results obtained in this study, it can be concluded that boiling has both positive and negative effect on water yam (*D. alata*). The positive effect will be derived from the reduction of the anti-nutritional factors while the negative effect is as a result of the reduction of nutritional factors. It is important to avoid overcooking since from the data obtained, it has been shown that the longer the cooking, the higher the loss in nutrients. A cooking time of between 30 and 60 min is recommended for *D. alata*.

REFERENCES

Akin-Idowu PE, Asiedu R, Maziya-Dixon B, Odunola A, Uwaifo A (2009). Effects of two processing methods on some nutrients and anti-nutritional factors in yellow yam (*Dioscorea cayenensis*). Afr. J. Food Sci. 3(1):022-025.

Akinmutimi AH (2007). Effect of cooking periods on the nutrient composition of velvet beans (Mucuna pruriens). Proceedings of the 32nd Annual Conference of PAT 2009; 5(1):92-102 ISSN: 0794-5213.

Alaise C, Linden G (1999). Food Biochemistry. Chapman and Hall, Food Science Book. Aspen Publishers Inc. Maryland. pp. 15-121.

Aletor VA, Adeogun OA (1995). Nutrient and antinutrient components of some tropical leafy vegetables. Food Chem. 53:375-379.

Association of Official Analytical Chemists (AOAC) (1990). Official methods of analysis (13th ed.) Washington, DC: Association of Official Analytical Chemists.

Balogun TF, Kaankuka FG, Bawa GS (2001): Effect of boiling full fat soyabeans on its amino acid profile and on performance of pigs. *Nigerian* J. Anim. Prod. 28(1): 45-51.

Boham AB, Kocipai DC (1994). Flavonoid and condensed tannins from leaves of *Hawaiian vaccinum vaticulum* and vicalycimum. Pracific Sci. 48:458-463.

Bourret-Cortadellas D (1973). Ethnobotanic Study of the Dioscorea, Yams New Caledonia. Doctorate 3rd cycle. Faculty of science, Paris, France. p. 135.

Coursey DG (1967). Yams: Tropical Agriculture series, Pp. 25-28. Longmans London.

Coursey DG (1973). Cassava as Food: Toxicity and Technology. In: Nestel, B. and R. MacIntyre, (Eds.), Chronic Cassava Toxicity, Ottawa, Canada, IDRC, IDRC-10e, pp. 27-36.

Ekpeyong TE (1984). Composition of some Tropical Tuberous Foods. Food Chem. 15:31-36.

Farris DG, Singh U (1990). Pigeon Pea. In: L. Nene et al. (eds). The Pigeon Pea, Pantancheru AP, 502324. India. ICRISAT. p. 467.

Farquer JN (1996). Plant Sterols. Their biological effects in humans, Handbook of lipids in human nutrition. BOCA Raton FL CRC Press pp. 101-105.

Hahn SK (1995). Yams: *Dioscorea* spp. (Dioscoreaceae) In J. Smaartt and Simmonds NW (Eds). Evolution of crop plants, Longman scientific and Technical, UK. pp. 112-120.

Higdon K, Scott A, Tucci M, Benghuzzi H, Tsao A, Puckett A, Cason Z, Hughes J (2001).The use of estrogen, DHEA, and diosgenin in a sustained delivery setting as a novel treatment approach for osteoporosis in the ovariectomized adult rat model. Biomed Sci Instrum.37:281-286.

Jude CI, Catherine CI, Ngozi MI (2009). Chemical Profile of *Tridax procumbens* Linn. Pak. J. Nutr. 8:548-550.

Kataria A, Chauhan BM, Punia D (1989). Antinutrients and protein digestibility (*in vitro*) of mungbean as affected by domestic processing and cooking. Food Chem. 3:9-17.

Lebot V, Malapa R, Molisale T, Marchand JL (2005). Physicochemical characterization of yam (Dioscorea alata L.) tubers from Vanuatu. Genet. Resour. Crop Evol. 53(6):1199-1208.

Longe OG (1986). Energy Content of some Tropical Starchy Crops. Nig. Agric. J. 21:136.

Martin F, Rubeste R (1976). The polyphenols of *Dioscorea alata* Yam tubers associated with oxidative browning. J. Agric Food Chem. 14:67-70.

Mc Williams M (1979). Food Fundamentals: John Wiley and sons Inc. New York U.S.A. pp. 125-130.

Mignouna HD, Abang MM, Asiedu R (2004). Harnessing Modern Biotechnology for Tropical tuber crop improvement: yam (Dioscorea spp) breeding. Afr. J. Biotechnol, 2(12):478-485.

Muzac-Tucker I, Asemota HN, Ahmad MH (1993). Biochemical composition and storage of Jamaica Yams (*Dioscorea* spp.). J. Sci. Food. Agric. 62:219-224.

Nzewi D, Egbuonu ACC (2011). Effect of boiling and roasting on the proximate properties of asparagus bean (*Vigna* Sesquipedalis). Afr. J. Biotechnol. 10(54):11239-11244,

Obadoni BO, Ochuko PO (2001) Phytochemical studies and comparative efficacy of the crude extracts of some homeostatic plants in Edo and Delta States of Nigeria. Global J. Pure Appl. Sci. 8:203-208.

Okaka JC, Enoch NJ, Okaka NC (1992). Human Nutrition. An Integrated Approach. Enugu State University of Technology Publ. Enugu 130-152.

Okwu DE (2004). Phytochemicals and vitamin content of indigenous species of South Eastern Nigeria. J. Sustain. Agric. Environ. 6:30-34.

Okwu DE, Omodamiro OD (2005). Effects of hexane extract and phytochemical content of *Xylopia aethiopica* and *Ocimum gratissimum* on the uterus of guinea pig. Bio-research 3 (In press).

Onu PN, Okongwu SN (2006): Performance characteristics and nutrient utilization of starter broilers fed raw and processed pigeon pea (Cajanus cajan) seed meal. Int. J. Poultry Sci. 5(7):693-697.

Onyeike EN, Oguike JU (2008). Influence of Heat Processing Methods on the Nutrient Composition and Lipid Characterization of Groundnut (Arachis hypogacaea) Seed Pastes. Biokemistri 15(1):34-43.

Onyenuga VA (1968). Nigeria's Food and Feeding stuffs, their Chemistry and Nutritive Value 3rd Edition. Ibadan University Press, Nigeria, p. 99.

Osuntogun BA, Adewusi SRA, Adewusi A, Ogundiwin JO, Nwasike CC (1989). Effect of cultivar, steeping, and malting on tannin, total polyphenol, and cyanide content of Nigerian sorghum. Cereal Chem. 66:87-89.

Shirahata S, Murakami H, Nishiyama K, Yamada K, Nonaka G,Nishioka I, Omura H (1998). J. Agric. Food Chem. 37:299-303.

Sopido OA, Ahiniyi JA, Ogunbanosu JU (2000). Studies on certain characteristics of extracts of barke of pansinystalia macruceras (K. Schem.) Pierve Exbeille. Global J. Pure Appl. Sci. 6:83-87.

Udensi EA, Oselebe HO, Onuoha AU (2010). Antinutritional Assessment of *D. alata* varieties. Pak. J. Nutr. 9(2):179-181.

Vernier P, Berthaud J, Bricas N, Marchand JL (1998). L'intensificationdes techniques de culture de l'igname. Acquis et contraintes. In: L'igname, plante séculaire et culture d'avenir. Actes du séminaire international Cirad-Inra-Orstom-Coraf, 3-6 juin 1997, Montpellier, France. Montpellier, France: CIRAD. pp. 93-101.

30

Influence of different concentrations of CO_2 in the postharvest conservation of peaches 'Aurora-1'

Joana Diniz Rosa Fernandes, João Emmanuel Ribeiro Guimarães, Josiane Pereira da Silva, Kelly Magalhães Marques and Ben-Hur Mattiuz

Laboratório de Tecnologia dos Produtos Agrícolas, Departamento de Tecnologia, Faculdade de Ciências Agrárias e Veterinárias, UNESP - UnivEstadualPaulista, Campus de Jaboticabal, Brazil.

The objective of this study was to evaluate the influence of different concentrations of carbon dioxide associated with a fixed concentration of oxygen in the postharvest conservation of peaches 'Aurora-1'. Experiment was based on a complete random split plot design, five atmosphere conditions, five analysis dates with three replications, each consisting of five fruits. The treatments were based on five atmospheres of CO_2, that is, 1, 3, 6 and 12% and a fixed concentration of 20% O_2, while control fruits were placed in an atmosphere of 21% O_2 and 0.03% CO_2. Fruits were stored at 12°C for 28 days and analyzed every 7 days for the following variables: external appearance, incidence of disease, accumulated loss of fresh mass, firmness and color of the pulp, soluble solid content, and titratable acidity The effects of the gas over time were analyzed by analysis of variance and mean comparison of the data by Duncan test at 5% probability. The atmosphere of 12% CO_2 + 20% O_2 allowed the peaches 'Aurora-1' keep the physicochemical characteristics for up to 21 days of storage at 12°C. The controlled atmosphere condition with 1% CO_2 + 20% O_2 was adequate to allow the absence of disease during the 28 days of storage without compromising the physical-chemical characteristics of fruits.

Key words: Controlled atmosphere, oxygen, *Monilinia* sp., *Prunus persica*.

INTRODUCTION

Production of peach cultivar 'Aurora-1' in the Jaboticabal – SP area in Brazil is viable, as this region shows a mild winter that enables the harvesting from September to November, since this cultivar has the need for less than 100 h of cold to break dormancy (Silva et al., 2013).

Peach is a perishable fruit, as postharvest physiological processes shorten its useful life, due to intense metabolism that favors disease incidence. Cold storage associated with atmosphere control is a choice for quality preservation of this fruit. Cunha Júnior et al. (2010) report

that peaches cultivar Aurora-1 'are not sensitive to chilling injury at temperatures of 2, 6 and 12°C, the fruits keep for up to 35 21 and 14 days.

Reduction of oxygen concentrations and increments on those of carbon dioxide in the storage atmosphere, combined with adequate temperatures, have shown to besuitable to increase the storage life and keep the quality of many horticultural products (Kader, 2003). Ferrer–Mairal et al. (2012) reported that a controlled atmosphere containing 10% O_2 + 5% CO_2 was more

Table 1. Occurrence of diseases, expressed as percentages, in peaches 'Aurora-1' submitted to controlled atmospheres containing different carbon dioxide concentrations, and stored at 12°C and 98% RH.

Days	Concentration of CO_2 stored at 12°C				
	Control	$1\%CO_2+20\%O_2$	$3\%CO_2+20\%O_2$	$6\%CO_2+20\%O_2$	$12\%CO_2+20\%O_2$
0	0.0	0.0	0.0	0.0	0.0
7	0.0	0.0	0.0	0.0	0.0
14	16.0	0.0	8.4	0.0	8.4
21	16.0	0.0	8.4	0.0	16.7
28	66.7	0.0	33.4	25.0	25.0

Obs. Results are given as the ratio of rot fruits to the total amount of fruits (n=15).

efficient than a 10% O_2 + 5% CO_2 – IC for the preservation of peaches 'Jesca' and 'Evaisa'. A controlled atmosphere where the oxygen and carbon dioxide concentrations are kept around 2 and 5 KPa, respectively, at lower temperatures (-0.5°C) reduced the occurrence of rots in peaches 'Eldorado' (Brackmann et al., 2005), while variations on the concentration of carbon dioxide (3, 5, and 10 kPa) associated with1.5 kPa O_2 have also shown to be efficient on rot prevention of peaches 'Douradão' (Santana et al., 2011).

Controlled atmosphere conditions have also kept the firmness of peaches 'Maciel' at 2kPa oxygen and 4 kPa carbon dioxide concentrations (Sestari et al., 2008),and at 5 and 10 kPa carbon dioxide associated with 1.5 kPa oxygen concentrations for peaches 'Douradão' (Santana et al., 2011), as does at 2 kPa oxygen and 5 kPa carbon dioxide concentrations for peaches 'Eldorado' (Brackmann et al., 2005).

The conditions of the controlled atmosphere are established based on the species, cultivar, temperature, and period of storage. Plums 'Laetitia' have shown a delayed ripening at a 2 kPa oxygen and 5 kPa carbon dioxide concentrations at 5°C (Alves et al., 2010) whereas peaches 'Maciel' have shown a better storage condition under 1 kPa oxygen and 3 kPa carbon dioxide at -0.5°C (Sestari et al., 2008).

This works aims to establish ideal storage conditions for peaches grown in tropical areas by the evaluation of results obtained in controlled atmosphere conditions containing different concentrations of carbon dioxide associated to a fixed concentration of oxygen in the postharvest conservation of peaches 'Aurora-1'.

MATERIALS AND METHODS

Peaches 'Aurora-1' collected at a private property located in Pirangi, State of São Paulo, Brazil (lat. 21°05'29"S, long. 48°39'28"W, alt. 538 m) were carefully and rapidly transported to the Laboratório de Tecnologia dos Produtos Agrícolas, at UNESP, Jaboticabal after harvesting. Firstly, fruits were selected discarding those showing lesions and/or poor color appearance in order to make a uniform batch before washing with potable water and treatment with sodium dichloro-S-triazinetrione dihydrate (Sumaveg®200 mg 100 L^{-1}) for 5 min. After drying at room

temperature, peaches were stored in 52.8 L capacity hermetic chambers. Gaseous mixtures humidified at 98% were injected through and inlet at the bottom part of the chamber and evacuated by an outlet at the upper part opposed to the entrance, under a 300 ml min^{-1} continuous flow.Controlled atmosphere was created using a flow board system, with pressure adjustment by a bottled gas cylinder valve, as described by Calbo (1989) that avoids the loss of gas mixture when a barostat is employed (Cerqueira et al., 2009).

Dried peaches were placed in experimental chambers and submitted to controlled atmospheres containing 1, 3, 6, and 12% CO_2, 20% O_2 and control fruits were kept at room atmosphere containing 21% O_2 and 0.03% CO_2. Fruits were kept for 28 days at 12°C and analyzed every 7 days. Different parameters were evaluated, external appearance was graded as 3=excellent; 2=good; 1=bad; 0=terrible (Cunha Júnior et al., 2010); hollow stem disorder was evaluated as A=absence of lesion and P=occurrence of lesion > 0.5 cm^2 and quantified as percentages; firmness was measured using a manual 0.8 cm tip penetrometer and results were expressed as N; loss of fresh mass accumulated in the period was obtained by measuring the weight of the fruit lot in an electronic scale and results were expressed as percentages; soluble solid content in the pulp was determined by a refractometer from drops obtained from the filtered fraction of homogenized materials and results were expressed as °Brix indices; titratable acidity was determined in a 10 ml sample, diluted in 50 ml distilled water, titrated with NaOH 0.1 M, and results were expressed in equivalent g of citric acid/100 g sample; color of pulp was determined by colorimetric readings using a MINOLTA-CR-400b colorimeter and results were expressed as luminosity, hue angle, and chromaticity values.

Experiment was based on a complete random split plot design, five atmosphere conditions, five analysis dates with three replications, each consisting of five fruits. The effects of the gas over time were analyzed by analysis of variance and mean comparison of the data by Duncan test at 5% probability.

RESULTS AND DISCUSSION

Diseases were caused mainly by *Monilinia* sp., and as it can be observed in Table 1, occurrences started at the 14th day of storage with a higher intensity (16%) in fruits under the control atmosphere, while the $3\%CO_2+20\%O_2$ or $12\%CO_2+20\%O_2$ treatments resulted in a 8.4% occurrence. The treatment in the atmosphere containing 1%CO_2 + 20%O_2 did not allow any fungi manifestation during the whole 28 days period of storage, and under the $6\%CO_2+20\%O_2$ atmosphere, symptoms of the

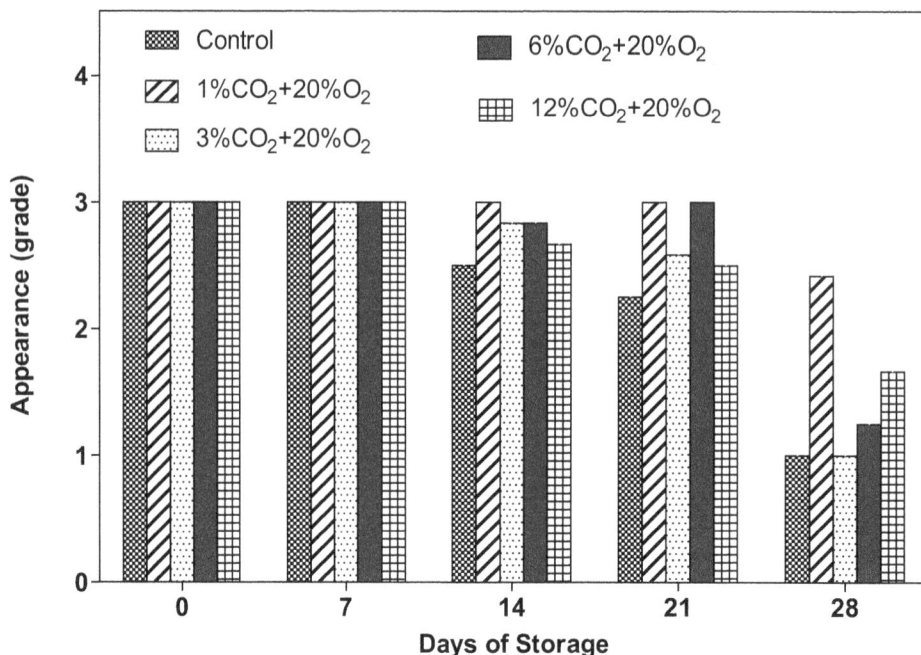

Figure 1. Appearance of peaches 'Aurora-1', submitted to controlled atmosphere containing different concentrations of CO_2, and stored at 12°C and 98% RH (Appearance: 3=excellent; 2=good; 1=bad; 0=terrible).

disease occurred only at the 28th day.

These results confirm the carbon dioxide fungistatic effect indicated by Kader and Ben-Yehoshua (2000), as fruits stored under CO_2 enriched atmospheres have shown a delayed growth of fungi or not growth at all in the period when compared to control fruits. According to Kader (1986), concentrations of carbon dioxide higher than 1% can minimize pathogen activity, as does observed at 1 kPa O_2 + 8 kPa CO_2 for peaches 'Eragil' (Brackmann et al., 2013),or 1 kPa O_2 + 3 kPa CO_2 for peaches 'Maciel' (Sestari et al., 2008), and 2 kPa O_2 + 5 kPa CO_2 for peaches 'Eldorado' (Brackmann et al., 2005).

Atmospheres containing 1% CO_2+20%O_2, besides not showing *Monilinia* sp., incidence, have also kept the external appearance of fruits (Figure 1) that showed excellent and good grades (3 and 2), thus preserving the quality during the whole period of storage. Atmospheres containing 6% CO_2 + 20% O_2 and 12%CO_2 + 20%O_2 were able to keep a good appearance of fruits up to the 21st day. At the 28th day, appearances of either control fruits and those stored at 3%CO_2 +20%O_2have shown a bad grade, thus making it inappropriate for commercialization. These results are in agreement with Santana et al. (2011) that have found a significant improvement in the appearance of peaches 'Douradão' stored for 28 days in controlled atmosphere conditions (5 and 10 kPa CO_2 + 1.5 kPa O_2).

During the storage period of 28 days, the highest loss of fresh mass was observed for the 3% CO_2 + 20% O_2

condition, while the lowest loss was observed for the 1% CO_2 + 20% O_2 condition, which kept the fresh mass of fruits as a possible result to the lack of disease. Atmospheres containing 6% CO_2 + 20% O_2 and 12% CO_2 + 20% O_2 have not shown any significant difference on the 21st day (2.08 and 2.12%, respectively). Nava and Brackmann (2002) did not observe a significant loss of fresh mass under controlled atmospheres for peaches 'Chiripá' during a storage period of 8 weeks, which was attributed to a reduced respiratory activity (Figure 2).

There was a higher maintenance of firmness (Figure 3) up to the 21st day for fruits stored at the atmosphere containing 12%CO_2 + 20%O_2, as does for the atmosphere containing 1% CO_2 + 20% O_2 that kept firmness of fruits and showed a reduced loss of fresh mass up to the 28th day. Cano-Salazar et al. (2003) have also reported that higher concentrations of CO_2 and O_2 (17 and 6 kPa) was advantageous for keeping the firmness of nectarines 'Big Top'. Lower concentrations of CO_2 (1 and 5 kPa) have also favored the maintenance of pulp firmness of apples 'Galaxy' (Brackmann et al., 2008), as does the atmosphere containing 2.0 kPa O_2 + 4.0 kPa CO_2 for peaches 'Maciel' (Sestari et al., 2008) and the atmosphere containing 2.0 kPa O_2 + 5.0 kPa CO_2 for peaches 'Eldorado' (Brackmannet al., 2005).

The maintenance of pulp firmness for fruits stored at controlled atmospheres can be attributed to the effect of those conditions on the decrease of biosynthesis and the action of ethylene, which reduces the activity of hydrolytic enzymes that are responsible for the degradation of cell

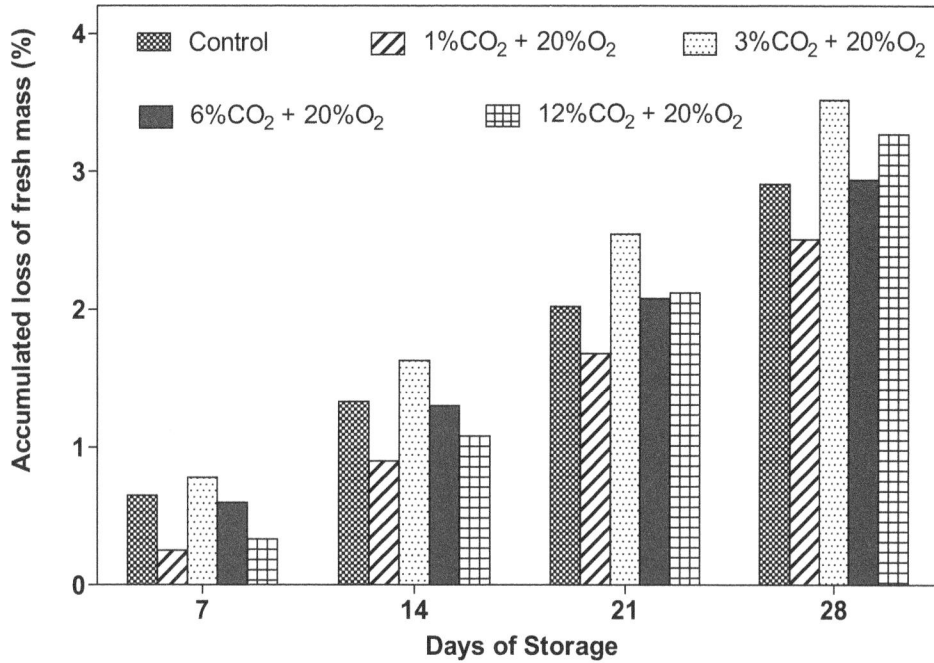

Figure 2. Accumulated loss of fresh mass for peaches 'Aurora-1' submitted to controlled atmospheres containing different concentrations of CO_2 and stored at 12°C and 98% RH. Vertical lines indicate standard deviation of the average value.

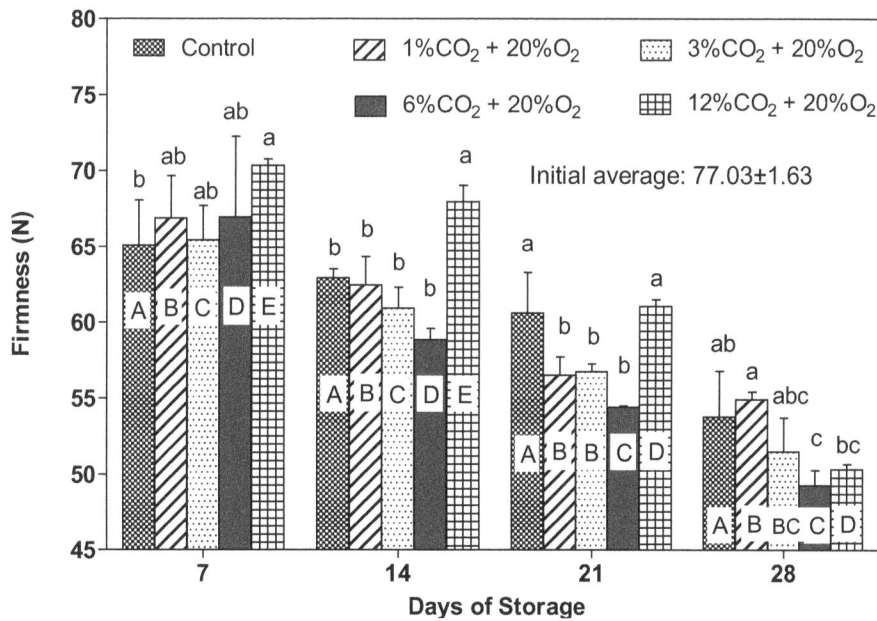

Figure 3. Pulp firmness of peaches 'Aurora-1' submitted to controlled atmospheres containing different concentrations of CO_2 and stored at 12°C and 98% RH. Vertical lines indicate standard deviation of the average value. Same lowercase letters within the group of columns and uppercase letters of columns between the groups did not differ by Duncan test (P <0.05).

wall components (Alves et al., 2010).

Pulp colour (Figure 4) of peaches stored at atmospheres containing $12\%CO_2 + 20\%O_2$ reached the final stage of maturity at the end of the storage period (28 days), followed by a change in the hue angle of the pulp color from yellow to orange, and darker fruits showing

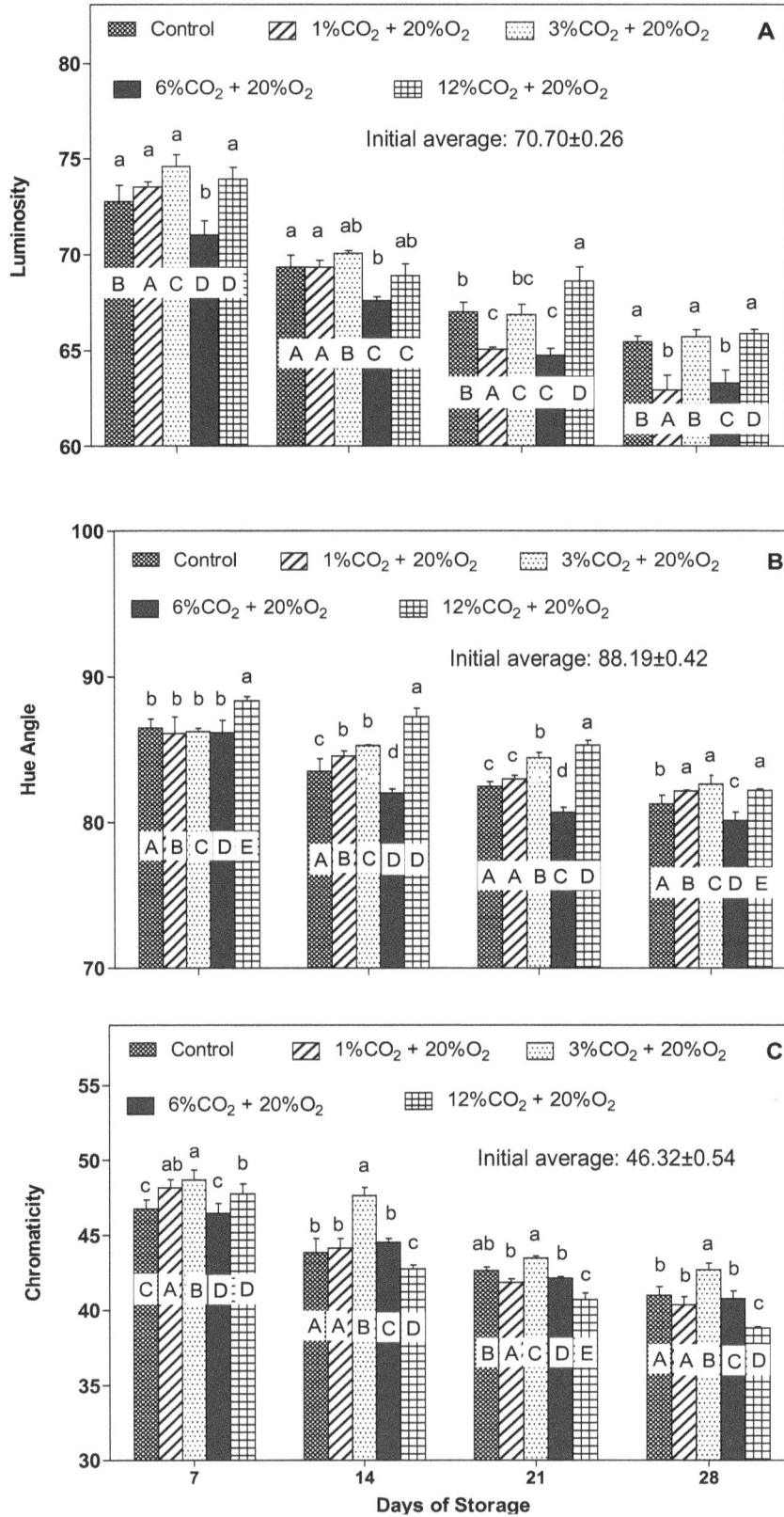

Figure 4. Luminosity, hue angle, and chromaticity values of pulp of peaches 'Aurora-1'submitted to controlled atmospheres containing different concentrations of CO_2 and stored at 12°C and 98% RH. Vertical lines indicate standard deviation of the average value. Same lowercase letters within the group of columns and uppercase letters of columns between the groups did not differ by Duncan test (P <0.05).

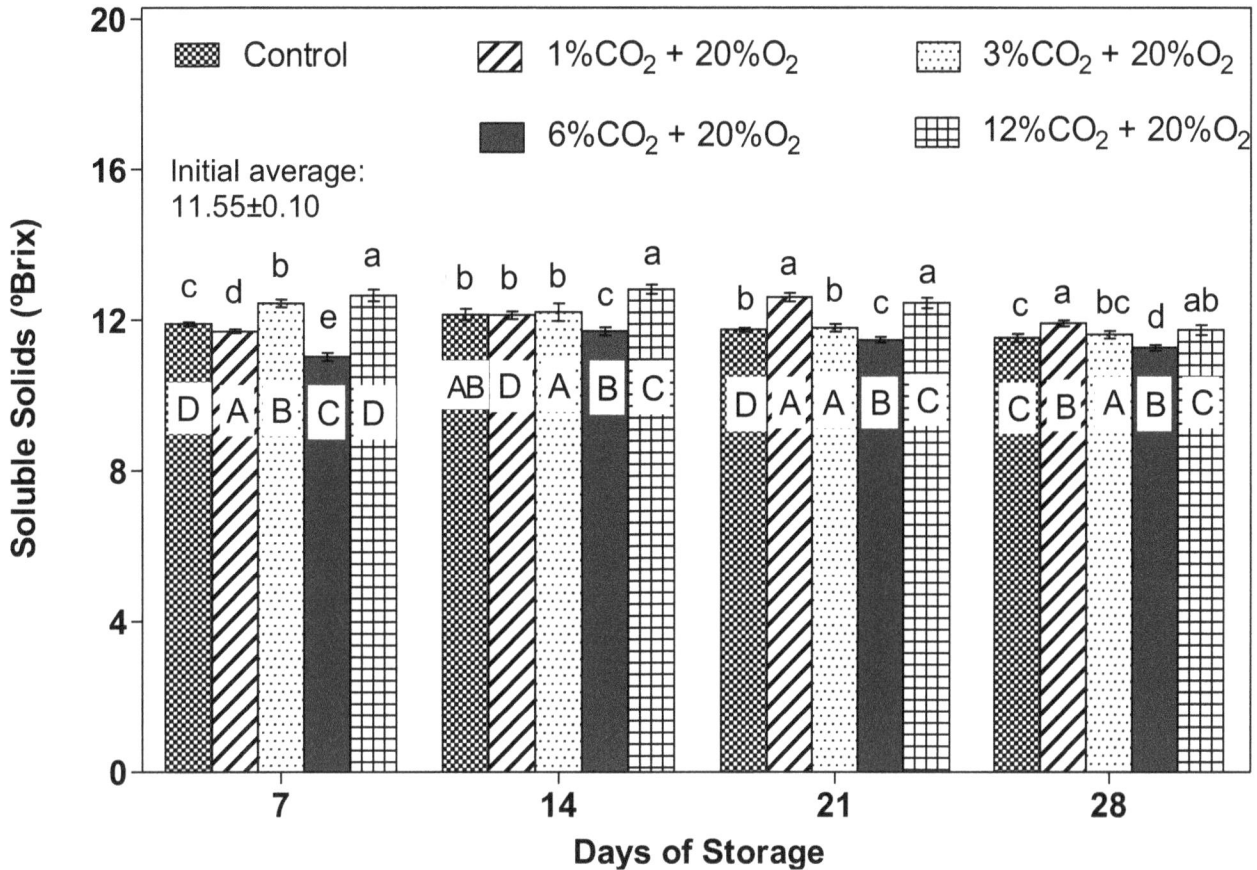

Figure 5. Content of soluble solids in peaches'Aurora-1'submitted to controlled atmospheres containing different concentrations of CO_2 and stored at 12°C and 98% RH. Vertical lines indicate standard deviation of the average value. Same lowercase letters within the group of columns and uppercase letters of columns between the groups did not differ by Duncan test (P <0.05).

either less luminosity and chromaticity values. A higher darkening of the pulp fruits was observed in the last two weeks under the 3%CO_2 + 20%O_2 (higher intensity), 12% CO_2 + 20%O_2 and control (higher intensity) atmospheres.

The lowest darkening and smallest hue angles of pulp color of fruits during the period of storage was observed under atmospheres containing 1% CO_2 + 20% O_2 and 6% CO_2 + 20% O_2, which can indicate a delayed maturity of these fruits. Brackmann et al. (2007) did not experience any efficiency of controlled atmospheres on the color of pulps of peaches 'Granada'. The lowest internal browning of peaches 'Maciel' was observed at atmospheres containing 1.0 kPa O_2 + 3.0 kPa CO_2 (Sestari et al., 2008) and at 1.0kPa CO_2 + 1.0kPa O_2 for plumps 'Laetitia' (Steffens et al., 2013).

Minor variations among the treatments during the period of storage were observed for soluble solids content (Figure 5). Atmospheres containing 1% CO_2 +20% O_2 and 12% CO_2 +20% O_2 have shown a small increase in the content of soluble solids at the 21st day of storage. The content of soluble solids can be associated to the stage of maturity at the harvesting of the fruits, the physiological maturity along the period of storage, and the loss of fresh mass, which favors the increase of soluble solids. Opposite results were reported by Cano-Salazar et al. (2013) that described a lower content of soluble solids and titratable acidity in peaches 'Early Rich' stored at atmospheres containing 6.0 kPaO_2 + 17.0 kPaCO_2. Weber et al. (2013) have also described the influence of different controlled atmosphere conditions on the content of soluble solids of apples 'Gala'. The soluble solids content in climacteric fruits, increases after harvest and the beginning of storage, depending on the hydrolysis of starch, and a decrease in long-term storage of sugars (Chitarra and Chitarra, 2005) occurs, which are metabolized by the respiratory process during ripening.

All treatments have shown a lowering on the titratable   acidity of the fruits. The highest values (Figure 6) were

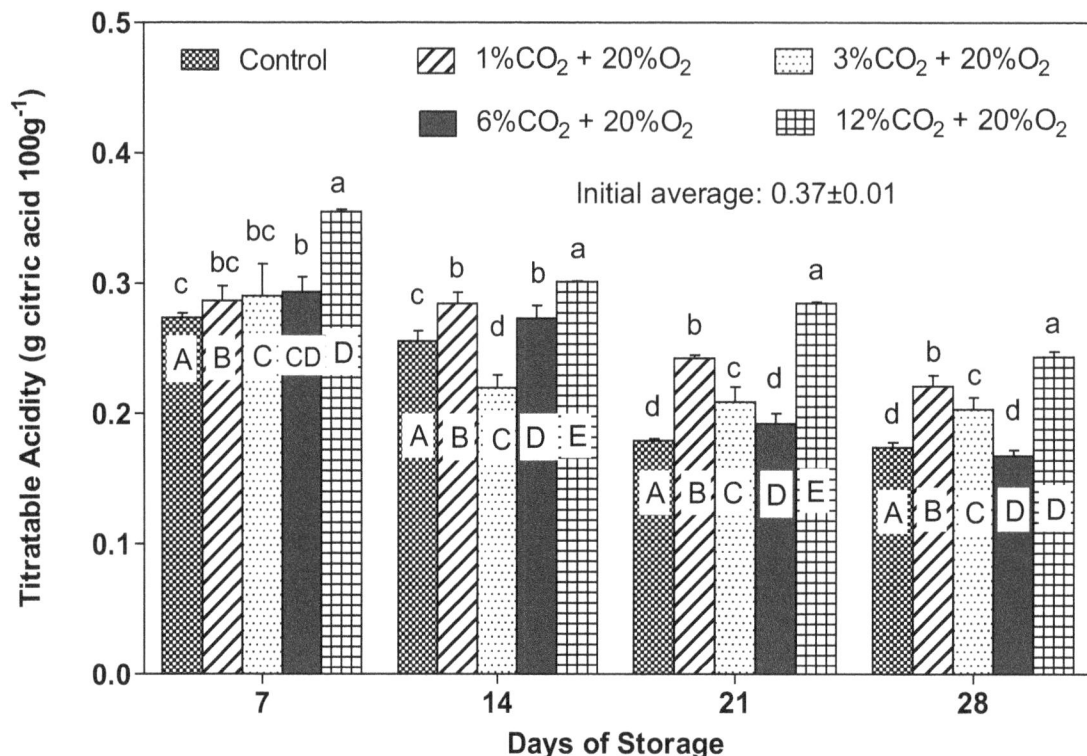

Figure 6. Titratable acidity in peaches'Aurora-1'submitted to controlled atmospheres containing different concentrations of CO_2 and stored at 12°C and 98% RH. Vertical lines indicate standard deviation of the average value. Same lowercase letters within the group of columns and uppercase letters of columns between the groups did not differ by Duncan test (P <0.05).

observed in fruits at atmospheres containing 12%CO_2 + 20%O_2, during the storage period of 28 days. The decreased titratable acidity values observed in the last two weeks of storage for the control and 6%CO_2 +20%O_2 atmospheres is assigned to ripening of fruits during the period, which is indicated by an increase in the concentration of soluble solids and a decrease in the firmness of the pulp. Rombaldi et al. (2002) have also demonstrated that peaches 'Chiripá' stored in a controlled atmosphere containing 1.5 kPa O_2 + 5.0kPa CO_2 showed a lower titratable acidity and a higher content of soluble solids during the first 30 days of storage. The decrease in the titratable acidity of the fruits is mostly due to levels of organic acids that, with a few expections, decrease along ripening as it constitutes excellent energetic reserves for the fruit (Kays, 1991).

Conclusions

The atmosphere of 12% CO_2 + 20% O_2 allowed the peaches 'Aurora-1' keep the physicochemical characteristics for up to 21 days of storage at 12°C. The controlled atmosphere condition with 1% CO_2 + 20% O_2 was adequate to allow the absence of disease during 28 days of storage without compromising the physical-chemical characteristics of fruits.

Conflict of Interest

The authors have not declared any conflict of interest.

ACKNOWLEDGEMENT

To FAPESP (Fundação de Amparo à Pesquisa do Estado de São Paulo) for financial support - process 2010/20514-0.

REFERENCES

Alves EO, Steffens CA, Amarante CVT, Brackmann A (2010). Quality of 'Laetitia' plums when affected by temperature and storage atmosphere. Rev. Bras. Frutic. 32:1018-1027.

Brackmann A, Bordignom BCS, Gieal RFH, Sestari I, Eisermann AC (2007). Storage of cv. "Granada" peach fruit in controlled atmosphere aiming the transport for long distances. Cienc. Rural. 37(3):676-681.

Brackmann A, Both V, Weber A, Pavanello EP, Anese RO, Santos JRA (2013). Controlled atmosphere, ethylene absorption and 1-MCP application during storage of 'Eragil' peaches. Científica. 41(2):156-163.

Brackmann A, Giehl RFH, Sestari I, Mello AM, Cuarientti AJW (2005). Use of controlled atmosphere storage of peaches for "Eldorado" harvested at two maturity stages. Brazilian J. Storage 30(2):209-214.

Brackmann A, Weber A, Pinto JAV, Neuwaldi DA, Steffens CA (2008). Maintaining postharvest quality of apples 'Royal Gala' and 'Galaxy' under controlled atmosphere storage. Cienc. Rural 38(9):2478-2484.

Calbo AG (1989). Adaptation of a fluxcentro for studies of gas exchange and a method for measuring capillary. Pesqui. Agropec. Bras. 24(6):733-739.

Cano-Salazara J, Lópeza ML, Echeverría G (2013). Relationships between the instrumental and sensory characteristics of four peach and nectarine cultivars stored under air and CA atmospheres. Postharvest Biol. Technol. 75:58-67.

Cerqueira TS, Cunha Júnior LC, Calbo AG, Jacomino AP (2009). Flowboard for postharvest gas mixtures applications to fruits and vegetables without waste of gas. In: International controlled and modified atmosphere research conference, Antalya. Leuven: ISHS. pp. 56-56.

Chitarra MIF, Chitarra AB (2005). Postharvest of fruits and vegetables: physiology and handling. Lavras: FAEPE P. 783.

Cunha Junior LC, Durigan MFB, Mattiuz BH (2010). Preservation of 'Aurora-1' peaches stored under refrigeration. Rev. Bras. Frutic. 32(2):386-396.

Ferrer-Mairal A, Remón S, Peiró JM, Oria R (2012). Effects of intermittent conditioning on the color and enzymatic activity of peaches during controlled atmosphere storage. J. Food Bioch. 36:129-138.

Kader AA (1986). Biochemical and physiological basis for effects of controlled and modified atmospheres on fruits and vegetables. Food Technol. 40(5):99-104.

Kader AA (2003). Physiology of CA treated produce. In: Oosterhaven J Peppelenbos HW (eds) International controlled atmosphere: research conference, Rotterdam, Netherland: Acta Hortic. 600:349-354.

Kader AA, Ben-Yehoshua S (2000). Effects of super atmospheric oxygen levels on postharvest physiology and quality of fresh fruits and vegetables. Postharvest Biol. Technol. 20:1-13.

Kays SJ (1991). Postharvest physiology of perishable plant products. AVI Book, P. 532.

Nava GA, Brackmann A (2002). The storaging of peaches cv. Chiripá, in controlled atmosphere. Rev. Bras. Frutic. 24(2):328-332.

Rombaldi CV, Silva JA, Machado LB, Parussolo A, Lucchetta L, Zanuzzo MR, Girardi CL, Cantillano RF (2002). Storage of chiripa peach (*Prunus persica* L.) in controlled atmosphere. Cienc. Rural 32(1): 43-47.

Santana LRR, Benedetti BC, Sigrist JMM, Sato HH, Anjos VDA (2011). Effect of controlled atmosphere on postharvest quality of 'Douradão' peaches. Food Sci. Technol. 31(1):231-237.

Sestari I, Giehl RFH, Pinto JAV, Brackmann A (2008). Controlled atmosphere storage for 'Maciel' peaches harvested in two ripening stages. Cienc. Rural, 38(5):1240-1245.

Silva DFP, Ribeiro MR, Silva JOC, Matias RGP, Bruckner CH (2013). Cold storage of peaches cv. Aurora grown in the Zona da Mata Mineira, Minas Gerais State, Brazil. Rev. Ceres, 60(6): 833 -841.

Steffens CA, Tanaka H, Amarante CVT, Brackmann A, Stanger MC, Hendgers MV (2013). Conditions of a controlled atmosphere for the storage of 'Laetitia' plums treated with 1-methylcyclopropene. Rev. Ciênc. Agron. 44(4):750-756.

Weber A, Brackmann A, Anese RO, Both V, Pavanello EP (2013) Controlled atmosphere to the storage of 'Maxi Gala' apples. Rev. Ciênc. Agron. 44(2):294-301.

Permissions

All chapters in this book were first published in JSPPR, by Academic Journals; hereby published with permission under the Creative Commons Attribution License or equivalent. Every chapter published in this book has been scrutinized by our experts. Their significance has been extensively debated. The topics covered herein carry significant findings which will fuel the growth of the discipline. They may even be implemented as practical applications or may be referred to as a beginning point for another development.

The contributors of this book come from diverse backgrounds, making this book a truly international effort. This book will bring forth new frontiers with its revolutionizing research information and detailed analysis of the nascent developments around the world.

We would like to thank all the contributing authors for lending their expertise to make the book truly unique. They have played a crucial role in the development of this book. Without their invaluable contributions this book wouldn't have been possible. They have made vital efforts to compile up to date information on the varied aspects of this subject to make this book a valuable addition to the collection of many professionals and students.

This book was conceptualized with the vision of imparting up-to-date information and advanced data in this field. To ensure the same, a matchless editorial board was set up. Every individual on the board went through rigorous rounds of assessment to prove their worth. After which they invested a large part of their time researching and compiling the most relevant data for our readers.

The editorial board has been involved in producing this book since its inception. They have spent rigorous hours researching and exploring the diverse topics which have resulted in the successful publishing of this book. They have passed on their knowledge of decades through this book. To expedite this challenging task, the publisher supported the team at every step. A small team of assistant editors was also appointed to further simplify the editing procedure and attain best results for the readers.

Apart from the editorial board, the designing team has also invested a significant amount of their time in understanding the subject and creating the most relevant covers. They scrutinized every image to scout for the most suitable representation of the subject and create an appropriate cover for the book.

The publishing team has been an ardent support to the editorial, designing and production team. Their endless efforts to recruit the best for this project, has resulted in the accomplishment of this book. They are a veteran in the field of academics and their pool of knowledge is as vast as their experience in printing. Their expertise and guidance has proved useful at every step. Their uncompromising quality standards have made this book an exceptional effort. Their encouragement from time to time has been an inspiration for everyone.

The publisher and the editorial board hope that this book will prove to be a valuable piece of knowledge for researchers, students, practitioners and scholars across the globe.

List of Contributors

G. Kandeepan
National Research Center on Yak (ICAR), Dirang, West Kameng District, Arunachal Pradesh 790101, India

S. Sangma
National Research Center on Yak (ICAR), Dirang, West Kameng District, Arunachal Pradesh 790101, India

Z. F. Bhat
Division of Livestock Products Technology Faculty of Veterinary Sciences and Animal Husbandry, Sher-e-Kashmir University of Agricultural Sciences and Technology of Jammu, R. S. Pura, Jammu, Jammu and Kashmir –181 102, India

V. Pathak
Division of Livestock Products Technology UP Pt. Deen Dayal Upadhyaya Veterinary Science University Mathura (UP) -281001, India

W. S. Dhillon
Department of Horticulture, Punjab Agricultural University, Ludhiana, India

B. V. C. Mahajan
Punjab Horticultural Postharvest Technology Centre, Punjab Agricultural University campus, Ludhiana, India

Devendra Kumar
Department of Livestock Products Technology, College of Veterinary and Animal Science, G.B. Pant University of Agriculture and Technology, Pantnagar-263145, India

V. K. Tanwar
Department of Livestock Products Technology, College of Veterinary and Animal Science, G.B. Pant University of Agriculture and Technology, Pantnagar-263145, India

Sanjay Kumar Bharti
Department of Livestock Products Technology, College of Veterinary and Animal Sciences, G. B. Pant University of Agriculture and Tech, Pantnagar– 263145, India

B. Anita
Department of Livestock Products Technology, College of Veterinary and Animal Sciences, G. B. Pant University of Agriculture and Tech, Pantnagar– 263145, India

Sudip Kumar Das
Department of Livestock Products Technology, College of Veterinary and Animal Sciences, G. B. Pant University of Agriculture and Tech, Pantnagar– 263145, India

Subhasish Biswas
Department of Livestock Products Technology, West Bengal University of Animal and Fishery Sciences, Kolkata – 700037, India

Anubha Upadhyay
Department of Crop and Herbal Physiology, College of Agriculture, Jawaharlal Nehru Krishi Vishwa Vidyalaya (JNKVV), Jabalpur (M.P.) 482004, India

Yogendra Chandel
Department of Crop and Herbal Physiology, College of Agriculture, Jawaharlal Nehru Krishi Vishwa Vidyalaya (JNKVV), Jabalpur (M.P.) 482004, India

Preeti Sagar Nayak
Department of Crop and Herbal Physiology, College of Agriculture, Jawaharlal Nehru Krishi Vishwa Vidyalaya (JNKVV), Jabalpur (M.P.) 482004, India

Noor Afshan Khan
Department of Crop and Herbal Physiology, College of Agriculture, Jawaharlal Nehru Krishi Vishwa Vidyalaya (JNKVV), Jabalpur (M.P.) 482004, India

P. Prabuthas
FST (Financial Services Technology) Laboratory, PHTC, Agricultural and Food Engineering Department, Indian Institute of Technology, Kharagpur, India – 721302

S. Majumdar
FST (Financial Services Technology) Laboratory, PHTC, Agricultural and Food Engineering Department, Indian Institute of Technology, Kharagpur, India – 721302

P. P. Srivastav
FST (Financial Services Technology) Laboratory, PHTC, Agricultural and Food Engineering Department, Indian Institute of Technology, Kharagpur, India – 721302

H. N. Mishra
FST (Financial Services Technology) Laboratory, PHTC, Agricultural and Food Engineering Department, Indian Institute of Technology, Kharagpur, India – 721302

Joel Bonales Valencia
Institute of Economic and Business Research, Universidad Michoacana de San Nicolás de Hidalgo Morelia, Michoacán, Mexico

Oscar Hugo Pedraza Rendón
Institute of Economic and Business Research, Universidad Michoacana de San Nicolás de Hidalgo Morelia, Michoacán, Mexico

José César Lenin Navarro Chávez
Institute of Economic and Business Research, Universidad Michoacana de San Nicolás de Hidalgo Morelia, Michoacán, Mexico

IB Oluwalana
Department of Food Science and Technology, Federal University of Technology, Akure, Ondo State, Nigeria

MO Oluwamukomi
Department of Food Science and Technology, Federal University of Technology, Akure, Ondo State, Nigeria

TN Fagbemi
Department of Food Science and Technology, Federal University of Technology, Akure, Ondo State, Nigeria

GI Oluwafemi
Department of Food Science and Technology, Federal University of Technology, Akure, Ondo State, Nigeria

V. P. Singh
Division of Livestock Products Technology, Indian Veterinary Research Institute, Izatnagar-243122, India

M. K. Sanyal
Division of Livestock Products Technology, Indian Veterinary Research Institute, Izatnagar-243122, India

P. C. Dubey
Division of Livestock Products Technology, Indian Veterinary Research Institute, Izatnagar-243122, India

S. K. Mendirtta
Division of Livestock Products Technology, Indian Veterinary Research Institute, Izatnagar-243122, India

L. E. Contreras
Chemical Research Center, Universidad Autonoma del Estado de Hidalgo. Carretera Pachuca-Tulancingo km 4.5, C. P. 42076, Pachuca, Hidalgo, Mexico

O. J. Jaimez
Chemical Research Center, Universidad Autonoma del Estado de Hidalgo. Carretera Pachuca-Tulancingo km 4.5, C. P. 42076, Pachuca, Hidalgo, Mexico

O. A. Castañeda
Chemical Research Center, Universidad Autonoma del Estado de Hidalgo. Carretera Pachuca-Tulancingo km 4.5, C. P. 42076, Pachuca, Hidalgo, Mexico

M. J. Añorve
Chemical Research Center, Universidad Autonoma del Estado de Hidalgo. Carretera Pachuca-Tulancingo km 4.5, C. P. 42076, Pachuca, Hidalgo, Mexico

R. S. Villanueva
Research and Assistance in Technology and Design of the State of Jalisco A. C. Normalista Avenue No. 800, Normal Hill, C. P. 44270, Guadalajara, Jalisco. Mexico

Hassan Abdalla Almahy
Department of Chemistry, Taif University, Faculty of Science and Education, Alkhurma, Kingdom of Saudi Arabia

Omaima Dahab Nasir
Department of Medical Laboratory, Taif University, Faculty of Applied Medical Sciences, Turaba, Kingdom of Saudi Arabia

A. O. Oyebanji
Nigerian Stored Products Research Institute, Onireke-Dugbe, P. M. B. 5044, Dugbe, Ibadan

M. H. Ibrahim
Nigerian Stored Products Research Institute, Onireke-Dugbe, P. M. B. 5044, Dugbe, Ibadan

S. O. Okanlawon
Nigerian Stored Products Research Institute, Onireke-Dugbe, P. M. B. 5044, Dugbe, Ibadan

S. O. Okunade
Nigerian Stored Products Research Institute, Onireke-Dugbe, P. M. B. 5044, Dugbe, Ibadan

E. F. Awagu
Nigerian Stored Products Research Institute, Onireke-Dugbe, P. M. B. 5044, Dugbe, Ibadan

VC Ezeocha
National Root Crops Research Institute, Umudike, Abia State, Nigeria

RM Omodamiro
National Root Crops Research Institute, Umudike, Abia State, Nigeria

E Oti
National Root Crops Research Institute, Umudike, Abia State, Nigeria

GO Chukwu
National Root Crops Research Institute, Umudike, Abia State, Nigeria

Ioan GONTARIU
Faculty of Food Engineering, Stefan cel Mare University, Street. Universitatii no. 13, 720229, Suceava, Romania

Ioan-Catalin ENEA
Agricultural Research and Development Station of Suceava, B-dul 1 Decembrie 1918, no. 15, 720246, Romania

Danela MURARIU
Suceava Genebank, B-dul 1 Mai no. 17, 720246, Suceava, Romania

Z. F. Bhat
Division of Livestock Products Technology, Faculty of Veterinary Sciences and Animal Husbandry, Shere-Kashmir University of Agricultural Sciences and Technology of Jammu, R. S. Pura, India – 181 102

Hina Bhat
Department of Biotechnology, University of Kashmir, Hazratbal, Srinagar, Jammu and Kashmir, India-190006

Xin Gao
College of Food Science and Engineering, Ocean University of China, Qingdao, Shandong 266003, China

Tian Yu
College of Food Science and Engineering, Ocean University of China, Qingdao, Shandong 266003, China

Zhao-hui Zhang
College of Food Science and Engineering, Ocean University of China, Qingdao, Shandong 266003, China

Jia-chao Xu
College of Food Science and Engineering, Ocean University of China, Qingdao, Shandong 266003, China

Xiao-ting Fu
College of Food Science and Engineering, Ocean University of China, Qingdao, Shandong 266003, China

M. A. Talpur
Department of Agricultural Mechanization, College of Engineering, Nanjing Agricultural University, Post Code 210031, Nanjing, Peoples Republic of China

J Changying
Department of Agricultural Mechanization, College of Engineering, Nanjing Agricultural University, Post Code 210031, Nanjing, Peoples Republic of China

F. A. Chandio
Department of Agricultural Mechanization, College of Engineering, Nanjing Agricultural University, Post Code 210031, Nanjing, Peoples Republic of China
Department of Farm Power and Machinery, Faculty of Agricultural Engineering, Sindh Agriculture University, Tandojam, Pakistan

S. A. Junejo
Department of Hydrology, School of Earth Sciences and Engineering, Nanjing University, P. R. China
Department of Geography, Sindh University Jamshoro, Sindh Pakistan

I. A. Mari
Department of Agricultural Mechanization, College of Engineering, Nanjing Agricultural University, Post Code 210031, Nanjing, Peoples Republic of China

C. C. Ojianwuna
Department of Animal and Environmental Biology, Delta State University, Abraka, Delta State, Nigeria

P. A. Umoru
Department of Zoology, Ambrose Alli University, Ekpoma, Edo State, Nigeria

Bendeddouche Badis
Ecole Nationale Supérieure Vétérinaire BP 161 El Harrach, Algiers, Algeria

Dahmani Kheira
Inspection vétérinaire de la wilaya d'Alger. BHCA El Harrach, rue Benyoucef Khettab. El Harrach, Algiers, Algeria

Priyanka Nayak
Manyavar Shri Kashi Ram Ji University of Agriculture and Technology, Banda, India

Dharmendra Kumar Shukla
Division of Post Harvest Management, CISH, Lucknow, India

Devendra Kumar Bhatt
Department of Food Technology, Bundelkhand University, Jhansi, India

Dileep Kumar Tandon
Division of Post Harvest Management, CISH, Lucknow, India

R. M. Omodamiro
National Root Crops Research Institute Umudike, P. M. B. 7006, Umuahia, Abia State, Nigeria

C. Aniedu
National Root Crops Research Institute Umudike, P. M. B. 7006, Umuahia, Abia State, Nigeria

U. Chijoke
National Root Crops Research Institute Umudike, P. M. B. 7006, Umuahia, Abia State, Nigeria

K. A. Arowora
Nigerian Stored Products Research Institute, P. M. B. 1489, Ilorin, Nigeria

B. A. Ogundele
Nigerian Stored Products Research Institute, P. M. B. 1489, Ilorin, Nigeria

A. O. Ajani
Nigerian Stored Products Research Institute, P. M. B. 5044, Ibadan, Nigeria

Lin-lin Cheng
College of Chemistry and Chemical Engineering, XinJiang University,Urumqi, 830046, China

Li-mei Xiao
College of Chemistry and Chemical Engineering, XinJiang University,Urumqi, 830046, China

Wei-Xin Chen
College of Horticulture, South China Agricultural University, Guangzhou 510642, China

Ji-de Wang
College of Chemistry and Chemical Engineering, XinJiang University,Urumqi, 830046, China

Feng-bin Che
Farm Product Storage and Freshening Institute, Xinjiang Academy of Agricultural Sciences, Urumqi, 830091, China

Bin Wu
College of Chemistry and Chemical Engineering, XinJiang University,Urumqi, 830046, China
Farm Product Storage and Freshening Institute, Xinjiang Academy of Agricultural Sciences, Urumqi, 830091, China

O. J. Olasoji
Institute of Agricultural Research and Training, Obafemi Awolowo University, Moor Plantation, Ibadan, Oyo State, Nigeria

O. F. Owolade
Institute of Agricultural Research and Training, Obafemi Awolowo University, Moor Plantation, Ibadan, Oyo State, Nigeria

R. A. Badmus
Institute of Agricultural Research and Training, Obafemi Awolowo University, Moor Plantation, Ibadan, Oyo State, Nigeria

A. A. Olosunde
National Centre for Genetic Resources and Biotechnology, Moor Plantation, Ibadan, Oyo State, Nigeria

O. J. Okoh
Department of Plant Breeding and Seed Science, University of Agriculture, P. M. B. 2373, Markurdi, Benue State

Chitra Pothiraj
ONIRIS - GEPEA – UMR CNRS 6144, BP 82225, F -44 322 Nantes Cedex 3, France

Ruben Zuñiga
ONIRIS - GEPEA – UMR CNRS 6144, BP 82225, F -44 322 Nantes Cedex 3, France

Helene Simonin
ONIRIS - GEPEA – UMR CNRS 6144, BP 82225, F -44 322 Nantes Cedex 3, France

Sylvie Chevallier
ONIRIS - GEPEA – UMR CNRS 6144, BP 82225, F -44 322 Nantes Cedex 3, France

Alain Le-Bail
ONIRIS - GEPEA – UMR CNRS 6144, BP 82225, F -44 322 Nantes Cedex 3, France

Fernande G. Honfo
Faculty of Agronomic Sciences, University of Abomey-Calavi, 01 BP 526, Cotonou, Benin Republic
Department of Agrotechnology and Food Sciences, Wageningen University, Wageningen, Netherlands

Kerstin Hell
International Potato Center (CIP), Cotonou, Bénin Republic

Noël Akissoé
Faculty of Agronomic Sciences, University of Abomey-Calavi, 01 BP 526, Cotonou, Benin Republic

Anita Linnemann
Department of Agrotechnology and Food Sciences, Wageningen University, Wageningen, Netherlands

Ousmane Coulibaly
International Institute of Tropical Agriculture (IITA-Cotonou), Cotonou, Bénin Republic

Ego U. Okonkwo
Nigerian Stored Products Research Institute Headquarters, Km. 3 Asa Dam Road, P. M. B. 1489, Ilorin, 240001, Kwara State, Nigeria

Kayode A. Arowora
Nigerian Stored Products Research Institute Headquarters, Km. 3 Asa Dam Road, P. M. B. 1489, Ilorin, 240001, Kwara State, Nigeria

Bukola A. Ogundele
Nigerian Stored Products Research Institute Headquarters, Km. 3 Asa Dam Road, P. M. B. 1489, Ilorin, 240001, Kwara State, Nigeria

Mike A. Omodara
Nigerian Stored Products Research Institute Headquarters, Km. 3 Asa Dam Road, P. M. B. 1489, Ilorin, 240001, Kwara State, Nigeria

Sunday S. Afolayan
Nigerian Stored Products Research Institute Headquarters, Km. 3 Asa Dam Road, P. M. B. 1489, Ilorin, 240001, Kwara State, Nigeria

ABANO Ernest Ekow
School of Food and Biological Engineering, Jiangsu University, 301 Xuefu Road, Zhenjiang 212013, China

Department of Agricultural Engineering, University of Cape Coast, Cape Coast, Ghana

M. A. Haile
School of Food and Biological Engineering, Jiangsu University, 301 Xuefu Road, Zhenjiang 212013, China

OWUSU John
School of Food and Biological Engineering, Jiangsu University, 301 Xuefu Road, Zhenjiang 212013, China
Department of Hospitality, School of Applied Science and Technology, Koforidua Polytechnic, Koforidua, Ghana

ENGMANN Felix Narku
School of Food and Biological Engineering, Jiangsu University, 301 Xuefu Road, Zhenjiang 212013, China
Department of Hotel, Catering and Institutional Management, Kumasi Polytechnic, Kumasi, Ghana

V. C. Ezeocha
Post Harvest Technology Programme, National Root Crops Research Institute, Umudike, Abia State, Nigeria

P. C. Ojimelukwe
Food Science and Technology Department, Michael Okpara University of Agriculture, Umudike, Abia State, Nigeria

Joana Diniz Rosa Fernandes
Laboratório de Tecnologia dos Produtos Agrícolas, Departamento de Tecnologia, Faculdade de Ciências Agrárias e Veterinárias, UNESP - UnivEstadualPaulista, Campus de Jaboticabal, Brazil

João Emmanuel Ribeiro Guimarães
Laboratório de Tecnologia dos Produtos Agrícolas, Departamento de Tecnologia, Faculdade de Ciências Agrárias e Veterinárias, UNESP - UnivEstadualPaulista, Campus de Jaboticabal, Brazil

Josiane Pereira da Silva
Laboratório de Tecnologia dos Produtos Agrícolas, Departamento de Tecnologia, Faculdade de Ciências Agrárias e Veterinárias, UNESP - UnivEstadualPaulista, Campus de Jaboticabal, Brazil

Kelly Magalhães Marques
Laboratório de Tecnologia dos Produtos Agrícolas, Departamento de Tecnologia, Faculdade de Ciências Agrárias e Veterinárias, UNESP - UnivEstadualPaulista, Campus de Jaboticabal, Brazil

Ben-Hur Mattiuz
Laboratório de Tecnologia dos Produtos Agrícolas, Departamento de Tecnologia, Faculdade de Ciências Agrárias e Veterinárias, UNESP - UnivEstadualPaulista, Campus de Jaboticabal, Brazil